DRY FARMING
IN THE NORTHERN
GREAT PLAINS
1920–1990

DEVELOPMENT OF WESTERN RESOURCES

The Development of Western Resources is an interdisciplinary series focusing on the use and misuse of resources in the American West. Written for a broad readership of humanists, social scientists, and resource specialists, the books in this series emphasize both historical and contemporary perspectives as they explore the interplay between resource exploitation and economic, social, and political experiences.

John G. Clark, University of Kansas, General Editor

DRY FARMING IN THE NORTHERN GREAT PLAINS

Years of Readjustment, 1920–1990

Mary W. M. Hargreaves

 University Press of Kansas

© 1993 by the University Press of Kansas

Published by the University Press of Kansas (Lawrence, Kansas 66049), which was organized by the Kansas Board of Regents and is operated and funded by Emporia State University, Fort Hays State University, Kansas State University, Pittsburg State University, the University of Kansas, and Wichita State University

Library of Congress Cataloging-in-Publication Data

Hargreaves, Mary W. M., 1914–
    Dry farming in the Northern great plains : years of readjustment,
    1920–1990 / Mary W. M. Hargreaves.
       p.   cm. — (Development of western resources)
    Includes bibliographical references and index.
    ISBN 0-7006-0553-3
    1. Dry farming—Great Plains—History—20th century. I. Title.
    II. Series.
    SB110.H37   1992
    338.1'0978'0904—dc20   92-8558

British Library Cataloguing in Publication Data is available.

Printed in the United States of America

10 9 8 7 6 5 4 3 2 1

CONTENTS

MAPS AND TABLES

Maps

Tables

ACRONYMS

AAA	Agriculture Adjustment Administration
AAM	American Agricultural Movement
ACP	Agricultural Conservation Program
ARS	Agricultural Research Service
ASCS	Agricultural Stabilization and Conservation Service
BAE	Bureau of Agricultural Economics
CCC	Commodity Credit Corporation or, in context, Civilian Conservation Corps
CNI	Conservation Needs Inventory
CRP	Conservation Reserve Program
CWA	Civil Works Administration
DRC	Federal Records Center, Denver, Colorado
EEC	European Economic Community
EI	Erodibility Index
EPIC	Erosion/Productivity Impact Calculator
ERS	Economic Research Service
FCA	Farm Credit Administration
FEI	Farm Equipment Institute, Chicago, Illinois
FERA	Federal Emergency Relief Administration
FHA	Federal Housing Administration or, in context, Farmers' Holiday Association
FmHA	Farmers' Home Administration
FSA	Farm Security Administration
GNDA	Greater North Dakota Association
GPC	Great Plains Agricultural Advisory Council
IWA	International Wheat Agreement
MAES	Montana Agricultural Experiment Station
MExtS	Montana Extension Service
MHS	Montana Historical Society
MSC	Montana State College, now Montana State University
NALS	National Agricultural Lands Study
NCA	National Cattlemen's Association
NDAES	North Dakota Agricultural Experiment Station
NDExtS	North Dakota Extension Service

NDFU	North Dakota Farmers' Union
NFHA	National Farmers' Holiday Association
NFU	National Farmers' Union
NRB	National Resources Board
NRC	National Resources Committee
NRI	National Resources Inventory
NRPB	National Resources Planning Board
NYA	National Youth Administration
OH	Oral History
PCA	Production Credit Association
PMA	Production and Marketing Administration
PWA	Public Works Administration
RA	Resettlement Administration
RCA	Resources Conservation Appraisal
REA	Rural Electrification Administration
RG	Record Group
SCD	Soil Conservation District
SCS	Soil Conservation Service
SDAES	South Dakota Agricultural Experiment Station
UFL	United Farmers League
USDA	United States Department of Agriculture
USGS	United States Geological Survey
USWD	United States War Department
WCU	Water Conservation and Utilization Program
WPA	Works Progress Administration

PREFACE

This research on the development of dry-land agriculture in the northern Plains was begun more than fifty years ago. During that period circumstances and my interpretative perceptions have changed considerably. As a student of Paul W. Gates and Frederick Merk, I was interested in frontier expansion and, specifically, in national land policy relating to that expansion. I began my study as a historical analysis of what Frederick Jackson Turner had projected as the apex of a vertical progression of stages in frontier development, agricultural settlement and its maturation into the commercial role for which it was regionally best adapted to contribute to the national growth.

In the volume *Dry Farming in the Northern Great Plains, 1900–1925*, published in 1957, I described the lack of a planned government program for land use in the region, the rapidity of the transition from agricultural settlement to commercial operations, the difficulties posed not only by the climatic and physiographic characteristics of the area but also by the exploitive interests, and the adjustments evolving, largely indigenously, in the process of adaptation to those circumstances. By 1925 the initial settlement phase had been accomplished, and the speculative froth had been largely dissipated. But the adjustments that would carry the development beyond mere self-sufficiency to lasting commercial significance were still exploratory. To review them and appraise their role in the broader context of the general welfare remained for further study.

I found as research extended over the succeeding decades that the Turnerian conception of frontiers freely developing in adaptation to their geographic potential was inhibited in relation to the northern Plains. This conclusion does not rest upon the argument of colonialism that has figured largely in accounts of regional history from New England to the Pacific Coast—dependence upon external investment of capital, labor, and carryover of traditional attitudes and institutions has been inherent to all frontier settlement. But while the general goal of such forces as related to earlier expansion had been contributory to development, the focus of extraregional programs on utilization of the Great Plains since the early twenties has been directed toward contracting the dry-farming effort. Why this should be the focus has become the basic concern of this study.

My perspective has not been intended as adversarial. The historian's function should be investigative, not hortatory. Presentation of data does, however, establish a framework for conclusions. In this case hyperbolic and mistaken assumptions have long prevailed. When one body of proponents would retire the cropland to a buffalo commons while another segment of analysts laments the threat of growing world hunger, the history of the dry-farming development affords applicable insight relative to the policy dilemmas of land utilization.

My research was initiated, long ago, with the help of a joint fellowship provided by Radcliffe College and the Brookings Institution, which afforded me guidance by Dr. Edward G. Nourse and brought me into contact with the great bibliographer of agricultural history, Dr. Everett E. Edwards. I am immensely grateful to both for the insight they gave during the period when land-use planning was in its formative stages. They made me aware of the large number of mimeographed, in-house studies under way in the Department of Agriculture during the thirties.

Since then I have owed much to the facilities of libraries and the services of their staffs—the Library of Congress, the library of the United States Department of Agriculture, the National Archives, the Federal Archives and Records Center in Denver, the New York Public Library, the library of the Montana Historical Society, the North Dakota State Library, and especially the libraries of the University of Kentucky. I could not have pursued this avocational research without the inestimable assistance of the last institution, a federal depository system with access to interlibrary loans, or without the particularly generous cooperation of the Montana and North Dakota historical societies and the North Dakota State University Library.

Professor Howard W. Ottoson, of the University of Nebraska and long active in the Great Plains Agricultural Advisory Council, critiqued my early plans for this volume and helped greatly in directing me to the *Proceedings* of the council. Some years ago Dr. Elmer E. Starch was kind enough to send me a lengthy manuscript on the dry-farming adjustment experience in relation to land retirement in Montana, and more recently Dr. Roy E. Huffman of Montana State University not only shared with me the insight of a participant long familiar with the local development but also gave me several locally published volumes, including the memoirs of Sherman E. Johnson, which, as the reader will note, proved particularly useful. To all these kind people—as well as to Montanans Richard B. Roeder, Thomas R. Wessel, and James Muhn—for their suggestions, many thanks.

I owe to Dr. John Clark, the series editor, profound gratitude for a detailed critique that forced me to clarify textual focus and linkages. My husband, Dr. Herbert Walter Hargreaves, the economist of the family, supplies not only critical challenge but also moral support—for which I am especially appreciative. I hasten to add that neither he nor any of the others mentioned are responsible for the interpretations here presented.

The University of Kentucky Research Committee has most generously on various occasions funded filming and acquisition of files from the National Archives and the Denver Records Center and, through the services of Mrs. June Carol Smith, provided for the transcription of the manuscript for the press. For such assistance, words are an insufficient expression of thanks.

Mary W. M. Hargreaves
Lexington, Kentucky
August 1992

INTRODUCTION

By definition dry farming relates to agriculture without irrigation in regions of semiaridity. That the zone of semiaridity fluctuates from season to season and between series of years occasions major instability in boundary delimitation. That semiaridity is itself effected by an amorphous combination of deviating patterns of precipitation, evaporation, wind velocity, exposure direction, terrain, and soil composition further complicates the problem. The setting of regional bounds becomes necessarily nominal. In this book, the northern Plains comprise the eastern two-thirds of Montana and the western half of the Dakotas, an area bounded on the west by the Rocky Mountains and on the east by the southward bend of the Missouri River. The eastern border is marked by an average, long-term precipitation approximating 15 inches annually, an amount narrowly marginal for successful agriculture. The southern boundary was chosen to permit local focus on the zone of predominant Northern Spring Wheat cultivation. The area was initially brought into production within a relatively compact period under the distinctive developmental impetus identified as the dry-farming movement.[1]

Semiaridity has inherently been a major consideration in the history of such land utilization, and because of its perceived limitation, agriculture was slow to develop on the Great Plains as a whole. During the opening half of the nineteenth century, travel accounts by people who traversed the region broadly categorized the zone west of Council Bluffs between the thirty-fifth and forty-ninth parallels of latitude as a "great desert," for the most part "sterile" and, except where capable of irrigation, "uncultivable and unproductive." By midcentury, however, it was discovered that livestock, which had been abandoned by overland settlers pushing on to the Far West, thrived on the native grasses of the Plains, and the development of large-scale range cattle operations after the Civil War marked a recognition that the region was a superb "grazing domain." To proponents of such land utilization, however, the very fact that grasses cured without rotting still indicated aridity too great for agriculture without irrigation. Meanwhile, miners and military personnel at Plains outposts began trying to grow vegetables, occasionally without supplementary watering. When the range-cattle industry was decimated in the mid-eighties as summer droughts were followed by severe winter blizzards, ranchers turned increasingly to cultivation of hay and forage crops in

1

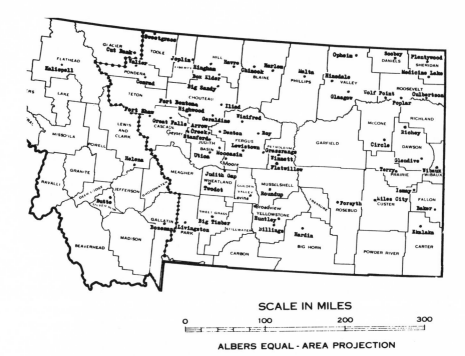

SCALE IN MILES

0   100   200   300

ALBERS EQUAL - AREA PROJECTION

Town and county maps of the northern Plains, Montana (above) and North and South Dakota (opposite). Dots outline the region included for the tabulations in this book. (From Bureau of the Census, *County Boundaries as of April 1, 1950* [Washington, D.C., 1953])

combination with stock production. In the early nineties an agent of the U.S. Geological Survey reported that irrigation was generally regarded as "indispensable" in Montana but a few farmers in the foothills of the Gallatin Valley and the Judith Basin were raising wheat successfully without it and irrigation was scanty and infrequent throughout most of the farming areas of the state. At the same time, as settlement expanded along railroad routes westward, eastern migrants who had traditionally practiced nonirrigated farming moved to the 101st meridian of longitude in the Dakotas and to sections of the Black Hills. Probing fingers of agricultural settlement had joined East to West, and the U.S. census announced the closing of the frontier in 1890.[2]

Widespread drought for much of the period from 1889 through 1895 provided warning of the climatic hazards of farming in the region. The agricultural advance was retarded, even pushed back for a time, but the drought also generated efforts to improve technological procedures. A variety of measures were proposed to counter the effects of drought—expanded irrigation, introduction of drought-resistant plant varieties, and specialized methods of

cultivation. Of the last approach, the most heavily publicized was the "Scientific Farming System" formulated by a Brown County, South Dakota, farmer, Hardy Webster Campbell. As fully developed and published in numerous editions of his *Soil Culture Manual*, first issued in 1902, Campbell's methodology required deep plowing, a packed subsoil, frequent surface cultivation, and cropping on land that had been kept cultivated as fallow throughout the previous growing season, a routine designed to retain the precipitation of two years for use during a single crop season by developing a reservoir of moisture in the subsoil and reducing surface evaporation. Supported by railway, milling, implement manufacturing, banking, real estate, and other commercial interests, this program was endorsed and promulgated by state agricultural leaders and an annually assembled national Dry Farming Congress. The effect was to offer assurance to prospective settlers that the hazards of semiaridity were readily surmountable. "It is altogether too common an idea that the quantity and quality of the crop depends upon the climatic conditions," Campbell wrote in 1902.[3]

In 1904 the U.S. Department of Agriculture began testing dry-land tillage technology. Departmental agricultural scientists were skeptical that a systematic formula could assure successful operations, but their test conclusions were long delayed. Meanwhile, regional proponents of Campbell's recommendations won congressional enactment of legislation in 1909 authorizing homestead entries of 320 acres on public lands in order to provide for farming in accordance with the fallow requirements of the Campbell program. A stock-farming homestead, authorized under legislation of 1916, permitted expansion of entries to 640 acres so that agriculture might be combined with stock operations.

Between 1900 and 1920 the rural population in the northern Plains increased more than 300 percent, from 194,000 to 620,000, and improved land in farms—acreage harvested, failed, idle, fallow, or plowable pasture—rose from a little over 1 million to over 27.3 million acres during the settlement boom that marked the quarter century ending in 1925. The area harvested in wheat, which became the principal cash crop, increased from 288,000 to 6,646,000 acres, but at the close of the period there were also about 995,000 acres of flax, 1,596,000 acres of oats, 456,000 acres of barley, 1,584,000 acres of corn, and 754,000 acres of alfalfa.[4]

In addition, the number of cattle increased from 1,334,000 to 2,170,000 head between 1900 and 1925, but the old style of range operations was in retreat. Cattle raising had become increasingly dependent upon supplementary feeding, and farming had achieved an integral status in the economy. Settlement of the region had counteracted the old notion that it was part of a great American desert or, at best, a domain to be restricted to cattle and sheep range.

But difficulties had also become apparent. The western Dakotas had experienced severe localized drought from 1910 to 1912. Amending the methodological focus of the Campbell program, other agricultural proponents for commercial interests—notably Thomas Shaw for the Hill railways and P. G. Holden for the International Harvester Company—then began to emphasize the importance of crop rotation, diversification, and incorporation of livestock in the operations. The North Dakota Bankers Association launched a "better-farming" campaign which helped bring about the legislation establishing the county extension service for agricultural education. Despite these adjustments, failure due to drought, which in some sections began as early as 1916 and continued as late as 1921, was widespread in the region from 1917 through 1920. Census data do not adequately recount the story of lost hopes and abandoned effort. One scholar has estimated that as many as three-fourths of the people who homesteaded Montana between 1909 and 1918 left their holdings by 1922. The rural population of western South Dakota showed little increase after 1910, and in the same decade the rural population of western North Dakota rose only 14 percent. Neither the dry-farming methodological system promulgated by Campbell nor the better-farming diversification programs had provided a panacea.[5]

Proponents of these measures could with considerable justification maintain that they had not been generally practiced. The intensive cultivation required under the Campbell system won few supporters outside of some of the foreign settlements, and even Russian and German colonists were found to have farmed with less than their traditional care. The slowness of sod deterioration on new breaking, high wheat yields on virgin soil, and heavy war-time demand for that grain had fostered continuous cropping under extensive farming practices. Letting fields lie fallow through the summer had never been a customary procedure in the Dakotas as it had in the early mining communities of western Montana. While use of cultivated corn ground as seedbed preparation for grain crops had gained some momentum in the Dakotas after the earlier drought years, by 1915 even that alternative as a technique for increasing moisture accumulation in the subsoil had been largely abandoned. The better-farming campaigns, which emphasized corn and feed crops combined with hogs and dairy cattle in a diversified program, survived principally in the form of increased livestock and hay production, stock farming, in southwestern South Dakota.

Regional agricultural leaders recognized that this was a land of peculiarly difficult climatic conditions with highly variable precipitation, high rates of evaporation, temperature extremes, short growing seasons, strong and persistent winds, and frequent summer hail storms but attributed much of the homestead abandonment to the fact that about half the settlers lacked any farming experience. Survey of the north-central Montana "triangle" from Great Falls north to Cut Bank and Havre revealed that of 550 homesteaders only 51 percent had farmed prior to their move into the area. Of the remainder, 11 had been railroad men; 10 had been clerks; 8, each, had been carpenters, miners, or unemployed widows; 7 had been schoolteachers; 6, each, were classified as merchants or "old maids" (which need not have disqualified them as farmers!), and the remainder ranged from ranch hands to people trained in various skills, household help, seafarers, "rovers," gamblers, or "professional homesteaders."[6]

A similar study of twelve townships scattered throughout western North Dakota showed the same wide gamut of prior occupational background. Of those settlers who had abandoned their homesteads before 1920, only 35 percent had been farm operators before filing their claims, and only 15 percent more had been raised on a farm or worked as farm hands. Almost 10 percent were women, mostly schoolteachers, stenographers, widows, daughters of settlers, and "elderly matrons." The analyst concluded that over half the former operators (54.3 percent) had entered the lands as a speculative venture, and nearly half had had other sources of income and did not depend on farming for a living. More than 20 percent had never farmed the homesteads.

Most settlers had merely sought to claim their share of the public lands before the supply was gone. Only 39 percent cited desire to establish a home as their reason for the undertaking. Many were not prepared or willing to accept the hardships of frontier settlement as a long-term commitment.[7]

The hardships of homesteading had been aggravated by the regional scarcities of wood for housing and fencing and water for domestic and agricultural uses. Milled lumber, barbed-wire fencing, and deep-drilled wells were costly when accessible. In some areas well water for household use could not be obtained at practicable depths; settlers in the early twenties were still relying on cisterns or hauling water by barrel from distant sources. Fuel, however, in the form of low-grade bituminous coal, lignite, was cheap and was found over much of the western Dakotas and eastern Montana.[8]

Agricultural operations also entailed special costs inherent to the regional technology. Land was cheap in comparison with that in more developed areas of the nation, but the summer fallow system required doubling the scale of cultivation, and the diversification program entailed use of more pasture. Public lands had been quickly engrossed in 160- to 320-acre homesteads, and pressure for expanded holdings had driven up prices on private offerings between 1910 and 1920, 50 to 100 percent in the "triangle," the northeastern, and the south-central districts of Montana, 100 to 150 percent in the favored Chouteau and Judith Basin counties. Speculative frenzy accounted for much of this demand in Montana; but in western North Dakota, where the availability of public lands was greatly reduced by 1910, pressure for increased holdings was also evident through the next ten years, most notably in the increased size of farms where additional land could be bought at relatively low prices from the Northern Pacific Railway grant. Farm management systems endorsed by agricultural scientists of the federal Bureau of Agricultural Economics in 1925 called for an average farm size of 640 acres in the great wheat district of northeastern Montana and 960 acres in the southeastern counties, where grain crops were generally coupled with livestock production.[9]

Larger enterprises required a heavier capital investment, for mechanization as well as for land purchase. Tractors and trucks were in widespread use by the early twenties, and where horses were still being used, they were teamed under multiple hitches, frequently as eight- to twelve-horse outfits to draw three- or four-bottom plows or two to three drills as a single operation. Most wheat was still cut with an 8-foot binder, but 12-foot push binders and headers were used for short grain stalks. Threshing by large custom outfits had been the general practice until the labor shortage of the war years, but by 1924 a considerable number of small cooperatively owned outfits had come into service. Introduced in 1918, 6- to 8-foot combined harvester-threshers powered by auxiliary engines had not been widely accepted; but by the mid-twenties sales of 8- to 10-foot combines, drawn and powered by medium-sized tractors, were increasing rapidly. While the number of farms in the northern Plains had increased by some 470 percent during the settlement period, the value of farm implements and machinery had mounted almost 1,600 percent.[10]

Credit for land and machinery purchases had become an important item in the regional costs of production. In most dry-land counties 50 to 80 percent of farm operators—over 82 percent in Jones County, South Dakota—carried mortgage debt in 1925. A survey by the U.S. Department of Agriculture showed average annual interest payments of $380 to $390 on owner-operated farms in the west north-central and western states, amounting to approximately one-fifth of the total cash expenses.[11] Farmers were heavily dependent upon the resources of local banks and mercantile establishments for such credit.

Under the Federal Reserve Act, national banks were permitted to make loans on farm lands for five-year terms and to discount agricultural paper for periods of six months. Administrative ruling during the war years had liberalized the policy by extending the definition of agricultural paper to cover farmers' and merchants' notes on tractor transactions. Although the reserves of local banks had increased during the settlement period, they were, during the early twenties, the lowest, next to those of the mountain states, of any banking district in the nation. They were far too low to meet the pressing demand for credit.[12]

Loans for agricultural purposes constituted from two-thirds to three-fourths of the business of commercial banks in the region. Mortgage records for three townships of Haakon County, western South Dakota, show that during the settlement period commercial banks were the principal institutional suppliers of funds for farm mortgages. In 1910 they provided 11.5 percent of the amount borrowed on first mortgages and another 1.2 percent of that held on second mortgages. By 1920 the proportion of their funding of first mortgages had changed little, but they held almost 63 percent of the amount backed by second mortgages.[13] The change reflected not only the increased demand for credit but also the declining role of other mortgage sources for the higher risk accounts. The localization of funding heightened the vulnerability to widespread disaster should regionally extensive drought develop for even a short time.

During much of the period individuals—chiefly external investors or former owners of the tracts sold—had been the most important suppliers of farm credit, which is evidence of the speculative nature of the activity. Such funding decreased from about 70 percent of the amount held on first mortgages and 61 percent of that on second mortgages in Haakon County in 1910 to about 46 percent and 28 percent in 1920. Mortgage companies, which were the second highest institutional lenders in Haakon County during the early settlement years, had held 10 percent of the amount of first mortgage loans and 14.4 percent of the amount lent on second mortgages. Local banks, real estate dealers, and loan agents, acting as brokers in placing mortgages with extraregional farm mortgage finance companies and private investors, had also played a large role. By 1920, however, such investment had also sharply declined. Insurance companies, another major source for mortgage lending, had taken little part in the development of the dry-farming areas.[14]

Credit management, generally, reflected a prevalent view of the regional development as a risky enterprise. Short terms, high interest rates, and multiple commission and administrative charges were characteristics of the lending operations. Long-term mortgages ran for only three to five years, with renewal contingent upon additional service fees. Interest rates at 12 percent on such loans had been common as late as 1913 and continued at about 10 percent through the war years. Commission charges varied from 2 to 10 percent additionally. Short-term bank loans, backed by note or chattel mortgages, carried basic rates of 10 to 12 percent, plus commission charges and various hidden costs such as prior deduction of interest. Store credit, ranging up to two years, carried similar rates, plus an additional cost of 5 to 7 percent in the price of the goods, which were discounted for cash purchases.[15]

The federal government had attempted to relieve the situation by passing the Farm Loan Act in 1916, but the direct benefits of the measure were meager through the settlement years. The joint-stock land banks, organized as one feature of this legislation, placed the major portion of their loans in older, better-established farming regions. While the operations of the federal farm land banks were more evenly distributed over the nation, they were largely suspended during judicial action from 1919 to 1921. None of these loans was extended in trans-Missouri Haakon County before 1930. When available, they were conservatively administered, averaging only 40 percent of the appraised value of the mortgaged land and buildings despite authorization for a 50 percent level.

Meanwhile, the state governments in the region had acted on the problem. By legislation dating from the 1890s, the Dakotas had authorized farm loans through investment of funds from the sale of state lands acquired as public-school and university grants from the federal government. Montana had enacted similar legislation in 1909 but required applicants to hold water rights to the pledged tracts, a restriction that had greatly limited use of the loans until it was repealed in 1917. Loans totaling some $4.3 million were made under the Montana program, but because of the heavy rate of foreclosure and quit-claim liquidation through the drought period, the state recovered the title on lands representing 95 percent of that sum, and no loans were made after the spring of 1922. Not all school-fund investment was confined to farm mortgages; during the dominance of the Nonpartisan League, North Dakota had had to act to require that at least one-third of the money be assigned to such use. As in Montana, the experience was not encouraging. By

1930 over $11 million of the North Dakota school funds had been invested in farm loans, but several hundred of those loans were delinquent, and the state had already foreclosed on over 102,000 acres. Further lending was discontinued in 1929 due to the lack of funds.[16]

Because of the scarcity of credit resources throughout the early settlement period, legislators instituted loan programs backed by sale of bonds or warrants. Such legislation in Montana, enacted in 1915, was emasculated by judicial ruling against its constitutionality. The North Dakota program, operated through the state bank established in 1919, was also hampered by court contest during its first three years. With the repudiation of the Nonpartisan political leadership that had sponsored the legislation, a generally conservative administration of loans prevailed, but the foreclosure rate was high, tax levies were necessary from the beginning to cover service costs on the outstanding bonds, and the Bank of North Dakota was finally forced into liquidation in 1934. The Rural Credit System of South Dakota, established in 1917, quickly became a major lending agency in the trans-Missouri district of the state. Its loans together with those from the state school fund amounted to almost half of the total first mortgage loans outstanding in Haakon County in 1925. By then, however, the state lending program had been forced into liquidation there, also.

The direct effects of such efforts were meager, but their value in moderating the terms of agricultural credit may have been considerable. Authorities in the Dakotas had attempted to channel a major share of the available funds into the western communities. The governmental enactments offered loans for longer terms under lower interest and more flexible repayment plans than those extended by private lenders. Interest rates declined about one percent between 1916 and 1920, and short-term personal and collateral bank loans in 1920 averaged about 10 percent, the lower range of the prewar rate scale. Collection of bonus or commission fees had been largely terminated, and optional advance-payment features were becoming more common.

Other factors generating unusually heavy regional costs involved problems of distance. Sparsity of settlement, normal under frontier conditions, here became a permanent situation as larger farms and extensive farming operations separated neighbors. Schools, churches, and recreational facilities were far apart and thinly supported, and roads, bridges, and other governmental services had to be provided for large areas of low population. Extension of telephone, electricity, and gas lines into rural districts lagged and, when available, entailed heavy installation charges. For settlers to obtain medical or dental care they had to travel long distances and pay for food and, perhaps, lodging. These circumstances lowered the quality of life and for some 14 percent of the settlers became the primary reason for abandoning their claims.[17] Social considerations aside, high taxes for public services, high transportation costs for farm supplies, long delays for machinery repairs, and heavy hauling expense for marketing crops greatly lessened the profit on crop production for a highly competitive world trade.

Drought and the related problems of scarce timber and water for domestic and farm use, capital requirements for large-scale land and machinery purchase, and limited popular support for social and governmental institutions formed an inherently regionalized cluster of difficulties in dry-land agricultural development. Most of the problems could be surmounted if the farming operations were profitable over an extended period, but profitability rested upon the market demand for the products. Regional dilemmas were involved in a much broader context of national and international commerce.

Wheat had been a prominent part of the exports of the United States since the early colonial period. International trading centers, rather than domestic terminals, dictated its price. Foreign production had been curtailed during World War I, and in the immediate aftermath the food shortage had continued. Returns on wheat to American farmers, indexed at 100 for the period 1910–1914, had held above 200 since 1916 and soared to 248.6 in 1919. By the summer of 1920, however, U.S. public reconstruction loans in Europe had been terminated, the War Finance Corporation had been suspended, and private lending for relief purposes was shifting to more remunerative enterprise. Europeans no longer had funds to supplement their shortages.[18]

Meanwhile, vast new areas of wheat production had been and were yet being developed, not only in the United States but also in Argentina, Australia, and Canada, competition for a dwindling market. Revised population studies showed declining birth rates both in the United States and Europe. Changes in dietary habits were reducing the use of cereal products, and the growth of motor transportation and mechanized farm equipment was decreasing the acreage needed for feed grains.

The carry-over of wheat in the United States by July 1, 1920, had become unusually large. As the new crop came on the market, prices collapsed. The farm price, which in June averaged $2.58 a bushel, dropped to $1.43 by December 1. A year later it stood at 90.5 cents a bushel. Moreover, farm expenses, also greatly increased during the war years, had reached a peak in 1919 and 1920 but remained high throughout the next decade. With the indices of prices received and paid by farmers scaled at 100 for the period 1910 through 1914, the parity ratio of farm receipts dropped from 118 in 1917 and 1918 to 104 in 1920 and to 75 in 1921. The purchasing power of wheat, specifically, during the last year stood at 78.[19]

High freight rates were also identified as a major factor in the crisis that developed. The problem for wheat growers of the northern Plains was enhanced by their inland location, which was far from domestic and foreign terminal markets. Between 1917 and 1920 railroad freight rates in the United States had been doubled, the last increase, a rise of 25 to 40 percent, coincided with the drop in wheat prices. Shortly afterward, the Canadian government, reacting to U.S. tariff legislation of 1920 and 1921, initiated several rate reductions on freight over dominion lines. By July 1922 charges on wheat shipped from Scobey, Montana, to Duluth, Minnesota, amounted to 22.5 cents a bushel, while the cost from Regina, Saskatchewan, to Port Arthur, Ontario, a distance 90 miles longer, was only 12 cents. An import duty of 30 cents a bushel more than offset the Canadian rate advantage in U.S. markets, but it could not protect the American export trade, and it could not prevent the decline of domestic prices as carry-over stocks developed.[20]

Surveying the areas of major dry-land development during the early twenties, L. F. Gieseker, soil analyst of the Montana Agricultural Experiment Station, repeatedly emphasized that the problems of agriculture in the region were due more to national economic changes than to protracted periods of drought. He noted, however, that the general situation had a heavier impact in the northern Plains because the type of operation there could not easily be altered to meet new conditions. In that northern latitude, wheat was the best adapted of all cash crops to withstand low rainfall. Since shortly after the turn of the century it had paid a higher return than livestock for the capital investment. Land once classified for farming could not be returned to grazing use without loss in property value for the individual and for the community tax base. Moreover, sod once broken could not be restored to range without a prolonged waiting period for re-establishment of grass. Even to institute stock farming, diversification of the farming activity with greater attention to livestock production, must be a gradual adjustment. Such a change would require new capital—better bred stock, more pasture, and supplementary implements; new experience; more labor; and increased community effort. Inflexibility consequently posed yet another problem of dry-land agricultural development.[21]

Counteracting the many difficulties, however, were regional advantages that might offset the limitations. The climate itself conveyed several benefits, as promoters had long advertised. The limited rainfall had reduced the leaching of soil nutrients, so that use of fertilizer had been unnecessary through the settlement period; where supplies of moisture were adequate, crop yields, even under continuous cultivation, remained unimpaired. From 50 to 75 percent of the precipitation that did fall was concentrated in the growing season, particularly in May and June, during the early stages of plant growth when most needed. The late summer tended to be a time of scant rainfall, which provided ideal conditions for the ripening grain and its harvesting, and the low humidity minimized the spread of plant diseases such as rust and smut. Warm, dry weather fostered maturation of the wheat berry with a high protein content that gave it special value in blending of flour. Baking tests showed increases in loaf volume amounting to 10 to 15 percent, depending upon the availability of protein that enhanced the elasticity of the dough. Since 1918 the Washburn-Crosby Company of Minneapolis had been using a protein map that rated wheats from various sections of the northern Plains, and during the early twenties public grain inspection laboratories throughout the region began to inform farmers of ratings that frequently entitled them to premium payments above the prices of standard marketing grades.[22]

Plains topography also afforded benefits as mechanization increased. Vast areas of level to gently rolling terrain provided an ideal surface for moving equipment. North of the Missouri River, in particular, glacial action had ground away surface projections and filled depressions with drift soils of good depth nearly as far west as the mountains. Broad benches extending back from the breaks of deeply entrenched stream beds were composed mostly of loams, friable, moderately good in water-holding capacity, and relatively free of large stones. The land was treeless except in the foothills and along streams and invited an extensive system of farm operation.[23]

"Some backing up will have to be done here," Dr. F. B. Linfield conceded in his report as director of the Montana Agricultural Experiment Station in 1923. He noted that many who had studied the problems of the region held more pessimistic views. Some were already contending that the public lands should never have been opened to settlement until they had been classified and that only those which offered "a reasonable opportunity" should have been released for homesteads and the remainder reserved as "community pasture." But in this and other reports of the early twenties, Linfield delineated his continuing conception of the northern Plains as an agricultural domain. Even the districts of low rainfall, some of which might "fall below sole dependence on grain growing, should not, however, be abandoned," he contended. Much of this country was capable of producing forage crops even in dry years, support for a livestock industry very different from the range-cattle operations of a quarter of a century earlier. In appraising the future course of Montana's agricultural development, he foresaw continued emphasis upon wheat: "Our natural conditions on the large areas of dry land are favorable to the production of wheat of high quality," he concluded.[24]

Readjustments had already been made in the region. In western South Dakota the greatest change had come between 1910 and 1920, when the total population had declined, the rural population had increased only slightly, and the number of farms had decreased, but there had been a large increase in the amount of cropland. By 1925 the number of farms had increased moderately and acres devoted to crops had more than doubled. Production data for the earlier period showed a large increase in the acreage of wheat and large increases in the acreages of corn and alfalfa, in the number of cattle, and in the number of swine. After 1920 the acreage in wheat had declined slightly and the number of cattle was reduced notably, but the other trends continued.[25]

In western North Dakota and eastern Montana the changes were marked only after 1920. The number of farms decreased by more than 2,300 enterprises in the former district and by over 9,000 in the latter, but in both there was a moderate increase in the amount of cropland. The acreage of wheat and flax had been expanded in both states, but so also had that in corn at the lower altitudes and in Montana that in alfalfa as well. The number of cattle had declined slightly in western North Dakota and increased only moderately in eastern Montana, reflecting the slow rate of recovery from drought losses; but in both areas the number of the more rapidly maturing swine and sheep had been expanded. Larger holdings and a combination of cash grain and crops for stock farming characterized the program trends.

Between 1920 and 1925 the average amount of cropland per farm increased from 187.56 to 306.83 acres in western North Dakota, from 174.56 to 330.74 acres in western South Dakota, and from 196.14 to 299.82 acres in eastern Montana.[26] Thousands of settlers had abandoned the region, and the number of farms had greatly diminished, but many of the holdings had been acquired by remaining operators, who thereby attained a more extensive and more economically profitable enterprise. This was not to be a development in the form of small, intensively worked farmsteads as originally projected under Campbell's dry-farming system or under Thomas Shaw's program for the rotational diversification more traditionally associated with midwestern agriculture. Promoters who had sought regional settlement as an extension of the "back-to-the-land" movement, popular as a Progressive cause in the early twentieth century, would find it necessary to postulate a commercial scale of agricultural undertaking, one in which the farmer must reckon the profitable margins of investing in additional land or additional machinery, of producing cash-crop wheat or supplemental feed for a larger herd of livestock.

Some "backing up" had been necessary. The Gieseker soil studies in Montana showed that the localities of heaviest abandonment were characterized by poorer soils and that they were not continued in production during the early years of the decade. In half a dozen counties, each, of western North Dakota and eastern Montana the amount of cropland declined between 1920 and 1925. In five counties of western South Dakota that change had occurred between 1910 and 1920, and in Stanley County it continued to 1925. A winnowing process had been under way that not only identified those settlers most committed to farming but also pointed to the need for revisions in land use and farm organization to sustain a viable long-term agricultural economy.

The readjustments in regional development thus embraced both climatic and commercial considerations. Neither aspect had figured very largely in the history of the formulation of American policy relating to disposition of the public lands. The most significant approach had been that encompassed in the work of Major John Wesley Powell. In 1878, at a time when the Plains were considered a grazing domain, Powell had published *Report on the Lands of the Arid Region*, calling, with some exceptions, for allotment of the lands as pasturage farms of 2,560 acres, with no more than 20 acres under irrigation, or as irrigated tracts with approximately 80 acres under ditch. He recognized that there might be exceptions to such a format—districts where the annual rainfall was concentrated in the growing season or foothill areas that received unusual amounts of moisture condensed by adjacent mountains. Appointed director of the Geological Survey in 1880, Powell undertook identification of such areas and classification of the tracts that might be developed under irrigation. The survey caused some delays and restrictions upon land entry which western congressmen bitterly assailed. Coupled with the opposition of eastern representatives, who resisted a program calling for federal expenditures for irrigation, the hostility led to rejection of Powell's approach.[27]

Popular protest against restraints on public land entry dated as early as the American colonial period. While the colonies surrendered their western land claims as a financial resource for the newly formed revolutionary federation, their citizens and their political representatives had promptly initiated a prolonged process of whittling away the limitations to free agricultural enterprise. As the semiarid regions were opened to settlement, those daring to undertake the venture and those seeking to promote it continued the pressure. In an earlier volume, this survey of dry-land development in the northern Plains pointed to the "general unwillingness to defer programs for land disposition" until an adequate experimental and educational basis existed for such expansion. "Dry-land settlers," it concluded, "provided the experimental background."[28]

Whether such an experimental approach should be continued; whether in the public interest further expansion onto semiarid lands was warranted; whether, indeed, even an increase of irrigated acreage could be justified were questions increasingly raised on commercial grounds by agricultural economists and governmental policy makers during the twenties. In his annual report for 1919, Secretary of Agriculture David F. Houston had introduced the discussion in the broad economic context of market demand when he questioned the opening of new lands:

> It is not in the interest of producers or consumers to have large fluctuations in agricultural production. . . . The aim rather should be to secure a steady flow of commodities of sufficient volume to supply an increasing demand at prices which will yield the farmer a decent wage and a fair profit on his investment. . . . There should be, and in the long run there will tend to be, no more farmers in the Nation than are needed to produce the quantity of products which can be disposed of at a profit.[29]

Secretary Houston's comment related specifically to the opening of new lands. The dilemma confronting dry-land farmers during the "readjustment years," however, was the dislocation for them and their communities when national planning agencies sought to reverse the development already under way. During the twenties, local leaders continued the indigenous, experimental process of adjusting the regional agriculture to the dual problems of drought and weak market demand. After an unstable recovery from these conditions during the mid-twenties, their efforts were overwhelmed by the severity of the difficulties that followed in the next decade. Much local planning and local leadership gave ground to external policy decisions. The technological adjustments of the twenties undergirded the dry-land agricultural program that resulted, but accompanying efforts to depopulate vast areas and resettle dry-land farmers on small irrigated tracts were generally neither popular nor successful. Alternative measures of the New Deal program that applied to agriculture in the region were chiefly beneficial due to the relief loans that permitted most operators to remain on their land.

While stratagems to cope with drought remained a continuing concern in dry-land agriculture, there was no other comparably widespread and prolonged stress from drought in the northern Plains during the period under review. Meanwhile, production soared. The loans of the thirties were quickly repaid, and except during brief periods production values on wheat were subsequently high, notably so until the late fifties, again in the early sixties, and from the early seventies until the mid-eighties.[30] Adjustments in farming programs were directed primarily to combat the problems of shifting market demand and very often to offset nationally established acreage limitations that penalized efforts to maximize the regional economies of scale.

Agricultural development in the northern Plains remained primarily centered in dry-land operations. In 1987 only about 10 percent of the harvested crop acreage in eastern Montana, about one-half of 1 percent of that in western North Dakota, and about 3 percent of that in western South Dakota were irrigated. Under the impact of national planning goals, large districts were taken out of agricultural use and other land was retired at least temporarily, yet after 1925 the amount of cropland in the region was increased 17 percent and the amount of harvested cropland, 12 percent. The northern Plains region produces about 12 percent of the national wheat crop, at less cost per bushel than the national average and for generally better returns than the national average. It also encompasses increased acreage in barley and alfalfa, to sustain a livestock industry over twice as large as that in 1925.[31] Dr. Linfield's belief that this would remain an agricultural domain was tested and justified.

National efforts to solve marketing problems have been directed to elimination of a large proportion of this production during much of the period here under review. Currently over one-third of the total cropland in the region has been declared eligible for retirement as "conservation reserve," although under 1982 management practices there was not as much as 1,000 acres of cropland in either Montana or North Dakota found to be losing productivity over 100 years at a rate greater than 5 percent, the level at which soil is considered renewable by natural forces. At a cost of more than $400 an acre over ten years, the expense to the U.S. government on rental contracts already negotiated for such retirement in the region will be over $2.3 billion, not to mention the loss of production. The contracts through 1988 covered 78 percent of the acreage sought for retirement in North Dakota and 60 percent of that in South Dakota, but only 28 percent of that in Montana. Yields on the land so far enrolled were notably lower than the state average. Presumably, the rental cost will need to be increased considerably to attract the remaining acreage.[32]

This is a high price to pay to reverse a development that has contributed so largely to the nation's capacity for alleviating hunger at home and starvation abroad, not to mention its diplomatic leverage and its balance of payments in international trade. That the policy was authorized in 1985 and renewed under legislation in 1990 represents less the failure of dry-land technological adjustments than the defects of national and international marketing arrangements. In the ensuing pages the record of dry farming during the readjustment years embraces periods of hope, protracted hardship, long-run success, and yet widespread repudiation. The achievements of this enterprise have been denigrated by lingering misconceptions from the past, generalized regionalization of local conditions, and politically dictated agricultural programming. As reports of the United Nations today warn of world population growth vastly larger than projections, posing unprecedented demands for greater resources, the history of this agricultural effort must remain open-ended.

PART ONE

The Twenties
Local Initiatives
for Revitalized Development

CHAPTER ONE

Unstable Recovery

The decade of the twenties brought an unstable recovery after the record-breaking drought of 1919. Not all areas in the northern Plains again simultaneously recorded severe lack of rainfall until 1929, but the condition was widespread in northern Montana during much of the period through 1922 and serious in southwestern North Dakota and northwestern South Dakota in 1921 and again in 1923. However, the Judith Basin district of central Montana, northwestern North Dakota, and southwestern South Dakota had favorable crop conditions from 1920 through 1923, and northeastern Montana had ample precipitation from 1922 through 1925. Drought recurred in north-central Montana, along the Canadian border, in 1924 and 1926 and in the Judith Basin area in 1924 and 1925, but the latter district had excellent years from 1926 through 1928. The western Dakotas fared well in 1924, and favorable moisture conditions continued in southwestern North Dakota and western South Dakota through 1925. While only central Montana had satisfactory rainfall in 1926, the situation improved generally in 1927, with so much precipitation that both spring planting and fall harvest were dangerously delayed. And despite dry weather in the early spring of 1928, the crop season proved nearly as favorable as the preceding one. Drought returned throughout the region for the next three years. It was a pattern of intermittent and scattered hardship that bridged troughs of prolonged disaster at both ends of the decade.

The deviations in rainfall scarcity had, however, lessened the distress and preserved an optimism that conditions would be better the next year. Carryover of moisture in some crop seasons made reasonably satisfactory yields possible despite alarming shortages of rainfall during the current growing season. During some years and in some sections winter wheat that had germinated under favorable precipitation of the previous autumn ripened before the onslaught of searing summer heat and drought. Other years when winter wheat failed to make a start, spring-sown grains supplanted it. If spring conditions were unfavorable, late-sown corn, oats, or barley might at least provide feed for livestock. Only 4 percent of the cropland acreage in eastern Montana failed to yield a harvest during the widespread dry weather of 1924, and only 11 percent during the general drought of 1929. The corresponding record for western North Dakota was 2 percent and 3 percent, respectively, for those years, and for western South Dakota, 7 percent and 4 percent.[1] In all areas some farmers reported successful enterprise, and everywhere others failed.

The production of wheat per acre had averaged only 4.79 bushels in the western North Dakota counties, only 3.80 bushels in eastern Montana, and 8.05 bushels in western South Dakota during 1919, the year of record drought, when many fields were never harvested but did afford some stock forage. In 1929, another dry year in all three areas, wheat yields averaged 9.74 bushels an acre in western North Dakota, 8.86 bushels in western South Dakota, and 10.74 bushels in eastern Montana. With such low yields operators were in deep trouble unless they had built up large reserves of cash and stock feed. In 1924, under more favorable conditions, in the Dakotas and into northeastern Montana, wheat averaged better than 14 bushels an acre in western North Dakota and eastern Montana and slightly over 11 bushels an acre in western South Dakota. At the prices prevalent during the decade, the latter returns approximated the margin between success and failure.[2]

Reports for localized districts and conditions showed great range, however, and individual farmers boasted of remarkable yields. Estimates ran at fifteen to fifty bushels an acre for winter wheat in central Montana in 1926, with spring wheat averaging about twenty bushels an acre in Fergus County and about fifteen bushels in the major producing areas generally. Carbon County that year led the state, with an average of twenty-seven bushels, and the south-central foothill counties all averaged well over twenty bushels an acre. But to the east, Prairie County then averaged only three bushels an acre; Dawson, five; and Richland, six. Hill, Liberty, and Toole counties, in the northern triangle district, also gave low yields.[3]

The western Dakotas, which had experienced hard times throughout much of the preceding decade, recovered so strongly during the early twenties that experiment station analysts reported the crop records from 1921 through 1925 to be as good as or better than those for the region east of the Missouri River. Although 1926 and 1929 were dry years throughout the area, yields in Hettinger and Adams counties in southwestern North Dakota and Perkins County in northwestern South Dakota exceeded those for the district adjacent, to the north, an area which generally ranked superior. There was greater variability in yield in this borderland between the Dakotas. Extending northeastward to encompass most of central North Dakota, the zone of high yield variability excepted the five northwestern counties of Divide, Williams, McKenzie, Burke, and Mountrail, where conditions ranked most consistently favorable.[4]

Nevertheless, wheat yields in these outstanding dry-land counties of western North Dakota were generally lower than those for the better districts in eastern Montana. In 1924 only McKenzie County, with an average of 16.6 bushels of wheat an acre, produced appreciably above 14.6 bushels among the northwestern North Dakota counties. Across the border in northeastern

Montana, the yields averaged 17.6 in Sheridan, 19.65 in Roosevelt, 18.63 in Richland, 17.77 in Daniels, and 15.20 in Valley, while in central Montana, Fergus County produced an average of 16.08 bushels, and to the south Big Horn County averaged 16.04 bushels. Yields for the wheat area of western South Dakota compared even less favorably. Perkins County averaged only 12.55 bushels; Corson, 11.42; Haakon, 9.36; and Jackson, 15.46. State averages for wheat yields in Montana from almost entirely dry-land cultivation exceeded 12 bushels an acre seven years from 1921 through 1930, while in North Dakota, including the rich Red River Valley, the yield was above that level only four years of the ten and in South Dakota only three years.[5]

Identification of such localized long-term differences and explanation for their occurrence were poorly researched. The common assumption was that newer areas of cultivation possessed a fertility of unexploited soil minerals that cropping would quickly diminish. It was a premise that underlay traditional agricultural science and the advocacy of diversification and crop rotation for the restoration and conservation of plant nutrients. Yet experimental plats in areas already cultivated in wheat for fifteen to twenty years had not so far evidenced consistently declining yields. The availability of soil moisture, not soil fertility, seemed the more significant consideration.[6]

Clearly, however, the explanation for localized yield differences was not dependent solely upon precipitation patterns. Only the Judith Basin district of eastern Montana and the central and southwestern counties of western South Dakota could claim long-term precipitation averaging 15 inches or more annually. Occurrence of less than that amount was 20 to 40 percent less frequent and average relative humidity in July was 5 to 10 percent greater in the western Dakotas than in eastern Montana. A researcher of the Montana Agricultural Experiment Station in 1927, noting superior yields in northeastern Montana in comparison with the struggling "triangle" district of the north-central High Line, found it depended on differences of little more than one inch in average precipitation during the growing season and a ten-day longer period of rainfall probability that resulted in greater frequency and intensity of precipitation during June and July. The higher relative humidity and slightly higher temperatures during that period that ripened crops a few days earlier were other factors more favorable to the eastward. The study failed to account for the divergence in yields between northeastern Montana and the adjacent district in North Dakota, but it evidenced a growing realization that very slight deviations in a variety of factors must be considered in predicting agricultural success under conditions of such unstable productivity.[7]

Studies that would coordinate the climatic influences with varying soils, notably in relation to their water-holding and drainage capacity, were evolving even more slowly. Soil science developed in this country amid profound disagreement. So-called pure scientists debated at length whether chemical or physical characteristics were the more important identifying qualities for

TABLE 1.1. Farms and Farm Production, Northern Plains, 1919–1987
(in thousands)

	Farms (number)	Land in Farms (acres)	Total Cropland (acres)	Harvested Cropland (acres)	Wheat Harvested (acres)	Wheat (bushels)	Flax Harvested (acres)
Eastern Montana							
1919	46	28,903	8,943	2,839	1,423	5,414	126
1924	37	26,178	10,871	5,366	2,881	40,852	244
1929	36	36,670	14,202	6,689	4,173	36,966	357
1939	30	38,256	13,206	4,697	2,840	35,816	66
1944	27	48,719	9,711	6,162	3,500	64,558	182
1949	25	49,295	11,469	6,382	4,370	43,263	57
1954	23	50,864	11,704	7,162	3,993	68,225	[a]
1959	20	52,320	13,005	6,898	3,498	64,728	16
1969	18	52,074	14,002	6,752	3,465	87,679	[a]
1974	17	51,876	13,508	7,019	4,497	103,561	3[b]
1982	16	50,152	14,482	8,088	4,930	147,975	[a]
1987	17	49,298	15,710	7,847	4,257	132,477	3[b]
Western North Dakota							
1919	34	16,388	8,908	6,767	3,184	15,245	243
1924	31	15,284	9,608	7,278	3,486	49,656	647
1929	33	18,174	11,390	8,608	4,689	45,693	627
1939	30	17,471	10,960	5,490	2,905	26,742	70
1944	27	19,544	10,013	8,206	4,411	73,906	357
1949	25	19,666	11,305	7,730	4,674	45,324	643
1954	23	20,172	11,366	7,512	3,514	29,592	1,148
1959	21	19,849	11,092	7,091	2,991	34,703	262
1969	18	21,319	12,199	6,314	3,232	89,988	262[c]
1974	17	20,803	12,075	5,918	4,087	83,450	160[c]
1982	15	19,711	11,578	7,001	3,890	104,659	93
1987	15	19,974	11,685	6,763	3,497	82,033	117
Western South Dakota							
1919	19	15,239	3,357	2,699	296	2,385	16
1924	21	13,182	6,784	3,024	279	3,077	104
1929	22	15,953	7,821	4,393	870	9,345	371
1939	17	19,185	7,043	2,030	414	3,116	1
1944	15	21,740	4,300	3,754	898	9,371	9
1949	13	23,403	4,979	3,980	1,191	9,434	47
1954	12	23,362	4,975	4,147	915	9,412	115
1959	11	23,349	4,716	3,142	718	8,658	23
1969	9	24,553	4,843	2,819	713	16,024	18[c]
1974	9	24,652	4,722	3,147	1,027	24,199	14[b]
1982	8	23,933	4,729	3,231	1,032	29,494	14[b]
1987	9	24,303	5,335	3,206	1,054	31,742	4[b]

Source: U.S. census of agriculture for related years.
[a]Not available.
[b]Some counties had too little production to count; record was aggregated.
[c]Commercial farms only. Where comparable data are available, the difference from total production is slight.

Corn (acres)	Sorghum (acres)	Barley Harvested (acres)	Oats Harvested (acres)	Rye Harvested (acres)	Alfalfa (acres)	Cattle (number)	Sheep (number)
78	3	17	139	71	249	896	1,426
394	3	72	470	33	399	966	1,583
132	1	203	172	63	497	938	2,612
141	9	139	258	41	317	701	2,309
156	4	494	399	11	466	1,309	2,300
126	2	368	172	4	402	1,235	971
155	[a]	1,026	186	[a]	616	1,929	1,362
108	[a]	1,037	187	18	651	1,752	1,427
55	1	1,325[c]	316[c]	7[c]	743[c]	2,151	836
83	2	958[c]	148[c]	4[c]	880[c]	2,416	529
65	[a]	1,345	144	[a]	974	2,082	474
61	[b]	1,646	112	24	817	1,859	502
98	3	180	530	889	30	568	106
396	1	285	838	473	30	564	128
283	3	804	543	468	80	516	288
346	58	369	433	255	12	456	284
393	39	829	830	41	31	816	277
365	17	277	478	46	84	699	115
464	4	741	683	133	387	956	214
558	5	1,040	434	89	477	832	272
176	10	404[c]	835[c]	104[c]	563[c]	969	197
139	4	336[c]	544[c]	22[c]	.876[c]	1,206	141
153	[a]	347	491	32	729	960	108
163	1	835	353	101	761	976	97
479	14	29	99	82	227	814	542
794	6	100	247	37	325	640	399
958	7	498	310	71	455	652	540
264	238	219	113	38	30	419	671
401	164	531	288	96	93	807	877
427	64	186	239	31	156	954	436
486	73	127	406	20	587	1,290	643
333	78	79	185	12	487	1,219	852
193	140	76[c]	274[c]	12[c]	701[c]	1,233	567
190	137	69[c]	170[c]	7[c]	850[c]	1,430	421[c]
173	123	48	245	5	878	1,317	362[c]
189	126	109	187	12	920	1,297	298

purposes of classification. Theoretical squabbling, expressed in charges of superficiality regarding opponents' investigations, confused legislators and their farm constituents, who sought an approach that primarily would afford guidance to agricultural practice. Milton Whitney, chief of the federal Bureau of Soils during its formative years, emphasized the structural properties of soils and, pragmatically, stressed investigation of the horizons in the top three feet as the level of crop root penetration. As other investigators urged more comprehensive surveys, governing authorities tended to question their cost. Such considerations had led to rejection in 1915 and 1917 of proposals for more than a general identification of the major soil types as described by the Bureau of Soils for western South Dakota in 1909 and to cessation in 1923 of county surveys initiated to elaborate similar reconnaissance work done by the Bureau of Soils in North Dakota in 1908. In the western district of the latter state only the report for Bottineau County had yet been prepared when the program was interrupted. Within the next decade, expanded studies on the soils of only two more western counties, McHenry and McKenzie, were added. A revived survey in South Dakota, begun in 1919, had been completed by the mid-twenties for a unit of only 250 square miles in the trans-Missouri country, in Lyman, Jones, and Stanley counties.[8]

A more extensive project was initiated for Montana in 1921, when the state agricultural experiment station in cooperation with the Bureau of Soils began mapping soils on a county basis along the High Line north of the Missouri. These were also designated as reconnaissance surveys, developed to complement the early work in the Dakotas. Soil classification was based on identification of the parent material, in this area primarily the result of glacial action, but the major series were subdivided according to the textural properties of the surface horizons. The depth and composition of subsurface strata were noted, but little reference was made to their agricultural implications.

The Montana studies were, nevertheless, developed as an announced effort to assist "the home-seeker in selecting a new location, . . . the investor in making loans, . . . the real estate man and land buyer in appraising land values, and . . . the farmer in determining adaptation and soil resources."[9] L. F. Gieseker, who prepared the publications, conceded that there was some question whether the approach reflected the crop-producing power of the soils, but he attempted to supplement the reports with recommendations on appropriate land utilization, observations on the crop and yield experience of local farmers, warnings where soil blowing had developed on light surfaces left without cover under the dry farmers' practice of summer fallowing, and estimates of the extent and degree of permanence of abandonment in problem areas. He also included data on temperature and precipitation, with particular reference to conditions during the growing season, the natural vegetation and its carrying capacity for livestock, availability of domestic water supplies, settlement history, transportation routes, marketing centers, unoccupied lands and their price range, and average farm indebtedness—in short, a wide range of detail designed to give "unbiased information" to prospective settlers.

Although the mapping provided only an impressionistic forecast of productive capacity, it was far superior to the best available prior to that time, the topographic surveys of the U.S. Geological Survey, the field notes by surveyors for the General Land Office, and taxing classifications of county assessors' records. By the end of the decade the Gieseker reports covered the northern tier of counties west to the foothills of the Rockies, and during the thirties they were to be continued southward and eastward through the dry-farming districts of the Judith Basin and the Musselshell country. More definitive studies, based on the historical records of the Agricultural Adjustment Administration and the farm surveys of the Soil Conservation Service, were still limited in availability half a century later. Even today, estimates of regional productive capacity rest largely on generalized summations from sampling of very small areas.[10]

Repeatedly during the twenties complaints were made that although weather conditions affected yields, the crucial concerns were marketing factors, high costs and low prices. The average price of wheat received by farmers nationally remained below $1.00 a bushel until 1924, when, because of a short crop in international trade, it increased to almost $1.25. It rose to nearly $1.44 the following year, held at around $1.20 during 1926 and 1927, but then again dropped below $1.00 in 1928. While it regained a few cents in reaction to the widespread drought of 1929, it collapsed at the close of the decade to a new low of $ .67. Meanwhile, transportation costs from the northern Plains to export markets so curtailed returns that most North Dakota farmers received nearly $ .10 a bushel less than the national average, as much as $ .29 less in 1921, and exceeded the national average only two years of the decade.[11]

For Montana farmers, on the other hand, and to a lesser extent for operators in western Dakota counties, low dockage on grain grown on relatively new farm land and the superior milling quality of dry-land production drew premiums over card prices which raised wheat returns above the national average five years of the ten. No. 1 Dark Northern wheat grown in the Scobey area of northeastern Montana, the northern terminus for a branch of the Great Northern Railroad, yielded 18 to 30 bushels an acre in 1921 and brought a $ .20 premium over the card price, raising the local return to $1.32 a bushel. That year almost 75 percent of Montana wheat graded no. 1, while only 14 percent of North Dakota wheat and 15.7 percent of that grown in South Dakota graded as high. The superior ranking of Montana wheat continued through most of the decade, dropping somewhat in 1927, when only 68 percent of the spring wheat graded no. 1, still well above that of South Dakota with 57 percent, North Dakota with 41 percent, and Minnesota with only 12 percent. During the banner year of 1928, over 87 percent of the Montana crop graded no. 1.[12]

Complaints mounted during the early twenties that the standard grading procedure failed adequately to compensate these dry-land farmers for the superiority of their wheat. It was customarily used by American millers for blending to raise the quality of lower grades; and since Canadian wheat, then the principal rival in quality production, bore heavy tariff charges in U.S. markets, wheat from the northern Plains commanded a strong domestic sale independent of world demand. Fear of mounting agrarian radicalism, spreading in the Nonpartisan movement from North Dakota into Montana during the election of 1922, led even conservative political and financial interests to support the call for change in the grading criteria.

An investigation by the U.S. Department of Agriculture during the harvest season of 1921 culminated in some revision of the regulations governing inspection at terminal centers and, in 1924, establishment of a new class, identified as "Hard Spring," as the highest official standard for quality in wheat. It required a test weight of at least sixty pounds to the bushel, with 85 percent of the kernels categorized as dark, hard, and vitreous. Dark Northern Spring, formerly the quality standard, now required only 75 to 85 percent of such kernels and a test weight of fifty-eight pounds to grade no. 1. Excessive moisture content, damaged kernels, mixed varieties, weeds, dirt, and other dockage lowered grades under all classes.[13]

Protein content, a major element determining the gluten quality for bread-making, was not, however, recognized as an official grading factor. Small farmers marketing through local elevators protested that middlemen, purchasing according to grade, sold the wheat at terminal markets on a protein basis. State agencies consequently moved to assemble data for better-informed marketing.

The North Dakota Agricultural Experiment Station initiated its first protein survey in 1922, with county extension agents gathering samples for test by the state cereal chemist. Montana established a grain inspection laboratory at Great Falls the following year, and the State Agricultural College began offering tests at Bozeman in 1925. Privately operated laboratories at Bainville, Montana, and Grand Forks, North Dakota, were also operating by mid-decade. The North Dakota Bankers Association emphasized the development at its annual meeting in 1925 and called for a wide education campaign to teach farmers the importance of such testing. Farmers were warned that those who sold to a local elevator on a "to arrive" basis would command no premium, but those who sold by sample could benefit according to the percentage of protein in the gluten. When only 13 percent of Montana farmers were found to have had their wheat tested in 1926, the governor called a conference to promote wider establishment of the necessary facilities. During 1928 and 1929 laboratories were operating at nine or ten marketing centers scattered over that state. In the latter year the *Wall Street Journal* contained an article about the high protein percentages of Montana grain and stated that this factor had become the principal determinant of wheat value.[14]

Compensation on the basis of protein content became an issue in South Dakota, too. Of twenty-one elevator managers interviewed during the winter of 1928–1929, only eight reported that they tried to reflect protein premiums in their payments to individual growers, and then primarily on lots of 1,000 bushels or more. Limited binning space, they argued, posed great difficulty in differentiating wheat at local markets on a protein basis separate from grade. A bulletin of the South Dakota Agricultural Experiment Station then advised farmers to hold back their high quality grain until they acquired information on both the protein level and the pressure of demand at terminal centers.[15]

Concerned that farmers were not being paid according to the protein content of grain for which millers were extending premiums to commission merchants, Representative Olger Burton Burtness of North Dakota and Senator Thomas Jones Walsh of Montana called for congressional action in 1928. Both men introduced bills seeking establishment of federal laboratories that would standardize the testing process as a basis for resolving disputes. A second Walsh measure would have incorporated protein content as a criterion in federal wheat grades. Neither measure was enacted then. It would be a half century later, in 1978, before the U.S. Department of Agriculture introduced an official test for protein content of hard wheat varieties and provided for recording the result as an entry on inspection certificates. While it has not yet been officially designated a grading factor, protein content remains a major consideration in wheat prices.[16]

The growing attention to protein measurement raised the returns to regional wheat producers. The five-year average for protein in Montana wheat was 13 percent in 1928. The annual average rose to 13.79 percent in 1929 and 14.23 percent in 1930. High Line counties averaged near the top of the spectrum. Western counties of North Dakota also produced high averages, 15.62 percent in Golden Valley in 1925, with samples ranging to 18.51 percent. Premiums on the 1924 crop amounted to ten to sixteen cents a bushel for 12 percent protein and twenty cents on 13 percent. The following year they ran somewhat lower, at six cents a bushel on wheat averaging 11 percent and twenty cents on that testing 15 percent. In 1926 they dropped still lower, to four cents at 13 percent in November, and almost disappeared the remainder of the season. But with a shortage of high protein grain in 1927 the premium rose as high as forty cents a bushel on 13 percent protein by midwinter. Through 1928 premiums ran on a sliding scale ranging from four cents a bushel at 12 percent protein to eighteen cents at 14 percent.[17]

That protein content was highest in the semiarid region, and frequently highest in years when meager precipitation caused low yields and shriveled grain of low test weight, seemed a particular compensation to dry-land cultivators. But market forces, not grading standards, dictated the amount of premiums. In 1929, when drought reduced yields in the northern Plains, recently expanded areas of cropping from western Nebraska and Kansas into eastern Colorado had not yet experienced declining precipitation, and supplies of high-protein wheat increased on the national market. Although protein percentages then ran high in Montana, "nearly three percent above last year's average," premiums amounted to only two cents for 13 percent protein, less than at any time since the factor had been measured. North Dakota farmers with protein measuring 14 to 19 percent in 1930 found themselves lucky to get three cents a bushel above card prices. At the highest return a bushel of wheat that August brought but seventy-five cents.[18]

Livestock operations did not cushion the farmers' reverses until late in the decade. Such production, the principal alternative to wheat cultivation on the semiarid lands of the northern Plains, had been decimated by the disastrous weather of 1919–1920, drought followed by a long, bitterly cold winter. Feed and forage crops and pasture and range had all been depleted. Prices had collapsed as animals were hurried to market, and livestock remaining in the area had died by the thousands. Between January 1920 and January 1921 the number of cattle declined in Montana by over 100,000 head and in South Dakota by 300,000. Herds were rebuilt over the next 5 years, but prices remained low. By 1926 another dry year again brought a sharp reduction in herds. Production remained low, despite improved prices, until 1928 and 1929. With the index of income from such operations set at 100 for the average of the years 1924–1926, the norm was exceeded only in 1925, 1928, and 1929 and fell below 80 four years of the decade. The index of crop income for the same base period stood at 102.93 and rose well above that level in 1924, 1925, 1927, and 1928, although it also fell below 80 four years of the decade.[19]

For farmers everywhere in the nation the twenties were a period of continuing depression. The parity ratio of prices received to prices paid reached as high as 92 (1910, 14=100) only in 1925 and ranged in the 80s most of the decade. Discussing the situation in 1924, Secretary of Agriculture Henry C. Wallace observed that in no year since the price decline of 1920 had income "sufficed to allow both a commercial return on capital and adequate rewards for the farmers' labor, risk, and management." Most farmers of the northern Plains had used cash reserves for seed and feed in 1919 and had been forced to resort to credit from then into 1922. Surveys showed that even with a good crop in 1922 about one-third of the farmers in northeastern Montana and about one-eighth of those in southeastern Montana and western South Dakota then had farm expenses in excess of receipts. A North Dakota study found that between 1922 and 1925 farmers in the western counties barely broke even, with decreased assets balancing decreased liabilities. With the norm (100) established as the average of 1924 through 1926, the index of Montana gross cash income sagged as low as 60.3 in 1921 and dipped even lower, to

55.9, in 1930. It was below 90 from 1920 through 1923, in 1926, and again from 1929 into the mid-thirties.[20]

Yet the parity ratio of operational income to expenditure for Montana farmers throughout the twenties ranged between 6 and 11 percentage points above the level nationally. The proportion of failure in cropland acreage was not greatly different between the semiarid and humid areas of the states of the northern Plains, as measured in 1919, 1924, or 1929. In 1922, after the initial speculative bubble had burst, 99 percent of the farmers surveyed in southeastern Montana, 94 percent of those in northeastern Montana, and 84 percent of those in western South Dakota had not only survived but had bettered their financial position through agricultural operations, as well as through appreciation of land values, since settlement. While the average labor return of sixty-two farmers surveyed in the "triangle" area of northern Montana amounted to a loss of $709 on their operations during the season of 1921, even then over one-third of them had received more than $500 after paying all expenses and allowing for depreciation on their buildings and machinery and 6 percent interest on their capital investment. The average net worth of farmers surveyed in northeastern Montana increased at the rate of 8 percent from agricultural activity during 1922. A study of farm income in western South Dakota through the remainder of the decade showed that the labor income, above farm expenses and allowing for 5 percent return on investment, exceeded $922 in 1924, 1925, 1927, and 1928, and ranged as high as $3,574 in the last of these years.[21]

At the close of the decade O. E. Baker, senior agricultural economist of the federal Bureau of Agricultural Economics, summarizing the recent agricultural development of the Great Plains, noted that four years of drought had been followed by three years of low farm prices and commented that it was the latter which had "broken the backs of so many settlers." After 1923, however, farmers with modern methods and mechanization had found that they could produce profitably with yields of only 10 to 12 bushels an acre on summer fallow, or as little as 6 to 8 bushels without the cost of fallow, he reported. Some operators with very large acreage had cleared as much as $50,000 during 1928, and most of the smaller farmers had then fared well. Land deemed "sub-marginal for crops a few years ago" had become "clearly super-marginal." "This Great Plains belt," Baker concluded, "is now the most dynamic agricultural region in North America."[22]

Notwithstanding the unstable record of the decade, new aspirations had been generated to transform grassland into grainland. Promotional activity had not been silenced by the early drought experience. Leaders of such effort argued with much justification that the conditions of 1919 had been the most severe on record, that a large proportion of the settlers had come as speculators, with insufficient capital and little experience in any kind of agriculture, and that every agricultural frontier had "passed through trials and tribulations in its infancy . . . and still these countries passed this period and safely arrived at a point where there was no further discussion as to whether or not they were really agricultural countries."[23]

State legislatures continued to authorize expenditures for exhibits of agricultural produce at county fairs and in state and national shows. That the intent of such displays was more to attract new settlers than to stimulate agricultural education was evident in the report of the South Dakota commissioner of immigration as he outlined his work for 1921, covering exhibits at county fairs in Iowa and Minnesota, state fairs in Iowa, Kansas, and Nebraska, and national stock and dairy shows. The North Dakota Department of Immigration announced in January 1920 that in an effort to attract 100,000 new farmers to the state it had contracted with an advertising firm to get its message before 10 million readers. Publicity departments of Montana and South Dakota issued new publicity brochures in the early twenties, and the Northern Pacific Railway reissued its bulletin, *Along the Scenic Highway through the Land of Fortune.* Homeseekers' rates were restored over the Great Northern Railway in 1921 and in the following year extended on all the major lines running through the region. Minneapolis financial leaders joined the promotional chorus, national journals picked up the theme, and audiences as far east as the New York Chautauqua heard lectures on the amazing opportunities "Out Where the West Begins."[24]

District and community development associations were also organized or revived. A South Dakota Development Association undertook to reorganize commercial clubs of communities in the west-river region during the summer of 1921 and found active support in a reactivated Kadoka Commercial Club. The following year business leaders of Minot and Williston joined efforts as the Northwestern North Dakota Development Association to raise funds for a publicity campaign to compete with Canada for settlers. The association promptly launched a "100,000 More" campaign, adopting the goal stressed earlier by the state immigration commissioner. At the same time a newly organized Montana Development Association announced plans for a huge advertising campaign by the northern transcontinental railway corporations in conjunction with local and state promotional organizations. As a step in this effort, the Northern Pacific Railway in December 1922 initiated a more lenient payment program for sale of its lands in the area, extending the loan period from ten to nineteen years. The Montana Development Association itself issued a bulletin featuring the agricultural possibilities of the state, and, perhaps more importantly, inaugurated a tilled-fallow contest to prove the success of the dry-farming technology. With less caution, the Billings commercial club urged that winter that the phrase "dry land" be dropped as applicable to Montana, on the basis that it was conducive to misinterpretation.[25]

The story of the gigantic farming venture of the Montana Farming Corporation under Thomas D. Campbell in Big Horn County was a widely featured aspect of this publicity. Begun as part of the war production effort, with backing by the U.S. government and the J. P. Morgan banking interests, Campbell's heavily mechanized operations in 1920 included 28,000 acres on the Crow Indian Reservation and 27,000 acres on the Ft. Peck Reservation in southern and northeastern Montana. Some 14,000 acres at that time were irrigated; but as Campbell concentrated his efforts in the Big Horn area after the war, he focused on dry-land wheat production. Despite disappointment in 1919, he enthusiastically announced the following spring that farmers who followed the advice of government agricultural educators and the old, experienced, successful farmers of each community "are as certain of a crop in Montana every second year as they are in the Red River Valley." He was "satisfied" that under such management dry farming over a decade would prove "a success, in spite of last year's drought."[26]

The Montana Farming Corporation was liquidated at a heavy loss to the backers after the crop season of 1921, but Campbell remained optimistic. He reorganized the enterprise as the Campbell Farming Corporation and continued to open additional acreage. By 1923 he had over 100,000 acres in cultivation, about half in crop and half as fallow. His venture was successful through the remainder of the decade; the reorganized corporation lost money for the first time in 1930. Success and Campbell's ebullient enthusiasm generated heavy publicity on the development and the area in which such a vast project could be conducted. By the end of the decade, accounts of his operations had appeared in *Everybody's Magazine, Current Opinion, Scientific American, Country Gentleman, American Magazine, World's Work, Magazine of Business, Fortune, Colliers, Outlook,* and *Independent,* and numerous newspapers of both local and extraregional circulation.[27]

Through the early twenties promotional leaders generally remained defensive in their efforts. Thomas D. Campbell's backers were not the only sponsors of regional development to question the usefulness of further investment. The South Dakota commissioner of immigration lamented in 1924 that his office had been unable to interest railway companies operating in that state to provide tours by agricultural exhibit cars. He had also reduced newspaper and fair advertisement because of lack of response, and the legislature had curtailed appropriations for that purpose during the next biennium. Within the region there remained livestock interests resentful that farmers had plowed up lands on which cattle and sheep had grazed freely for decades. Outside the Plains, farmers to the east had long deplored the opening of new and cheaper lands to cultivation, a complaint that grew louder as surpluses mounted. Reviewing the losses on wheat production forecast in July 1923, even the Montana commissioner of agriculture, labor, and industry, who directed the state's promotional activity, commented that the economic problem should be solved before more land was reclaimed or settlers drawn into the area.[28]

Efforts to enact legislation providing for licensing of realtors had proved fruitless in the northern Plains during the early settlement years, but such a law was approved in Montana in 1921. The Montana commissioner of agriculture was at that time also empowered to examine proposed colonization projects. The Gieseker soil surveys, the first of which was published in 1923, were undertaken, as previously noted, with specific intent to gather and publish reliable data for prospective development. M. L. Wilson's extensive report on conditions in the northern Montana "triangle" area, which was one of the major centers of disastrous drought, appeared the same year. That November, Commissioner Chester C. Davis called a Land Settlement Congress to meet the following spring to set forth a land development policy for the state.

Representatives of the American Farm Bureau, the National Farmers' Union, the Montana Stockgrowers' Association, the Montana Development Association, the transcontinental railway companies, the State Bankers' Association, the Real Estate Dealers' Association, the U.S. Reclamation Service, the State Agricultural College, and the boards of county commissioners met and recommended consolidation of the efforts of public and private agencies for a state-wide land settlement program to operate under a state advisory board. This board, in turn, called for appointment of county committees to investigate and report on all local settlement projects and to make reliable information available to prospective settlers. All work was to be channeled through the state Department of Agriculture, which in cooperation with the federal Crop Reporting Service published annually a review of statistical data covering the period for which records were available.[29]

At best, however, such regulatory efforts were limited. They commanded local attention more as a medium for promotion of advertisement than for its regulation. The functions of the Montana Land Settlement Board and of the commissioner of agriculture were limited to advice and information. The weakness was manifest in 1929, when three projects were continued in spite of disapproval by the commissioner.[30]

South Dakota's more limited conceptualization of a land settlement board proved similarly ineffective. Hoping to attract veterans of World War I as settlers for the trans-Missouri region, the state established a board in 1919 that selected and acquired 10 quarter sections in Todd County for development as a group settlement project. The colony was never formed. Instead, the land was offered for sale as part of a general veterans' credit program. The terms were exceedingly generous, allowing a first mortgage for up to 70 percent of the valuation of the property and a second mortgage for up to 90 percent, to be amortized over thirty years at 6 percent interest. In 1925 the settlement board was abolished and the loans transferred to the state Rural Credit Board for management. By 1930, 187 of the 347 farms approved for credit under the program had reverted to the state and 10 more were in process of foreclosure. A survey conducted for the U.S. Department of Agriculture in 1930 reported that at least thirty-two of the states were devoting public funds to attract settlers but not one was then regularly attempting to direct them to the areas best suited for agriculture or to help them get established.[31]

Although governing authorities might attempt to act cautiously, they frequently succumbed to over-optimistic visions. They were subject to enormous pressures for local development and the associated distortions in advertising. States were not only involved indirectly as they were called upon to serve the promotional interests of constituents, but they were also directly concerned in selling land holdings and in the revenue increase as areas were shifted from grazing to agricultural utilization. Lack of long-range data on localized conditions, yield experience, and rapidly changing agricultural technology further weakened regulatory capabilities.

Agricultural educators, called upon to provide guidance on such matters, were often identified with irresponsible developmental activities. Representatives of the state agricultural colleges sitting on regulatory boards were heavily outnumbered by a range of banking, mortgage company, real estate, and other commercial spokesmen who also comprised such agencies. At meetings for instruction of farmers, agricultural extension workers customarily shared the platform with development agents for the western railroads. The relationship with such interests was often even closer. Dr. John H. Worst, former director of the North Dakota Agricultural College, had become the state commissioner of immigration. Carl Peterson, extension agent of Fergus County, Montana, for thirteen years, resigned in 1926 to assume charge of the land operations of a realty company. His successor, Dan Noble, resigned three years later to become an agricultural adviser for the Milwaukee Railway. At the close of the crop season of 1926, Alfred Atkinson, long a proponent of dry farming at Montana State Agricultural College and then president of the college, wrote euphorically of Montana's "unique record of quality production" in a letter published by the *National Bank Review*, an organ of the Northwestern National Bank of Minneapolis. Citing the high protein and grading percentages of Montana wheat, he added that her potatoes commanded a premium over those of any other state, her apples brought the highest prices in the New York market, her registered seed was recognized as the nation's finest, and her peas and beans commanded premium prices. Conceding that distance from markets was an offsetting concern, he concluded: "Yet on the whole our agriculture is just now situated perhaps better economically than that of most other agricultural states of the Union."[32]

The crop season of 1926 had not, in fact, been a particularly good one in much of Montana; but those for the next two years were outstanding. The promotional effort was accordingly intensified. In the spring of 1926 the governor and commissioner of agriculture had called a Land Settlement Congress to work out a program. The Milwaukee, Northern Pacific, and Great Northern railway companies all published new booklets to advertise the state—indeed, the Milwaukee issued five and the Northern Pacific six in that single year. The Greater North Dakota Association, reorganized in 1925, also conducted a general advertising campaign through the succeeding year and long continued a focal force in such activity. State and county governments began to liquidate their holdings of overdue land contracts and tax claims at that time, by collection of the debts, by abrogating a portion of the obligation in return for payments that covered the remainder and met the current assessments, or by resale of the abandoned property. In the spring of 1927, the Bank of North Dakota initiated a highly publicized offering of lands that it had taken over during the crisis period, and investors, generally, renewed efforts to sell foreclosed property. Later in the year the Northwest Land Finance Corporation, capitalized at $25 million was organized at Minneapolis, with Louis W. Hill of the Great Northern Railway Company heading the directors, for the purpose of buying up such lands, maintaining a field sales and immigration service to attract settlers, and underwriting purchases of lands through other agencies of agricultural finance. Through the winter of 1928–1929, a Custer Battlefield Highway Association, spanning the district from northwestern South Dakota into eastern Montana, maintained a staff of some one hundred contact agents from Ohio westward into Wisconsin and Minnesota to disseminate data on land opportunities. Also that winter the Northwestern Agricultural Foundation was organized at Fargo, on call from the Greater North Dakota Association, to correlate developmental activities throughout the northern Plains and Minnesota.[33]

Recurrent drought in 1929 stimulated the effort. The Great Northern Railway Company sponsored a series of programs on the network of the National Broadcasting Company featuring the "up-an-coming Northwest," to supplement newspaper and magazine advertisement already under way in eastern publications. Through the spring of 1930, agents of the agriculture and development department of the Milwaukee line met with farmers and businessmen in western North Dakota to urge their support for such promotional effort, and the following September the company circulated nearly 100,000 copies of its publication describing the district covered by its routes. Meanwhile, the Montana governor and commissioner of agriculture called a Development Congress to meet in February 1930 after which once again a commission was appointed to advise on policy for land advertisement, settlement, and utilization. The congress also proposed organization of a new Montana association "to induce desirable immigration here and to keep those now here satisfied and contented." Plans called for division of the state into twelve district bodies, each to be comprised of at least 150 members who would be charged a membership fee of ten dollars to promote the work. The following winter the association was organized, and by January 1931 the structure for a "working" body committed to "inside development and outside advertising" was in place.[34]

How influential such efforts were in attracting new settlers into the region is difficult to measure. Public lands available for entry had been scarce in North Dakota from the beginning of the decade, and the dwindling supply of such lands in South Dakota was evident as entries fell below 100,000 acres annually after 1924. In eastern Montana, however, from 350,000 to 500,000 acres of public land were entered annually through the end of the decade. About 250,000 acres of state land, partially repossessed contracts, were sold in Montana in the four years ending in 1929. Other blocks of sod land were brought under plow as ranches in previously developed areas were subdivided for sale. The Northern Pacific Railway Company reported sale of some 257,000 acres in the single year 1928, an increase of 84 percent in acreage sold and, even more remarkably, an increase of 179 percent in sales receipts over the returns of the previous year. Much of this development came in the Redwater Valley district of eastern Montana, where a 62-mile branch of that railroad was opened through Dawson and McCone counties to the town of Circle. In 1926 a branch of the Great Northern Railway was extended some 30 miles from Scobey to Opheim and Glentana, tapping another vast district in western Daniels and Valley counties of northeastern Montana.[35]

In part, the increased land sales represented an effort by ranchers to acquire lands necessary to grow hay as range areas on public domain were occupied, but the increase in acreage of harvested wheat—some 45 percent in eastern Montana, 35 percent in western North Dakota, and 212 percent in western South Dakota between 1924 and 1929—indicated that the development was primarily agricultural. Very large increases in wheat acreage, amounting to more than 50,000 acres each, occurred during that period in northern and eastern Fergus County, north through Chouteau and Hill, and northeast through McCone, Valley, Dawson, Richland, Roosevelt, and Sheridan counties of Montana, in McHenry, Ward, Mountrail, Williams, McLean, Dunn, McKenzie, Stark, and Hettinger counties of North Dakota, and in Corson and Perkins counties of South Dakota. The trend was only slightly less marked in Phillips County of Montana, western Bottineau, Renville, Slope, and Adams counties of North Dakota, and Dewey, Meade, and Bennett counties of South Dakota.[36]

Like earlier agricultural settlement in the region, much of the activity was accompanied by speculative investment, inexperience in locating favorable farm sites, and heavy costs of establishing new communities. Again land prices began to rise: in Montana during 1927–1928 from $23.30 to $24.20 an acre for improved farm land, generally; from $16.80 to $17.90 an acre for cultivable dry land; and from $6.40 to $6.80 an acre for improved grazing land. Again the individual units of sale tended to be small, a maximum of 160 acres of improved land under sales by the Bank of North Dakota, an average of about 263 acres under the contracts of the Northern Pacific Railway. Even the lack of farm background, so frequently cited as a cause of the collapse of the settlement effort a decade earlier, was repeated. An immigration agent for the Milwaukee Railway Company, selecting small tracts around Harlowton, in Wheatland County, Montana, for promotion during the spring of 1928, commented that the immigration was now by factory men, miners, and industrial workers generally, seeking a place "with a better chance for winning a livelihood."[37]

The new settlement did not, however, constitute a great increase in the total farm population of the region. During the first half of the decade, the number of farms declined from 45,594 to 36,591 in eastern Montana; there had also been a loss of 2,337 farms in western North Dakota; and the increase had amounted to only 1,281 in western South Dakota. The number remained relatively stable after 1925, continuing to decline to 36,324 by 1930 in eastern Montana, but increasing from 31,296 to 32,968 in western North Dakota and from 20,511 to 22,484 in western South Dakota. The last of these areas, with a 17 percent cumulative increase during the decade, provided the most striking development of the period. Elsewhere, promotional groups had been struggling merely to replace the tremendous losses that had marked the disaster of the immediate postwar years.[38]

The expansion of farming acreage represented, instead, a major adjustment in the scale of individual enterprises during the decade.[39] This transition was undertaken on lands of increasing marginality for long-term yields, with need for far greater expenditure on capital investment than had characterized the earlier settlement period, and it occurred at a time of mounting crop surpluses and declining agricultural prices in the world market. The struggle to formulate a program of dry-land agriculture that could withstand recurrence of drought now assumed an additional dimension centered on the increasing hazards of a drastically changing economic environment. Although further approaches to this range of difficulties evolved in later years, the groundwork is apparent in the emerging adjustments of the 1920s.

CHAPTER TWO

Emerging Adjustments

Agriculture in the northern Plains continued to be centered upon dry-land wheat cultivation during the twenties. Prolonged drought through the early years of the decade brought revived interest in a variety of long-standing proposals for alternative adjustments—irrigation, diversified cropping, and increased reliance upon livestock rather than cash-grain production—but such measures failed to offset the trend toward expansion of specialized wheat cropping by dry farming. Instead, efforts were undertaken to improve the technology through use of better-adapted crop varieties, refinement of methodology, greater attention to farm management that would make the most of cheap land and mechanized operations, and more involvement of farmers themselves to help achieve these goals. Contrary to the premises of the early settlement effort, these adjustments rested upon widespread recognition of the hazards of climatic conditions and marketing constraints, yet laid a groundwork for continued development, not retreat.

Irrigation was featured only intermittently, as a response to periods of drought or as an accompaniment of efforts to support restructuring the debt on established projects. Pleas for construction of irrigation systems were notable in Montana from 1919 to 1921 and again in 1923 and 1926. Prices for lands with and without irrigation were usually quoted in advertising booklets, and a publication of the Montana Department of Agriculture, Labor, and Industry in 1921 even commented that homeseekers would "at present" find their "biggest opportunity" on irrigated lands or by a combination of livestock raising with farming. The legislature that year also adopted a joint memorial calling upon Congress to provide funds necessary for completion of long-suspended federal reclamation projects.[1]

The annual conference of Montana county agents in January 1920 had urged use of irrigation wherever feasible in conjunction with dry-land tillage, and during the winter of 1923, M. L. Wilson's advocacy of similar measures in reference to the "triangle" district was cited by state extension workers as generally applicable in dry farming. Blaine County represented a model of this effort to develop small reservoirs for salvage of flood waters, with the county agent helping to lay out some forty such systems supplementary to dry-land cultivation. But the Gieseker soil survey in 1923–1924 found that by then most of the works had already been abandoned. The Montana commissioner of agriculture as early as 1924 deplored "misdirected settlement" efforts to "develop new land through drainage or irrigation projects where the costs made such projects economically unsound."[2]

Some interest in expanding the use of irrigation was also reflected in northwestern North Dakota during the early twenties. Irrigation, suspended for four years under the Williston pumping project of the federal Reclamation Service, was resumed in 1919. For several years, about 27 percent of the irrigable acreage was watered; but in 1922 the percentage dropped to 21, and during the next two years, to about 15. During the winter of 1925–1926, the government sold the works to private interests, who agreed to give the irrigation district free use of the pumps, to cancel all prior indebtedness, and to provide power for irrigation at two cents a kilowatt hour, a rate of about one-third that previously charged by the government. Although some thirty farmers and landowners had expressed interest in resuming irrigation and the local commercial club worked diligently to enroll an acreage sufficient to meet the minimum requisite for operation, only one operator agreed to water a substantial area. The local editor reported that "most all others fell far short of the number which was possible, some refusing to use water on any of their land."[3]

Although Slope County settlers of western North Dakota also requested further consideration of development along the Heart River in 1922 and again in 1925, the Reclamation Service still found too little support in the area to warrant construction. Meanwhile, outright hostility to such development, expressed in part by the state's leading newspaper, blocked expansion of such works in western South Dakota.[4]

Throughout the region, the federal reclamation projects, with heavy construction, operating, and maintenance charges, were under-populated, poorly developed, and financially insolvent. By 1924, in Montana alone, the federal government had expended $15,634,000 for construction costs and an additional $2,891,000 for operation and maintenance and had received reimbursement of only $1,650,000. National delinquency rates on such contracts were so great that in 1924 Congress enacted legislation to permit restructuring the debt. Construction charges, originally prorated at an annual payment per irrigable acre over ten years, under the new legislation were to be based on the productive power of the land, at a rate of 5 percent of the average gross annual acre income over the ten preceding years or, if in operation less than that period, on the basis of earnings for all the years of record. Two years later, after extensive surveys, the law was further revised to exclude from the projects lands found to be unproductive, nonirrigable, or incorporated in railway rights-of-way and various federal reservations. The secretary of the interior was to require execution of contracts by the water users' associations or irrigation districts calling for repayment of the entire charges against all productive lands remaining in the projects within a period not to exceed forty years from the date of notification when water was available. The secretary was also authorized to suspend payment on operation and maintenance charges at his discretion for periods not exceeding five years upon payment of interest at 6 percent annually. The arrangements thus wrote off overdue repayments amounting to some $14 million and deferred repayment of another $12 million.[5]

Comprehensive studies of the economic development of individual projects at this time indicated that in most cases holdings were considerably in excess of the authorized 40- to 160-acre limits, requiring eventual relinquishment of the excess and an expansion of the population if anticipated operating and long-term costs were ever to be met. Alfalfa and small grain were the principal crops raised. Neither had proved profitable, in large part because yields were too low and freight charges too high on the bulky, low-value products to cover the added costs of irrigation. Cold winters prevented production of more profitable specialty fruits and vegetables. Sugar beet cultivation, desirable both for sugar and stock feed, was restricted because of a lack of local manufacturing facilities. Creamery development, which would provide a market for irrigated alfalfa, also lagged as slow town growth provided limited local markets. Moreover, because most settlers did not possess sufficient capital or credit, dairy herds were of low quality. Beef cattle, which grazed on adjacent range grasses and were fattened on irrigated hay, afforded the principal marketing outlet, but until late in the decade brought lower prices than dry-farmed wheat.

Whenever precipitation appeared at least marginally adequate for dryland cultivation, settlers were reluctant to pay for crop watering. In 1927 only 64 percent of the irrigable acreage on the Huntley, Montana, project was being irrigated, only 19.4 percent of that on the Milk River project, 23.4 percent of that on the Sun River project, 26.8 percent of that on the Lower Yellowstone project, extending from eastern Montana into western North Dakota, and 35.7 percent of that on the Belle Fourche project in western South Dakota. At that time Director Linfield of the Montana Agricultural Experiment Station estimated that with current irrigating costs the crop returns under irrigation would need to be three or four times those from the average dry farm. While he did not believe it was impossible to attain such yields, he emphasized that they required "the highest type of farm practice." Questioned about the morale of the settlers, project directors described it as "Poor,—practically all settlers desire to sell."[6]

By 1927 the federal Reclamation Service was strongly defensive of its program. Midwestern farmers were emerging as strong proponents of a more active governmental effort to cope with problems of agricultural surplus and were critical of public expenditure to expand areas of crop production. The repeated demands for legislation to refinance irrigation contracts were bringing heavy congressional attack. In his annual report to the president for 1927, Secretary of Agriculture William M. Jardine described the policy of giving settlers on federal reclamation projects extended periods to repay construction charges without interest as an "extensive subsidy to agricultural expansion" that worked "a hardship on other farmers." He questioned the justification for federal participation in land reclamation where agricultural expansion was the primary concern. When farming was sufficiently profitable, he argued, there would be plenty of land from cut-over and semiarid districts "not requiring extensive reclamation, much of it already within the boundaries of farms." During the next few months, the Association of Agriultural Colleges and Experiment Stations, the American Farm Bureau Federation, and officials of the federal Bureau of Agricultural Economics expressed the view that construction of works by the U.S. government to irrigate new land should be suspended until there was need for additional crop production.[7]

Proponents of the federal reclamation program protested that such criticism represented mere interdepartmental rivalry, but by November 1928 Elwood Mead, commissioner of the Reclamation Bureau, conceded that "some of the most influential civic organizations of the country" had reached the conclusion that new construction of irrigation works by the federal government should be halted until there was need for more crops. An editorial in the *Chicago Tribune* in March 1928 had described "the record of reclamation in the United States . . . [as] one of chronic failure and loss." A survey by the agricultural committee of the U.S. Chamber of Commerce had shown that its members generally called for at least a temporary halt to the program. Perhaps most damaging, a committee of the American Society of Civil Engineers, reporting to the annual meeting of that body in October 1928 after a two-year study, reached the same conclusion: "Agricultural conditions due to overproduction are such at present that it is undesirable for the Federal Government, except in the case of commitments already made, to bring new areas under cultivation."[8]

Meanwhile, the Reclamation Bureau struggled to attract new settlers onto its projects, particularly those in Montana and western South Dakota. Major railroads were called upon to advertise the available lands, and with the cooperation of the Great Northern and the Northern Pacific companies a booklet was specially prepared featuring the Lower Yellowstone project. The publicity was circulated through some thirty farm journals and newspapers and garnered 27 settlers for the Belle Fourche project and 24 for the Lower Yellowstone, but by autumn Mead confessed that the results had been "discouraging." The newcomers had merely taken over developed "farms with houses on them"; unimproved lands could be neither sold nor rented. A year later, the chief engineer of the Reclamation Service advised Mead that the Belle Fourche and Lower Yellowstone projects needed 800 settlers "and during the past year . . . secured three or four at too great a cost to the railroads and projects." The projects, he concluded, were "not yet ripe for an extensive settlement program and results . . . [would] continue to be disappointing until there is a change in agricultural conditions."[9]

Mead, himself, reported to the secretary of the interior in 1928 that a new approach to irrigation development was necessary. The plan of building works to irrigate unimproved, privately owned land, with payment of construction costs dependent solely on contract with an irrigation district, had not brought about the desired improvement. Farmers needed "more capital and more science and skill in farming" to meet the increased costs of irrigation. Until sufficient funding could be made available to make farms "habitable and productive," as well as to provide canals, further construction "*should cease*," he concluded.[10]

Local newspapers occasionally urged expansion of irrigation as drought returned in 1929 and 1930, but the impetus for such development remained transitory. Irrigation had not represented a significant adjustment in the regional agriculture. In eastern Montana only 10 percent of the cropland acreage was irrigable in 1929, and only 53 percent of that area was irrigated. In western North Dakota the comparable percentages were .2 of a percent and 38 percent, respectively, and in western South Dakota, 1.6 percent and 55 percent. In each state the irrigated acreage was markedly less at the end of the decade than at the beginning.[11]

Diversification as an adjustment had more widespread and more persistent advocacy. State immigration departments, developmental associations, railway and implement companies, bankers associations, and other commercial interests joined in promoting and featuring the production of corn, alfalfa, feed grains, dairy cattle, hogs, poultry, and even bees. The governor of North Dakota proclaimed "Diversified Farming Week" in the spring of 1922 and again in 1923. During the early twenties, J. G. Haney, successor to P. G. Holden in the educational work of the International Harvester Company, conducted farmers' meetings and wrote an extended newspaper feature series on the importance of corn and alfalfa as an accompaniment for "More Dairying for Montana." Toward the end of the decade, he added sweet clover as a featured introduction. Evan W. Hall and, later, Dan Noble, as agricultural development agents for the Milwaukee Railway Company, also promoted diversification. The Great Northern and Northern Pacific railway companies ran demonstration trains emphasizing the importance of more highly bred beef and dairy cattle, poultry raising, and use of pure seed. These companies also joined in the activities of the Agricultural Credit Corporation, a Minnesota-based, privately financed, nonprofit concern organized in 1924 to lend money to country banks for livestock loans. Besides encouraging farmers to make such investments on favorable terms, bankers' groups were particularly active in assisting boys' and girls' clubs in corn-, pig-, or calf-raising projects.[12]

Such efforts were intended to be educational and were part of a national campaign to counteract the postwar agricultural depression, but the promotional value of featuring crop diversification in regional publicity material was also important. Like the work of Dr. Thomas Shaw and P. G. Holden, the program pointed to developmental potential for the semiarid lands in accordance with traditional eastern conceptions of agricultural operations. Promotional groups tended to attribute the recovery of agriculture in the region at mid-decade to the progress of such educational efforts, with little recognition that weather conditions had also improved. In a publicity feature during the spring of 1928, the *Minneapolis Tribune* opened its account with the theme: "Diversification in agriculture has brought prosperity with a capital 'P' to Montana."[13]

State and county agricultural extension workers generally supported the diversification campaign. Agricultural scientists had long taught that crop rotation and use of livestock manure were necessary to maintain soil quality, and already soil blowing under dry-land summer tillage techniques was indicating deterioration of the tilth of the lighter soils. Many also presented the argument that variety in farm production offered security against single-crop failure or low prices. While a few local leaders, like F. B. Linfield and M. L. Wilson, saw little role for large-scale dairy, hog, or sheep production in dryland districts, they encouraged raising at least enough farm animals to provide self-sufficiency in domestic use. Believing, too, that corn held little promise as a grain crop in the northern sections, they valued it as a rough-forage crop. Linfield, in particular, had endorsed the dry-farming movement initially for the contribution it might afford in support of the regional beef-cattle industry.[14]

Several publications of the U.S. Department of Agriculture written during the early twenties reiterated the view and added that wheat in the cropping system should be considered a speculative venture, merely supplementary to the basic livestock economy. By that time, few local leaders, including Linfield, held so restricted a conception, but one of the notable features of the dry-land development throughout the settlement years had been the large number of cattle maintained in conjunction with wheat cultivation. Between 1900 and 1920 the number of cattle in the dry-land districts of the northern Plains had been increased by 59 percent, and in western North Dakota, by nearly 150 percent. "Combination farming" had, in fact, long constituted a regional form of diversification in the northern Plains.[15]

However, such operations had depended upon use of native range, and grassland had become scarce in settled areas. As early as 1923, M. L. Wilson noted that both in tax assessment and in the returns under competitive cropping, tillable land in the north-central "triangle" district of Montana was valued too high to be used for grazing. The Gieseker surveys indicated that the carrying capacity of native grass, at 20 to 40 acres a head of cattle for a growing season, was low, at best, across northern Montana. Even farm pasture required an average of between 9 and 18 acres a head in the area south and west of the Missouri River in North Dakota. Moreover, the drought years had shown that pastures as well as croplands were vulnerable to prolonged rainfall deficiency. Dry weather that deterred the growth of spring crops also forestalled the growth of grasses and dried water holes. The distress of suffering livestock had preceded the losses from diminished crop harvests in 1919, and rebuilding of foundation herds required greater time and capital than the planting of crops the next season. Regional proponents of diversification identified a great need for production of more tame grasses, forage, and feed crops.[16]

Corn acreage was increased greatly during the early twenties, more than quadrupled in western North Dakota and eastern Montana and expanded by 66 percent in western South Dakota, which had already undergone a marked shift toward such cropping during the preceding decade. But the trend toward corn production slackened notably in all three areas between 1924 and 1929 and actually declined during the latter period in western North Dakota and eastern Montana. At its peak level in 1924, corn was grown on only 4 percent of the total cropland in western North Dakota and on only 11 percent as much acreage as was devoted to wheat. In the northwestern wheat district of the state, corn occupied as little as 2 percent of the total crop acreage. In eastern Montana, the acreage in corn amounted to less than 4 percent of the crop acreage in 1924 and about 14 percent of that devoted to wheat.[17]

For winter feeding, livestock growers primarily depended on wild hay, but also used alfalfa and grain hays. The acreage of alfalfa was increased moderately in eastern Montana and western south Dakota, but only slightly in western North Dakota. It proved profitable in some dry-land districts as a seed crop, but for hay yields it was commercially noncompetitive without irrigation. Sweet clover had proved even less popular because it was a biennial with limited yield during the first year. Without irrigation it rarely provided more than one cutting, and seed production was often unsuccessful if the first growth was cut for hay. After advocating use of these legumes for fourteen years, J. G. Haney lamented in 1928 that he could find only one and one-quarter acres of either alfalfa or sweet clover in all of Williams County, North Dakota.[18]

Farmers in wheat-producing areas of the northern Plains were more likely to feed their livestock grain hays during the period when pasturage was not available. Because the labor requirements for feed grains were competitive with those needed for wheat, however, in most years price differentials were a seriously limiting factor against their production. When wheat prices fell sharply at the end of the decade, the federal farm board urged the feeding of even wheat, but such use for the cash crop was not usually practicable. Regional analysts also contended that wheat was less palatable than either barley or oats. While barley yielded more than both wheat and oats in most years and was second only to corn in feeding value, the most productive variety then available was White Smyrna, a bearded form which was undesirable for hogs and was short-strawed and difficult to harvest. In 1928 the Montana Experiment Station emphasized the need for improvement of all standard barleys and particularly for a hull-less variety of satisfactory yield.[19]

Early in the decade, the acreage in oats had exceeded that in barley throughout the region, but as the number of horses declined in the latter half of the period, barley superseded oats. Rye was unimportant in any district through this period, as also were sorghum, tame hays other than alfalfa, and other small grains cut for hay. In 1929 the acreage in the principal feed and forage crops (corn, barley, oats, and rye) in western North Dakota amounted to only 43 percent of that in wheat, and in eastern Montana, less than 14 percent. Only in western South Dakota was the agricultural program oriented more strongly toward the requirements of livestock than to cash-wheat production, with the acreage in wheat there amounting to only 48 percent of that in stock feeds.

By the end of the decade, however, the number of cattle had been increased in both western North Dakota and eastern Montana, even beyond the high levels of 1919. Under favorable prices prevailing through the latter half of the decade, the number of sheep was also growing in the foothill districts of eastern Montana, although elsewhere in the region sheep were not a major factor in the economy. While the number of swine rose during the early half of the decade, it declined sharply in western North Dakota and eastern Montana after 1924. Even in western South Dakota, where the greater availability of corn contributed to a larger production of swine, the number of such animals remained low, less than 15 percent of the state's total production. Despite some development of turkey raising in central and northern Montana during the latter half of the decade, the industry was limited by the brevity of the holiday marketing season and the need for close proximity to shipping facilities. Poultry enterprise was not commercially significant in the dry-land districts. Despite the promotional emphasis upon traditional diversification, the economy of the region remained centered in livestock and cash-grain production, with the former still largely dependent upon access to steadily dwindling native pasture areas.

The situation neared a crisis even before prolonged drought returned to the region at the end of the decade. Congressional enactment of legislation in 1928 permitted organization of an area of 108,804 acres between the Mizpah and Pumpkin creeks in southeastern Montana as a grazing district, closing the public lands to entry for other than mineral development and authorizing grazing leases for up to ten years. It was the first of similar arrangements during the early thirties covering 192,320 acres of public land in Rosebud and Custer counties and 175,000 acres in Garfield County of Montana. Such action, however, related to areas generally denominated as grazing land, which had been rejected or abandoned for homestead development and committed to range-livestock utilization. Provision for maintaining range areas in districts where a combination of ranching and farming was practicable was another matter. In his annual report for 1929, Director Linfield of the Montana Agricultural Experiment Station questioned whether hay and forage for livestock could be raised competitively on cultivable dry lands. Moreover, a survey of successful farm operations in southeastern Montana during the years 1926 and 1929 indicated that when weather conditions were unfavorable for the cash crop, wheat, the yields of feed crops also tended to be low. As he noted the scale of expanded livestock production, Linfield voiced concern that prices in that industry might shortly be as depressed as were those for wheat raising. "This must be given particular consideration before advising dry-land farmers on low-grade farming land to turn their activities toward livestock production," he warned.[20]

Flax was another cash grain. It was commonly seeded on newly broken sod land, but because of the development of wilt, had not retained a place in the cropping system during the early settlement years. Following World War I, that situation changed. Increased demand for linseed oil, extracted from flax seed, had raised prices; and tariff protection, amounting to thirty to forty cents a bushel under legislation of 1921 and 1922, and increased to as much as fifty-six cents a bushel in 1929, continued the marketing advantage through the twenties. Meanwhile, H. L. Bolley and his assistant, O. A. Heggeness, at the North Dakota Agricultural Experiment Station, developed the wilt-resistant varieties Buda and Bison, which were distributed for general cultivation during the decade. Newland, a variety developed by Charles H. Clark, of the Division of Cereal Crops and Diseases of the U.S. Department of Agriculture, was tested at the Judith Basin, Montana, branch station, and also distributed at this time, offering superior yield and drought resistance, although continued susceptibility to wilt. Under such stimuli, the acreage in flax was increased over two and a half times in the northern Plains between 1919 and 1929. At its peak, however, the acreage in flax was slight compared to that devoted to wheat, about 1.3 million acres compared to nearly 9.75 million acres, respectively, in 1929.[21]

The wheat yields in Montana set new records during 1922 through 1924, 1927, and 1928, those of the last two years being over 25,000,000 bushels greater, each, than the amount in any previous year. The yields in North Dakota were also outstanding in 1922, 1924, 1927, and 1928, and the portion of it produced in the western counties was steadily expanding, while that to the eastward was being curtailed. South Dakota recorded peak yields in 1922, 1927, and 1930. Secretary of Agriculture William M. Jardine, who was a strong proponent of diversification, conceded that in the semiarid region it had "not been found practicable to accomplish much through diversification of crops." Wheat prices after 1924 had "permitted or encouraged an expansion of wheat production in some regions," specifically, he added, "in the new lands of the

Great Plains region." As prices fell precipitately at the end of the decade and governmental authorities were urging drastic limitation of production, Director Linfield of the Montana Agricultural Experiment Station emphasized that there was "no pronounced surplus" of high protein wheat. "It is entirely possible," he wrote in 1930, "that a readjustment may take place in the United States in which the soft wheats will be largely fed to livestock and the other wheats, such as are raised in Montana, be used for milling purposes."[22]

The recognition accorded protein measurement in 1927 and 1928 had contributed to this optimistic viewpoint, but changes in dry-land farm management and technology developed during the decade had been equally important. The long period of plant exploration, testing, selection, and crossing of strains for adaptation to the semiarid region was beginning to yield significant results. The northern Plains produced mostly hard red spring wheat, even to the southern border of western South Dakota and the foothill districts of central Montana, where farmers tried persistently to raise wheat as a fall-sown grain. Winter-kill often occurred, but higher long-term yields and an earlier harvest that minimized the dangers of summer heat and drought made winter wheat desirable in restricted areas. Karmont, a variety of winter wheat developed by selection from a Kharkof strain by E. L. Adams at the Judith Basin, Montana, branch experiment station in 1911, was first distributed for commercial growing in 1920 and in a decade's testing outyielded Marquis, the leading spring wheat, by an average of 3.6 bushels an acre. Newturk, another winter wheat, selected at the Moro, Oregon, branch station in 1916, was placed in field tests in the Judith Basin in 1922 and distributed for general use in 1927. It proved even more promising. Since these new varieties appeared to be equal to standard varieties in milling and baking quality, agricultural scientists urged that some acreage be diverted to them as a safeguard in moisture utilization. They conceded, however, that they should not supplant the spring wheats, which normally survived better and brought higher prices.[23]

Improved spring wheats were also being developed. The durums, a class used primarily for semolina in pasta, were benefited by the introduction of Nodak and Mondak, selections from the old standard, Kubanka, made by R. W. Smith, a cerealist of the U.S. Department of Agriculture working at the Dickinson, North Dakota, branch experiment station. Higher yielding and more rust resistant than other durums, they won high acclaim in areas of such specialization, chiefly north-central North Dakota. Durums were not, however, as popular farther west, because under dry-land cultivation their yield was only a little better than that of standard spring varieties but usually brought considerably lower prices.[24]

In the northern Plains, Marquis, a standard spring wheat developed as a Canadian cross and tested in the United States since 1912, had attained preeminence by 1919 and continued to be dominant through the next five years.

But by 1924, Kota, a North Dakota selection from a Russian variety, was found to be more resistant to rust than Marquis and on that basis achieved temporary popularity. While Kota had produced higher yields than Marquis under test during the dry years of the early twenties, its long-run returns were less favorable, and it also gave way to more promising innovations.[25]

Ceres was the product of a Marquis-Kota cross developed by L. R. Waldron of the Dickinson branch station and distributed in North Dakota in 1925. Resistant to both rust and drought and superior to Marquis and Kota in yield, it quickly became the leading variety in the Dakotas. The Adams County, North Dakota, extension agent reported that as late as 1927 only two farmers in that county had produced Ceres, but by 1930 more than one-half of the crop was of that variety. Meanwhile, Supreme, a Canadian selection from a cross of Preston and Red Bobs, an Australian strain, won strong favor in central Montana, where it ripened somewhat earlier than Marquis, gave better yield, and was judged equally satisfactory for milling. Supreme was also strongly resistant to smut, a serious problem for Montana wheat producers at mid-decade. It won favor so quickly that, although almost none of the new variety had been available for general planting in 1926, a Fergus County planning conference estimated that one-half of the local spring wheat crop would be of Supreme in 1928. Under the moister conditions of that season, however, Supreme's susceptibility to rust became evident, and yields declined. By the following spring, enthusiasm for Ceres as an alternative swept eastern Montana, as it already prevailed in the Dakotas. Newspapers featured the Dickinson introduction, and settlers so eagerly awaited shipments of the scarce new seed that it was dispensed by ration through the county agents. Ceres was to remain a leading variety throughout the region over the next decade.[26]

Maturing experimentation in dry-farm methodology was an equally significant development. As Hardy Webster Campbell had originally formulated that methodology, he had emphasized the need for plowing deeply and packing the subsoil to create a reservoir for moisture accumulation, to be maintained by frequent and intensive cultivation of a surface mulch. At a time when broadcast seeding was still prevalent, he had recommended seeding in rows, so that young grain could be cultivated to preserve the protective mulch. He adopted the idea of alternating crop and fallow several years later, and related it to his earlier view by terming it "summer tillage," emphasizing that fallow land should be cultivated throughout the summer to maintain the mulch. The practice of frequently leaving fields lying fallow had a long tradition in agriculture of the region; the attention to intensive surface cultivation was the novelty of Campbell's system.[27]

As easterners occupied the semiarid lands, few had been willing to forego regular cropping on their newly broken holdings. Still fewer had been prepared to follow Campbell's intensive tillage requirements. Many agricultural scientists, particularly those of the federal Department of Agriculture, questioned whether either phase of the program assured success, but their efforts to test the system had run little more than a decade when the severity of the drought culminating in 1919 required immediate conclusions. If, as seemed indicated, grain planted on fallow land did yield better than continuous cropping, how was this to be explained? Campbell had contended that the principal function of the fallow was to provide a moisture reservoir by capturing and retaining precipitation and by holding underground water attracted to the surface through hygroscopic action under the pressure of subsurface packing. His critics doubted whether the latter view was tenable, in view of the depth to ground-water strata in the region. If moisture retention, alone, was the goal, would cropping with corn or potatoes, which required frequent cultivation, not prove as effective as the fallow? Comparative tests were under way.

While it could be argued with accuracy that the early settlers had failed to pursue recommended methods, researchers at branch experiment stations and demonstration farms were not able to provide cumulative test data on the value of those methods until the early twenties. Relatively little support emerged for deep plowing, packing, or cultivation more than was necessary to keep down weeds. A circular of the Montana Agricultural Experiment Station published in 1920 still warned that land broken in the spring should be packed and disked immediately afterward to prevent the soil from drying out, but added that fallow and fall-plowed land should be left rough. Other bulletins in the twenties specifically rejected the "dust mulch," as a deterrent to penetration of water and contributor to soil blowing. "Subsoiling," or plowing to a depth of more than eight inches, had not been found sufficiently useful to warrant the greatly increased cost of production. Seeding in rows wide enough for cultivation had also proved to be too expensive. Harrowing winter wheat in the spring reduced weeds, but it damaged the young grain and reduced yields. Other studies reinforced Campbell's emphasis upon early seeding and plowing of the fallow, but his call for seedbed preparation as a fall operation was tempered by strong warnings against work when soil conditions were unfavorable.[28]

What remained of the Campbell system centered primarily in the practice of maintaining land in cultivated fallow during alternate years. In a speech before the Williston Chamber of Commerce in April 1921, Commissioner Greenfield, as development agent of the Great Northern Railway, stressed summer fallowing for successful dry-land agriculture. In reports at their annual meeting the following August, Montana bankers emphasized the same theme. That autumn the Montana Development Association, an organization of merchants and business leaders centered at Billings, pressed a major campaign to organize "summer fallow," or "summer tillage," clubs, in which farmers would pledge to commit part of their acreage to the procedure. In November the Montana Farm and Loan Bankers Association announced that it was distributing 30,000 pamphlets written by Campbell on proper tillage methods for the fallow and that conformity with the recommendations would be a requirement for those seeking seed loans in 1922. The Development Association brought Campbell, himself, then more than seventy years old, to deliver lectures in some thirty-five counties during that month, and Campbell provided a series of letters further explaining his program to members of the tillage clubs. The following spring, the Development Association hired "field men" to supplement the effort locally. A statewide contest was held, with prizes for the clubs most successful in collecting pledged acreage. By the autumn of 1922, a total of 10,000 farmers and businessmen were enrolled in the program, 4,400 of them as farmer members of tillage clubs who had pledged a total of more than 225,000 acres to the system. The following winter, the publicity was extended eastward into northwestern North Dakota. A bulletin published by the Northwestern National Bank of Minneapolis had recommended that the practice be extended throughout western North Dakota and parts of western South Dakota.[29]

The promotional focus in this campaign was clearly stated. Proponents emphasized that the publicity associated with the movement would "help to offset false impressions left in the past by half-truths." Montana had lost many residents during the drought period, one local editor noted, and the farms must be peopled before the state could prosper. The president of the Montana Development Association announced that the program had been undertaken, not as "a cure-all," "but for the single purpose of furnishing ample proof that with proper methods applied, Montana's soil would produce consistently." The organization planned to keep detailed records to be published in agricultural, banking, and financial journals throughout the nation. "Unquestionably," a Judith Basin banker explained, "these records will prove that farming, even in the driest districts of Montana, is profitable if proper methods are used."[30]

Agricultural educators—experimental and extension personnel, alike—supported the campaign. Research tests consistently showed that wheat on cultivated fallow gave more yield than that on any other seedbed, although the cause was still uncertain. In most reports, the result was attributed to weed control, but tests at the Huntley, Montana, branch experiment station in 1921 indicated that the results were most impressive under drought conditions, that moisture retention under tilled fallow had been the critical factor. A Montana station *Bulletin* in 1929 cited both explanations and added a third, that nitrogen development was enhanced in fallow soil. Since nitrogen was a critical factor in formation of protein content, it appeared that dry-land technology itself contributed to the superiority of wheat produced in the region.[31]

While data on the use of fallow are lacking before this period, the expense of breaking sod, the high prices for wheat, and the appeals for increased food production as a patriotic obligation had undoubtedly fostered continuous cropping. Consequently, the census report of 1924 that over 1.5 million acres of cropland in eastern Montana were idle or in cultivated fallow suggests that the summer tillage campaign had considerable effect. Only about 500,000 acres were so categorized at that time in western North Dakota and less than 150,000 in western South Dakota. That the educational impact extended beyond the promotional excitement and the pressure of immediate drought experience was also evident as after the bumper wheat crop of 1927 a larger fallow acreage than ever before was reported in Montana, with increases in nearly all non-irrigated sections. By 1929 almost 2.5 million acres of cropland in fallow were reported in eastern Montana, nearly 1 million in western North Dakota, and slightly over 200,000 in western South Dakota.[32]

After 1923, however, the promotional campaign in support of the practice had virtually ceased. Many of those who had pressed it most ardently became engrossed in the publicity directed to stimulate diversification. Somewhat defensively, the president of the Montana Development Association explained that the effort to restore confidence in their ability to produce had been a necessary preliminary to "diversification in any form. . . ." Arguing that the procedure caused soil drifting and was conducive to a one-crop system that robbed soil of productive capacity, J. G. Haney had rejected the practice of fallow even at the height of the campaign.[33]

Agricultural educators also recognized early that there were problems associated with the practice. Clyde McKee of the Montana Experiment Station warned in 1922 that the effectiveness of the system varied with conditions, that where the soil was shallow and had a rock layer near the surface, the practice of fallowing would do no more good than any other system. Gieseker advised against leaving the lighter soils fallow, and M. L. Wilson called for maintenance of a clod mulch through the summer until mid-August, when a surface crust was to be permitted to develop. A. Osenbrug, superintendent of the Judith Basin branch station, even suggested that winter wheat might need to be eliminated from the cropping system because of the seriousness of soil drifting during the long period when cultivated land was left uncovered.[34]

From the early twenties, a variety of techniques was introduced to cope with the difficulty—listing, furrow drilling, and seeding into grain stubble. As early as 1923, Wilson suggested "strip farming," alternating crop and fallow in strips; "strip-fallow cropping," or planting two to four rows of corn or sunflowers across the fallow in strips perpendicular to the wind; and scattering straw on the higher knolls. A study of Montana wheat farms at the end of the decade reported that "strip farming" was rare, but that a few operators had been using the procedure for three or four years, based upon its extensive use north of the border, in Canada.[35]

About 1922 the "duckfoot" cultivator was introduced. By a series of flat, shoe-like blades attached to rods suspended from a transverse bar the implement cut under the stubble, stirring the subsoil while leaving the surface rough. Serving to control weeds and protect the surface from blowing, it proved particularly effective in maintenance of fallow. By 1926 it was used almost exclusively in lieu of the older disk and spike-tooth harrows for cultivation. In the late twenties, such cultivation was being proposed as a form of "plowless fallow," eliminating the cost of initial plowing of the fallow and reducing that for its overall maintenance by 40 to 65 percent.[36]

The cost of sacrificing a year's crop productivity was a growing concern as wheat prices declined. Experimentation with a variety of crop rotations showed few alternatives in preparation for wheat. The grain yielded poorly after use of "green manures," field peas, rye, or sod crops such as alfalfa, brome grass, or sweet clover that had been turned under rather than harvested. The moisture reserve had been so exhausted that even the fertilizing effects of the legumes did not offset the deficiency. Other small grains, such as barley and oats, had similar moisture requirements and caused a similar reduction in the yield of subsequent crops. They could not compete with wheat commercially or with corn as stock feed and were planted primarily as "catch crops" that could mature if planted after a wheat crop failed to germinate or as feed grains where corn could not be grown. Corn as a cultivated crop served to conserve moisture for the succeeding crop better than other grains or grasses. While corn was not as effective as fallow in developing a moisture reservoir, and subsequent wheat yields were accordingly lower, cost analysis indicated that the total value of the corn-wheat sequence surpassed that of wheat on fallow, provided that sufficient livestock were maintained to utilize the crop.[37]

In the latter situation problems still existed. Apart from the growing scarcity of range and pasture for livestock, corn rarely matured for grain in northern latitudes and at high altitudes. When grown on dry land, it was frequently even poor silage. Its principal value in the northern Plains was as coarse forage and fodder. Moreover, the labor requirements of corn-growing interfered with other farm activities—for fallow preparation in the late spring, unless that practice were to be abandoned entirely, and for threshing of grain in the autumn. As a substitute for fallow, corn also imposed severe limitation on the acreage prepared for subsequent cropping, since a single operator could handle only about 40 acres of corn with the equipment in use during the early twenties.[38]

In the late twenties, experiments pointed to the development of large-scale, low-cost corn production. Multiple-row listers drawn by tractors had been introduced for plowing and seeding, and sled cutters, corn binders, and mechanical pickers had become available for harvesting. Agricultural analysts urged the use of trench silos, introduced from Canada, and extension agents in southwestern North Dakota organized a series of "trench-silo picnics" to support the idea during the summer of 1927. But corn harvesting equipment was not at that time in general use; 80 percent of the crop was still harvested as pasture. Judith Basin, Montana, farmers, discussing problems of dairying at this time, dismissed the use of silos as "not practical." Meanwhile, use of the duckfoot cultivator had enabled the individual operator to cultivate ten times the acreage as fallow that could be handled under older methods of corn cultivation.[39]

Use of new equipment and mechanization in wheat operations were the most notable adjustments in dry-land technology during the decade. Farmers quickly responded to the economic advantage of "plowless" fallow in preparation of the seedbed. The duckfoot cultivator, the most widely used implement for this practice, was found on seventy-four of eighty-five farms surveyed in Montana in 1930. A one-way disk, a combination of plow and harrow, was introduced later in the decade and offered advantages in areas where stubble or weeds were particularly heavy. The disk blades on the one-way disk were somewhat larger in diameter than those on the older tandem disk cultivator, and were placed to throw all the dirt in one direction in a stirring action that cut trash and stubble into the soil and left a ragged surface. The disks required less power than either a moldboard plow or a duckfoot cultivator and could be adjusted to cut to varying depths as the season advanced. Use of the one-way disk was most advantageous where continuous cropping was practiced, and North Dakota Experiment Station personnel criticized its use as tending to promote that procedure. In Montana, where it won some favor in winter-wheat districts, it was found on only twenty-eight of eighty-five farms surveyed in 1930.[40]

Rod weeders, introduced in the northern Plains still later, also left a cloddy surface. They were less effective than the older implements in controlling weeds, but cultivated as much ground more cheaply. As long as heavy growth had not been permitted to develop, they appeared to offer considerable promise by the close of the decade. The Montana Experiment Station estimated at that time that the cost of fallow preparation had been reduced from about $3.25 an acre at the beginning of the decade to about $1.53 an acre with the new implements.[41]

There had been no great change in seeding equipment—the furrow drill was not considered advantageous in spring-wheat districts—but reorganization of the operation through use of larger power units saved costs. Even before tractors were available in three- and four-plow sizes, use of larger teams of horses had become very general in Montana. Tractor development from the war years into the mid-twenties tended toward very light two- and three-plow machines, too light for heavy duty and, although relatively inexpensive, at $500 to $1,000 in purchase price, still too costly for mass sale during the financial depression of the period. Horses were used for heavy work; and where teams were kept, it was usually cheaper to add one or two than to buy a tractor of equivalent power. Agricultural scientists in the region at the time began to use twelve-, fourteen-, or sixteen-horse units and to expand the cultivated acreage accordingly. With a four-horse team, an operator could handle approximately 200 acres, half in fallow and half in crop, but with a twelve-horse team the cultivable area was increased to 640 acres, and with sixteen horses, to 800 to 900 acres.[42]

About 1924, medium-sized, three- and four-plow tractors appeared on the market, at an average purchase price of $1,300. Since the weather was more favorable and grain prices somewhat improved through the next five years, sales of the machines increased rapidly, in Montana from several hundred in 1924 to 1,160 in 1925, 1,749 in 1926, 3,607 in 1927, 5,251 in 1928, and 4,340 in 1919. Farmers welcomed the introduction of tractors because they permitted timely operations with relative independence from hired labor. The machines could work continuously at heavy loads, were not affected by hot weather, and required small storage space and no attention when not in use. While the cost of machinery rose from 10 to 37 percent of the total production expense, the labor requirement for producing an acre of wheat was reduced about two-thirds, to about two and one-half to three hours.[43]

Tractors were being improved rapidly. They were made more efficient in fuel use through reduction in weight relative to drawbar horsepower and better adjustment of the ratio between drawbar and belt horsepower. Anti-friction bearings, better protected from dust, brought greater efficiency, less susceptibility to breakdown, and economy in operation. Before the end of the decade, headlights were available so that night work might be done in rush seasons, and by the early thirties, introduction of rubber tires reduced the amount of soil packing, provided greatly increased fuel economy, and made the machines more maneuverable. The last factor became particularly important as farming operations were extended to scattered and more distant locations.[44]

Power take-off features brought improvements in complementary farming implements—better grain drills, power lift of cultivator gangs, operation of grain binders and combined harvester-threshers, and a variety of machines for routine farm chores. Introduction of the general-purpose tractor in 1923 permitted extension of tractor use to row crops with adaptation of specialized implements such as automatic plow guides, two- and four-row cultivators, corn pickers, mechanical huskers, silo fillers, and potato diggers. So rapid and sweeping were the changes that obsolescence rather than depreciation became the relevant cost consideration.

Elaboration of hitches had become an important development with the focus on large teams during the early twenties, and it remained a key factor in the efficient utilization of tractor power. Most farmers preferred machines with a double shift for reserve power and used machines with much greater capacity than was needed for individual operations. Studies by the Montana Experiment Station in the late twenties indicated that this practice, together with poor carburetor adjustment, was requiring at least one-fourth more fuel than necessary and costing close to $2.00 a day for each tractor. One solution to the problem was building economical loads by multiple hitching, linking from two to as many as five or six implements in one unit. Under tandem hitches, implements might be combined to cultivate and seed, perhaps even pack, in a single operation. On level land, a four-plow tractor could handle a four-bottom plow, two 7-foot tandem disk cultivators, two 8-foot duckfoot cultivators, two 10-foot drills, or a 16-foot combine, and provide wide coverage on a single passage of the equipment.[45]

Introduction of a prairie-type combined harvester-thresher during the mid-twenties was another important development for dry-land agriculture. Huge combines, drawn by thirty or more horses and powered by a large ground driving wheel, adapted from West Coast implements of the late nineteenth century, had been used on some of the larger farms of central Montana throughout the settlement period, but their size, cumbersome power arrangement, and heavy purchase cost had delayed widespread acceptance. By 1916 small 6- to 8-foot machines, to be drawn by six to eight horses and powered by auxiliary engines, had been introduced in the intermountain region, and during World War I 12-foot combines for use with larger teams or tractors began to appear in Montana. In 1926 a new model became available, with a cut of 8 to 10 feet, to be operated by power take-off from a tractor. It could be used economically on comparatively small farms, harvesting an average of 275 to 450 acres of wheat. Combine sales, which had amounted to a total of only 51 for the entire state of Montana in 1924, rose to 144 in 1925, 284 in 1926, 925 in 1927, and 1,685 in 1928. Custom combining, at a charge ranging from $2.47 to $3.32 an acre, was also frequently performed with such machines but more commonly with those of 12- to 20-foot width, still operated by auxiliary engine.[46]

Combines increased the speed of harvest, reducing the labor hours per acre approximately from 4.6 hours for the old binder and stationary thresher to 45 minutes with a combine, not including hauling time. Even more important to farm operators and their wives, who cooked for the harvest gangs, was the reduction of labor required. Most 8-foot combines were operated by one man, apart from grain haulers, and two men could handle up to a 16-foot machine. In a study conducted in 1926, over half the labor was found to be performed by the owner or his family, some by regularly employed hired hands, and the remainder largely by exchange with neighbors. The hiring of a large force of migrant workers was no longer necessary. While, as with tractors, the initial cost and the rate of obsolescence were high, the 10-foot combine proved more economical than the older binder when 60 or more acres were to be harvested and more economical than the old header at 100 acres. In view of prevailing custom rates, however, ownership of a combine was economical only when about double those acreages were to be harvested. After extended study of the costs of production in the area, the Montana State

College announced that power farming, especially use of the combine, had reduced the cost of harvest from $1.75 to $ .03 a bushel.[47]

Acceptance of combines had not, however, been without prolonged criticism. Threshing rigs still handled most of the crop in Fergus County, Montana, in 1927, when a farm without a tractor was considered "very rare" in that area. By the following year, three-fourths of the crop was reportedly combined. Only about 50 of the machines were operating in the northwestern North Dakota wheat district in 1927, and only 1,172 in the whole state. By 1929 the state total had risen to between 2,500 and 3,000, and in many localities as much as three-fourths of the wheat was then combined.[49]

The principal concern had centered, as with the use of headers, in the danger that grain would not be evenly ripened and that moisture levels in the binned wheat might exceed the grading standard of 14 percent. Since the grain had to be left standing seven to ten days longer before it could be cut, the danger of damage by weather, disease, or insects was increased. While proponents of combining argued that the reduction in handling lowered grain losses in the harvest process, others contended that the delay for even ripening also led to increased shattering of over-mature heads.

In the long run, improvements in varietal development and emphasis upon use of pure seed alleviated the problems. Marquis, Supreme, and Ceres were all wheats strongly resistant to shattering. But in the late twenties, extension of the practice of windrowing became widespread in the region as an intervening step between cutting and threshing to permit better drying of the grain. At first headers or binders were adapted for this swathing process, but in 1927 the International Harvester Company introduced a windrower, which cut the grain and let it slide onto the stubble, where it remained suspended so that air circulated around it. Several days to a week later, the grain was gathered and threshed. By 1929 all companies operating in the spring-wheat region were offering windrowers and pick-up attachments for their combines. The process added about fifty cents an acre to the cost and somewhat lengthened the time required for harvesting but the expense still remained considerably less than one-half and the time about one-quarter that necessary for the work by binder and threshing machine.[49]

Windrowing contributed to the spread of combining in the Dakotas, where the problem of uneven ripening was more marked than in the dryer areas to the west. Experience in Montana showed that in most years the practice was unnecessary, and by the end of the decade the combine, alone, was used there almost entirely. As depression developed during the thirties, the more expensive practice of windrowing was also largely abandoned in the western Dakotas.[50]

Before the end of the twenties, other complementary farm equipment was designed to facilitate the harvest process. Combine models were usually equipped with straw spreaders to scatter the straw, rather than leave it in narrow strips that interfered with later cultivation. Because combines could generally thresh a wider strip or heavier yield than could be cut in the semi-arid region, extenders were provided for the header mechanism, thus equalizing the working capacity. Screens were added to reduce loss of grain passing from the header to the thresher, and recleaners were introduced for both combines and separators to reduce the mixing of weeds in the grain. Tanks were also attached to combines, so that grain could be run directly into trucks without shoveling.[51]

The rapid increase in use of trucks, both for farm hauling and for marketing, ranked with the introduction of tractors and combines as a major feature in the agricultural development during the decade. In 1920 only 774 trucks were to be found in the whole state of North Dakota, 4,353 in South Dakota, and 1,225 in Montana. By 1924 the number had increased to 2,000, 9,000, and 3,000, respectively, and by 1930, to 16,990, 14,816, and 14,615. Trucks were then used primarily for local hauling, and they expanded both the capacity of the load and the distance that could be traversed practicably to markets. The Montana Agricultural Experiment Station estimated at the end of the decade that movement of a wheat crop from a 500-acre farm 25 miles from an elevator, with half the acreage lying fallow and a yield of twenty bushels to the cropped acre, would have required fifty two-day trips for a farmer hauling 200 bushels a trip with a six-horse team, or a total of 100 days. With a truck he could move 400 bushels on each of four trips a day and complete the task in 25 days.[52]

A South Dakota study providing detail at the county level showed that the rate of truck increase was greatest in the trans-Missouri country, amounting to 291 percent between 1923 and 1930, the period of most rapid expansion, while that for the state as a whole was only 138 percent. Counties newly developed—Tripp, Todd, Butte, and Perkins—had particularly large numbers of trucks, a circumstance explained by the limited railway development in the contributory districts. The largest proportion of the hauls, 22.7 percent, were from points distant 31 to 35 miles to the local terminal. Indeed, 7.5 percent of the farmers hauling to the town of Philip, in Haakon County, traveled more than 40 miles, one way; eight farmers hauled their crop more than 55 miles; and one, 82 miles. At the state-authorized rate of twenty-four cents a hundredweight for commercial hauling, a distance beyond 45 miles would have been economically impracticable; but only 22.6 percent of the trucking was done by commercial operators. Almost 60 percent was done by farmers individually, and neighbors frequently cooperated for the longer hauls.[53]

Meanwhile, the number of automobiles also increased, from 47,711 in 1920 to 78,798 in 1930 in North Dakota, from 58,202 to 81,923 during the same period in South Dakota, and from 22,072 to 38,166 in Montana. During 1927, 10,000 vehicles were sold in Montana. By the end of the decade, 90 percent of the farmers in that state had at least one car, and almost one-half had more than one. The changes brought about by this development revolutionized the social aspects of life in a region of sparse settlement and vast distances from schools and churches, medical facilities, and communal recreational activities. Through the improved access to town services and enlarged scope of communication, as well as the enhanced mobility in conduct of farming operations, the prevalence of motorized vehicles also significantly altered agricultural enterprise during the decade.[54]

Road improvement was a necessary concomitant of truck and automobile traffic. The availability of matching federal-state funds under the federal Aid Road Act of 1916 was supplemented by legislation in 1921 that authorized increases in the federal government's share by one-half the percentage that unappropriated public land bore to the total area of the states. Yet progress in road building was limited in sparsely settled dry-land districts by the inability of county commissioners to raise the matching funds. Montana was unable to use over $10,000 allotted as its share for the biennium 1923 to 1925, funds that were redistributed to North Dakota. Between 1919 and 1927, Montana had constructed only 1,153 miles of about 5,000 projected as a main trunk system. North Dakota by mid-decade had built some 1,630 miles under the arrangement and estimated that it had thereby reduced hauling costs by 50 percent. Less than one-half the work provided even gravel surfacing; the remainder was graded earth. A process of oiling the gravel before spreading it, to be packed by travel, was just coming under test in the area in the summer of 1929. The new process was approved by all three states for use the following spring.[55]

Mobility meant that "suitcase" and "sidewalk" farming came to the northern Plains. The wheat grower need no longer live on the land he farmed; he could reside 5, 10, or 15 miles away in town. Throughout the region, farmers were seeking to expand their operations to utilize more fully the productive power now available to them. The development in eastern Montana was evident as average farm size rose from 633.91 acres in 1920 to 715 acres in 1925 and to 1,009.51 acres in 1930. The number of farms of less than 500 acres about halved during the decade, while the number with over 1,000 acres nearly doubled, most of the increase coming in the period from 1925 to 1930. The trend was less marked in the western Dakotas during the first half of the decade, with average farm size even declining somewhat during that period in western South Dakota; but here, too, expansion of agricultural operations followed the introduction of mechanized equipment. Average farm size increased between 1925 and 1930 from 488.36 to 551.25 acres in western North Dakota and from 642.67 to 709.54 acres in western South Dakota. While the number of farms remained relatively stable in all three areas during the latter half of the decade, declining slightly in eastern Montana, total farm acreage was being increased by 40 percent, 19 percent, and 7 percent in eastern Montana and western North and South Dakota, respectively.[56]

In areas already largely developed prior to 1920, much of the subsequent farm expansion was achieved by consolidation of small, scattered holdings, the abandoned and foreclosed property of nonresidents. Analysts conducting a Montana survey of 100 "successful" farms in 1929 noted "with surprise" the large proportion of the enterprises that represented a combination of three, or sometimes as many as eight or ten, farms or units of land formerly operated separately. Research studies in North Dakota indicated that while the few comparatively large farms in western districts were becoming smaller, as ranches were broken up, small farms throughout the western two-thirds of the state had been consistently expanded since homestead settlement. Data showing that 62.3 percent of the operators and tenants on new holdings in western North Dakota between 1919 and 1926 came from the same community in which the acquired property was located and that another 9.4 percent were from other localities north and west of the Missouri River further reflected the enlargement of existing enterprises. Less than one in four of the land buyers during that period had moved into the region from outside the state.[57]

Initially, accelerated rates of tenancy marked the trend. By the latter half of the decade, mounting land sales became evident. Emphasizing that the sales were primarily to local tenants and owners who were adding to their holdings, the Greater North Dakota Association estimated that 500,000 to 600,000 acres had been sold in that state in 1927 and that the rate was doubling the following spring. The clerk of Fergus County, Montana, reported that some 75,330 acres in that one county alone changed hands in 1928.[58]

To organize farm programs so that the new resources of mechanized equipment were best integrated with available labor and land for greatest levels of productivity required managerial adjustments not encountered before. Regional agricultural leaders realized that the key to success was close accounting, careful planning, and maximizing the advantages of their climate and terrain. Scant data had been published on costs of production in the region before that time. Early estimates made by the U.S. Bureau of Statistics, based upon correspondents' reports, had been published in 1909 merely as state averages and related to areas developed before settlement of the major dry-farming regions. A North Dakota study based on accounts of 350 farms in that state for the years 1912–1914 had not been published until 1919. The Montana Agricultural Experiment Station, with brief reports on surveys of costs of wheat production in the Gallatin Valley and the Judith Basin, exceptionally favorable areas, had provided in 1914 and 1918 the earliest estimates applicable to dry-farming practice. Interrupted through World War I, such investigations were revived by both the federal government and the states in the early twenties.[59]

Establishment of the Bureau of Agricultural Economics in the U.S. Department of Agriculture in 1922 led to detailed surveys of the progress of farmers and recommendations for organization of their operations in eastern Montana and western South Dakota in 1923 and 1924. This information was circulated, however, only as mimeographed reports. M. L. Wilson's lengthy study of dry farming in the north-central Montana "triangle" district, published by the Montana Extension Service in 1923, included recommendations for improving farm organization. Meanwhile, Rex Willard, author of the earlier North Dakota survey, continued his collection of farm accounts through 1920 to 1925 and began to analyze the basis for localized differences. Passage of the Purnell Act in 1925, which authorized grants to the state experiment stations, scaling to a maximum of $60,000 annually by 1930, for research in agricultural and home economics and rural sociology, made publication of Willard's work possible and opened the way for numerous similar reports on the operations of practicing farmers, with emphasis on developments in farm size, land utilization, agricultural methodology, and mechanization.

By the end of the decade the Montana Agricultural Experiment Station was strongly emphasizing research conclusions that pointed to the importance of expanding the scale of grain production. M. L. Wilson, head of the department of agricultural economics, noted that there were still few "large-scale capitalistic farms" in the state. While he conceded that the three- and four-plow tractor farms were more "flexible" than large-scale farms, he warned that "those farmers who . . . fooled with a few old cows and hens, thereby missing their chance to farm 800 acres with a power unit, . . . [were] rapidly losing out." Montana studies showed a difference of $ .72 cents an acre in the cost of wheat production between a three-plow 800-acre farm and a four-plow 1,100-acre farm, with both organized at maximum efficiency. The cost could be reduced by $1.60 an acre with a six-plow tractor on 1,800 acres.[60]

Lands that consistently yielded over 15 bushels could, it was believed, yield a fair margin of profit over the long run. At the same time, however, it was estimated that a cash grain farm would need to produce at least 2,000 to 3,000 bushels of wheat annually to provide a family income comparable to that of a livestock ranch with 100 to 150 head of cattle. Since approximately one-third of the farms under dry-land operation in Montana still ranged between 260 and 599 acres, most of them the 320-acre units of the Enlarged Homestead legislation, they needed expanded holdings if they were to maintain summer fallow. When the price of wheat fell to seventy-five cents a bushel in 1930, the analyst concluded that a comfortable standard of living would require 900 to 1,200 acres with yields ranging between 15 and 20 bushels an acre on summer fallow; 1,600 to 1,800 acres, if yields ran 12.5 to 15 bushels to the acre; 2,400 acres, if yields fell to 10 to 12.5 bushels.

Wilson's work underlay the focus upon such findings in this period, and his role, both as a reflection of changing perceptions and an influence on the adjustment process, was to be significant regionally and nationally in the thirties. Born in the corn belt of Iowa, he had homesteaded as a wheat farmer near

Fallon, Montana, in 1910. He had served as an assistant in agronomy at the Montana Agricultural Experiment Station for two years, and in 1914 had become the state's first county extension agent, in Dawson County. He had directed the state's nascent extension service for nearly a decade while pursuing graduate studies in agricultural economics at the University of Wisconsin, where he was awarded the M.S. degree in 1920. Through Henry C. Taylor, one of his Wisconsin mentors who, as chief of the newly formed Federal Bureau of Agricultural Economics, visited Montana in the fall of 1923, Wilson and his friend Chester C. Davis, then the Montana commissioner of agriculture and publicity, were drawn to Washington the following year. Davis was to remain as a lobbyist for George N. Peek's proposals for agricultural equalization and as head of the New Deal Agricultural Adjustment Administration. Wilson worked for a year and a half in the Office of Farm Management of the Bureau of Agricultural Economics, but returned to Montana in 1926 as head of the department of agricultural economics at the State Agricultural College.[61]

Wilson's Washington contacts, notably Dr. W. J. Spillman, a colleague in the Office of Farm Management, had greatly influenced him. Spillman had come to reject diversification as a solution for depression in wheat farming and had recognized the need for stronger ties with farmers as a basis for more effective communication toward agricultural reform. Both approaches were to characterize Wilson's outlook. On one point, however, he did not then share the views of his Washington colleagues: He still looked to increased efficiency, rather than production control, as a solution to farmers' problems.

Shortly before going to Washington, Wilson had assumed direction of Fairway Farms, Incorporated, a project financed by loans from the Rockefeller Foundation to buy up abandoned farms, equip them, and enable tenants to purchase them over time through operations under the managerial guidance of the Agricultural College. The project was conceived as a means of rehabilitating regional farmers who had lost their holdings during the early drought and was designed to demonstrate that carefully selected tenants operating on properly equipped farms, with mechanization and approved technology, could succeed in family enterprises on the semiarid plains. Eight farms were selected from the poorer dry-land districts of eastern Montana. One was partially irrigated; the remainder were dry farmed. The operations stressed allocation of large power units per worker, either with a twelve-horse team or a medium-sized tractor; low cost practices, such as duckfoot cultivation rather than plowing of summer fallow, and use of a header or small combine instead of stationary thresher; supplemental livestock, such as dual-purpose cows, sheep, or hogs, to supply family living and some alternative income; and maintenance of feed reserves sufficient for at least a year's supply. The farms were large, ranging from 640 to 2,640 acres. As described in 1926, two were operated under very diversified cropping programs, another had about 400 acres in wheat, and a fourth had approximately 1,000 acres in wheat, with the wheat acreage in each case balanced by a like amount in summer fallow or corn. The remaining farms were not organized at that time. Farmers leased the property under tenant-purchase contracts designed to amortize the costs, with initial charges scaled to cover taxes, insurance, interest on debt, operating expenses, family living, and payment for farm or livestock equipment before land purchase.[62]

Although conditions in neither 1925 nor 1926 produced outstanding crops in the region, Wilson viewed the accomplishments as promising. When he returned from Washington, he resumed active direction of the program, paying particular attention to its experimental aspects related to farm management. He saw advantages in large-scale operations, not so much because of differences in labor output as because large-scale operators tended to give greater attention to managerial considerations such as cropping methods, marketing arrangements, and accumulation of reserves. By 1929 he concluded that serious error had been made in deciding on farm size in the Fairway project; farms of three to four sections, 1,920 to 2,560 acres, had proved more economical than those of one to two sections. "If we had the whole settlement of Montana to do over and were faced with the question of recommendation as to whether we should have large-scale or small-scale tractor farms, we think we should be inclined to go as far with large-scale tractor farms as there was available managerial ability to operate them," he wrote.[63]

Wilson's Fairway Farms project was based on the experiences of individual operators in a variety of peculiarly difficult settings, very different from the carefully nurtured plats of the experimental substations. His earlier survey of farming in the north-central Montana triangle had been optimistic as he reviewed the experiences of so-called "successful" farmers in the area, and he had tried to draw conclusions from them pointing to improved farm practice. In 1927 this approach was adopted again as an educational device when the Montana Extension Service initiated a "fact-finding survey," channeling the program's annual "short courses" into a series of six district economic conferences, involving from 1,500 to 2,000 participants, to draw up programs for agricultural development in various localities. The first of these meetings was held at Bozeman in February, when over sixty operators representing the southeastern counties met in specialized groups to consider problems on livestock; grain, forage, and potato production; canning; bee-keeping; dairying; poultry raising; and land utilization.[64]

In Fergus County, as preparation for a similar gathering called to meet the first week in March, County Agent Dan Noble had held a series of seven committee meetings on the same topics, together with farm credit and home living. Committee members for the most part had been selected by chairmen of the local community clubs and were chosen "on their merits as having had some success with the particular commodity which they represented." The meetings had, however, been conducted with careful effort to encourage free discussion and expression of the views of the participants. Their recommendations were brought before the March assembly for consideration and led to advisory endorsement of such practices as use of summer fallow on one-third of the acreage committed to wheat, control of soil blowing "by recommended methods," attention to protein measurement, careful weighing of the relative advantages of combines or binders, increasing capacity of native pastures by planting of sweet clover, brome, or western rye grass, incorporation of livestock on wheat farms where feed was not being fully utilized, but reduction of the number of livestock where reserves of feed were not available. Maintaining reserves was emphasized, both as liquid cash or negotiable paper and as livestock with appropriate feed stocks. The conferees agreed that a yield of 15 bushels an acre was necessary to "break even" when wheat was priced at a dollar a bushel, but offered no recommendation on utilization of lands which failed to meet that standard.[65]

The variety of committees, with their emphasis upon diversification, and the skepticism of the economic benefits of mechanization were more indicative of local programs for farmer guidance than of the newer developments being emphasized at the Agricultural College. Moreover, the publicity generated by the activity was quickly associated with advertisement of the area. The proceedings of the conference, subsequently published as a fifty-page "Program for the Development of Agriculture in Fergus, Montana," were widely distributed as a promotional release by the Cook-Reynolds real estate company of Lewistown. Such journals as the *Country Gentleman, Nation's Business,* and the *Minneapolis Tribune* acclaimed it as "the Fergus County plan of self-study, of searching out tried farming practices that were working successfully and were adapted to local use. . . ." Nevertheless, the emphasis given to public expression and local planning held significance as the beginning of a type of assembly that was to be institutionalized in governmental programs a decade later.[66]

Such sessions were repeated in Fergus County through the end of the decade and extended into southwestern North Dakota in 1928, again as representative gatherings of the "leading" farmers of the region. In 1929 a farmers' "economic" conference, the first ever held in the area, met at Williston, in northwestern North Dakota, with twelve group meetings called to plan operations for the district. Similar meetings were held in ten counties of western North Dakota the following spring, with an average attendance of about 250 "farm residents."[67]

From January to March 1929, Wilson carried the message of low-cost production through mechanization to fifty more traditional farmers' gatherings in conjunction with the Great Northern Railway Company's operation of a "Low Cost Wheat Train" across northern and central Montana. Featuring fifteen carloads of the most advanced machinery for wheat, corn, and hay operations, Wilson noted that in the past five years farming had changed from diversified to commercial production, "from farming to make a living to farming with the idea of making money." He emphasized that such agriculture required heavy capital outlays and good managerial ability, but he contended that it had elevated the standard of living. He warned, however, that there was always the possibility of another lean year and that farms must be run like a business, with accumulation of reserves. Some 12,000 wheat farmers, as many as 700 at the little northeastern Montana town of Plentywood despite below zero weather, visited the train and heard the message.[68]

From April to September 1929, Wilson toured Russia on invitation from AMTORG, the Soviet trade agency, to provide advice on large-scale wheat farming. He returned to the United States with renewed conviction that American wheat farmers must produce at lower cost to stay competitive in the world market. As the Montana State College planned its annual farmers' week program for February 1930, Wilson proposed to chair a session on the topic, "Can wheat production be curtailed or production re-adjusted?" His prefatory remarks would pose two alternatives: a general "domestic allotment plan," in accordance with proposals recently published by Spillman and John D. Black, or a program of "Regional farm adjustments based upon comparative costs and advantages of different regions in the United States." Wilson's preference, which grew out of the survey conducted in 1929 by the Montana Agricultural Experiment Station concerning the operations of a hundred "successful" tractor wheat farmers and the accumulating data from the Fairway Farms project, pointed to the latter solution. Director Linfield's forecast of Montana's primacy in wheat production echoed the same viewpoint in the annual report of the station later that spring.[69]

Reacting to public criticism of impracticality in experimental findings, Wilson and his colleagues hoped that their research on the economics of production might be expanded to test the effectiveness of mechanized equipment under a variety of farming systems in "actual farm surroundings." For this purpose, a ninth farm was to be established under the Fairway Farms project, at Brockton, in Roosevelt County, a district considered representative of average Montana conditions and typical production requirements. But when low prices slowed the progress of the operators on Fairway Farms, the Rockefeller Foundation terminated support for that project altogether beyond February 1930. As the deadline approached, Wilson was trying to get a $15,000 grant to fund the proposed three-year research project in low-cost production.[70]

In December 1929 the Experiment Station leaders directed their appeal to the federal government. The approach was made through Elmer Starch, assistant professor of agricultural economics at Montana State College, who was a friend and former college classmate of Representative Victor Christgau of Minnesota. Attempting to whet the congressman's interest in research on commercial, as distinct from production, problems in farming, Starch lamented that extension leaders in the northwest had concentrated on the importance of diversification for too long. "I haven't seen a real honest to gosh paying diversified farm west of the Red River and surely not west of the Missouri," he wrote. That the proposed legislation was focused on the regionalized conceptualization of the Montana program was evident when on June 30, 1930, Christgau introduced a bill, "To aid farmers in making regional readjustments in agricultural production to assist in preventing undesirable surpluses." In his accompanying remarks Christgau alluded to the recent changes through mechanization that had made wheat production in Montana possible with one to three hours of labor per acre, while it required seven to nine hours per acre in the Red River Valley of eastern North Dakota and twelve to fourteen hours per acre in southern Minnesota. "That some regions are better adapted for the production of certain farm commodities than others has become more apparent during the last few years," he commented. "One of the purposes of the bill is to aid the farmers in certain regions to determine what commodities they can produce to their greatest advantage in view of actual and potential competition with other regions in the country and also in competition with producers of agricultural commodities in other countries."[71]

The Christgau proposal was, however, a two-sided coin. If specific districts were to be encouraged to exploit their regional advantages, other areas must accordingly change their production or succumb to the competition. The text of the bill did not refer to compulsory land retirement. It assumed that farmers would seek "to make the best economic use of the land which they operate" and that "programs of readjustments in the light of controlling undesirable surpluses . . . [would] have to be built up on a reorganization of farms and substitutions of lines of production in such manner that the net productivity of the farms . . . [was] not decreased thereby." The debate, nevertheless, quickly shifted to the broader theme of nationally planned land use. As drought returned to the northern Plains in 1929 and continued through much of the thirties, discussion carried beyond the old issues of public land distribution to concern for the public good in all land use. The Christgau bill never emerged from committee. Conditions so radically changed in the dryland districts that the obverse side of the proposal confronted its regional proponents with the prospect that farming in vast areas would be terminated.[72]

The twenties had begun in a period of adversity for farmers and farm-related businessmen and ended upon the threshold of even greater hardship. The severity of the latter circumstances and their comparability to the earlier events have tended to obliterate the record of intervening progress, yet the decade largely defined the pattern of subsequent adjustment in the region. Improvement of crop varieties and modification of earlier tillage procedures had introduced better dry-farm technology. Adaptation of farm machinery to the requirements of large-scale operations and a developing appreciation of the importance of fine-tuned cost accounting had pointed up regional assets in commercial cropping. A strong start had been made in the accumulation of data on localized farming conditions and practices. Even the incorporation of resident operators in a community-based approach to planned agricultural development had been initiated.

Such operations still remained an experimental venture. The problem of recurrent drought was unresolved. Neither irrigation nor diversification had proved economically viable. The adjustments in dry-farm methodology remained to be tested over time. Promotional optimism was still too rampant, and too many inexperienced farmers expanded operations into fringe districts. Even for many established settlers, the period of prosperity had been too brief to build capital reserves, whether for livestock or grain production, after the drought and depression of the early twenties. Enthusiastically reporting the achievement of Montana's mechanized, large-scale operations, M. L. Wilson conceded that they depended heavily upon use of credit. "If we should have a bad crop we would begin to have some pain in the northwest," he had commented. That time came with the onslaught of renewed drought in 1929.[73]

PART TWO

The Thirties
Introduction of Federal Programs

CHAPTER THREE

A Decade of Disaster

The early settlement of the northern Plains had evolved from a background of questioning, even outright denial, that agriculture was possible in the region. Despite the achievements of the dry-farming movement, it had suffered major reverses from drought in the western Dakotas between 1910 and 1912 and more generally over the entire area from 1919 through the early twenties. Almost as serious as climatic difficulties during the latter period had been the decline of prices and increase of competition in marketing of wheat. For local supporters of agricultural enterprise, the combination of these circumstances had caused serious reflection concerning the future course of dry-land cultivation. By the latter half of the decade, however, improvements in the methodology of such operations and growing recognition of the regional advantages for production of superior quality wheat with economies of cheap land and mechanized operations had rekindled hope for continued expansion.

These adjustments had evolved primarily through local initiative. Little national planning had entered into the opening and agricultural settlement of the region, and little had been extended to consideration of the problems encountered. State agricultural leaders, with strong pressure and support by interested commercial bodies, had been called upon to provide guidance for trial-and-error development without long-run data on either climatic and soil conditions or cost-of-production and marketing trends. Such studies were barely beginning in the twenties, when farmers were already on the land.

The organization in 1922 of the Bureau of Agricultural Economics as a federal agency to assemble information for assessment of the nation's agricultural production and Secretary of Agriculture Henry C. Wallace's coinciding introduction of annual reports on the "Agricultural Outlook" marked a new assertion of federal leadership in such concerns. Significantly it was confined to an informational rather than a directive role. Nearly ten years later the Christgau proposal, as envisioned by M. L. Wilson and Elmer Starch, with its focus on differentiating regional specializations, still carried the assumption of individual choice among farm management programs according to rational assessment of competing alternatives. With the severity of conditions during the thirties, however, came a leverage for centralized planning as individual, community, and state farm relief programs were drawn into externally conceptualized roles through dependence upon federal funding.

Throughout the northern Plains precipitation was below the long-term average for at least nine years from 1929 through 1940. During half that period, the deficiency amounted to three or more inches. Where the normal amount ranged at only 13 to 15 inches annually, such a decline carried the supply of moisture below that needed to grow vegetation. Those conditions prevailed over much of the region consecutively in 1930 and 1931 and from 1934 through 1936. In central and north-central Montana, the situation improved during 1932 and 1933 and still permitted fair returns on winter wheat in 1934, but to the eastward across Montana and over the western Dakotas, the drought resumed as early as 1933. In western South Dakota the conditions were worst in 1931, 1933, and 1935, because much of the rainfall occurred outside the growing season. At Dickinson, in southwestern North Dakota, the precipitation was 73 percent of normal in 1933, dropped to 50 percent in 1934, improved slightly in 1935, but sank to 42 percent in 1936. After that, conditions eased in the Dakotas, but the drought during 1937 in northeastern Montana and northwestern North Dakota was described by a state meteorologist as the "most devasting" ever known in the region. In the Judith Basin district, the heartland of successful dry-land agriculture in Montana, moisture remained three to five inches below normal from 1934 through 1940, with only a minor break in 1938, and even then it remained below the long-term average.[1]

By 1933 dust storms described as the "most spectacular . . . in the history of the Upper Missouri country" swept the Dakotas eastward as far as the Twin Cities and Chicago, and during the spring of 1934 they were generated from the Rockies to the Red River Valley, centering in the Plentywood area of northeastern Montana. Skies were dark; travel and outdoor activity were halted; housewives lamented the futility of spring house-cleaning; and farmers, their young crops destroyed, looked hopelessly at layers of sand drifted high against buildings and fences. Most commonly a springtime phenomenon, dust storms became increasingly numerous throughout the year as the long drought continued. In 1937, Williston, North Dakota, reported a total of thirty-two storms by the end of August, five of which occurred during that harvest month. Analysts have found little evidence of serious wind erosion on cropland in the northern Plains, but structural damage to the soils was clearly a threatening problem.[2]

Extreme temperatures compounded the effects of drought. Spring was late in 1929, with freezing and frost extended through mid-May, so vegetation was about two weeks behind normal growth by June, when the drought set in. The summer months that followed provided the longest period of hot, dry weather recorded locally to that time. The heat set a new high in August, but the first week in September brought a killing frost, with temperatures falling to 24 degrees. Another hot summer in 1930 was marked by the hottest July known to that time, and the following year was described as warmer than any of the past four. Again in 1933 and 1936 temperature records were broken, while in 1934 "almost continuous hot winds," "withering winds," along with rainfall deficiency destroyed vegetation. Under sharp shifts of temperature, there were occasional violent storms with high winds, blowing dust, and damaging hail, but little precipitation.[3]

The concomitant invasion of grasshoppers and locusts during the decade was not generated by the drought—grasshoppers had first appeared on the Rosebud Indian Reservation, west of the Missouri River in South Dakota, at a time when crops looked promising in July 1931, and had spread in great numbers east of the river into Iowa and Nebraska by the end of the summer. But as discouraged farmers abandoned drought-ravaged fields, the insects found increased nesting areas northward and westward over the Plains. Nearly 50 percent of the crop acreage in southwestern North Dakota was infested by mid-April 1933, and nearly all counties of Montana reported serious incursions. The damage was particularly heavy in north-central Montana in 1934 and in southeastern Montana in 1935, reducing crop and pasture yields in 1935 when some recovery had been anticipated as July rains alleviated the drought. The following year, Wibaux and Fallon counties of eastern Montana reported grasshoppers "flying in by the millions . . . ," and large numbers had spread northward into Valley County on the Montana Highline. The "Mormon Cricket problem" had become so serious in eastern Montana in 1937 that nearly 1,500 relief workers were employed to dust the fields.

The insects remained a devouring horde over the path of their previous migration. The South Dakota Experiment Station reported that during the hot, dry summer months of 1936 grasshoppers "in such a torpid condition that it was possible to approach them without causing them to hop or fly" were to be found "clinging to objects of all kinds such as trees, bushes, weeds and other plants, fence posts, telephone posts, farm machinery and farm buildings," often "in such large numbers . . . that the weight . . . caused the plants to bend so that the tops . . . touched the earth beneath." Despite the availability of public funds for baiting or dusting and the use of work-relief labor to assist in the eradication effort, the insects were a serious problem in the region throughout the remainder of the decade.[4]

Unlike the situation during the twenties, when the production record showed some variability, the persistent drought of the thirties depleted moisture reserves so much that grain yields failed to reflect the slight gradations of improved precipitation. In Renville County, of northwestern North Dakota, where long-term production per seeded acre of wheat, from 1926 through 1948, averaged approximately 10 bushels to the acre, the yield did not rise above 8 bushels between 1930 and 1939, inclusive. In five years of the ten it amounted to less than 3.2 bushels. The situation was similar in adjacent Ward County, North Dakota, and in Ziebach County, of west-central South Dakota. It was little better in northeastern Montana, where returns were somewhat higher during the early thirties but fell to a disastrous level in 1936 and fell even lower in 1937. Although production in central and north-central Montana continued to be "relatively favorable" through mid-decade, there, too, it dropped to averages of but 4 and 6 bushels an acre in 1936 and 1937. The crop in this primary wheat district of the state did not return to the levels of the mid-twenties until 1938, when Fergus County yielded a soaring average of 19.8 bushels to an acre and other counties in the district produced at an average of 9 to 13 bushels. Montana, which had averaged production of only 35,217,000 bushels of wheat annually from 1928 through 1937, harvested some 72,349,000 bushels in 1938, 54,240,000 in 1939, and 60,811,000 in 1940.[5]

In spite of sharply declining wheat yields during the early thirties, prices for the crop dropped to new lows. Montana's annual production from 1924 through 1928, averaging 48,104,000 bushels, had been valued at an average of $61,881,000 annually. The crop for 1929, totaling 40,688,000 bushels, brought only $38,741,000; and that for 1930, 33,698,000 bushels, sold for only $16,332,000. The farm price per bushel on December 1, 1930, was $ .48, compared to $ .95 a year earlier and the five-year average of $1.05. By 1932 it had dropped to a low point of about $ .30. After that the yearly average rose, approximately doubling as more severe drought and the New Deal crop adjustment program reduced national supplies; but the price of wheat remained below $1.00 a bushel through the decade, except briefly in 1937. During 1938 and 1939 it again dropped to about $ .60.[6]

Prices for other small grains, corn, hay, and livestock closely mirrored those for wheat. The value of beef steers declined steadily from $13.42 per hundredweight at Chicago in 1929 to $5.42 in 1933. This market, too, reflected scarcities induced by drought, as beef prices rose to $10.26 in 1935 and $11.47 in 1937 after heavy selling of cattle during the periods of feed shortage, but thereafter the level of prices declined to about $9.50 in 1938 and 1939.[7]

Stock-raisers and grain-growers who were able to hold reserves for sale during the brief times of drought-induced increase in prices were exceptional. A study by the Montana Agricultural Experiment Station, covering dry-land farms in seventeen counties from 1928 to 1935, found that 65 percent of them had gross receipts of $2,000 or less, which it was estimated amounted to a net income of not over $100, annually, to cover living expenses and taxes. Thirty-eight percent had gross returns of less than $1,000. The U.S. Bureau of Agricultural Economics, basing its report on records running approximately a decade and a half, estimated that farmers in the Scobey-Plentywood and Divide County sections of Montana and North Dakota had averaged a gross income of $1,352, and those in the vicinity of Hettinger, North Dakota, $1,583, including the value of products consumed on the farm. In 1934 the Scobey-Plentywood farmers averaged gross receipts of only $792 and those around Hettinger, $1,054, most of that income being payments under the Agricultural Adjustment program, federal livestock purchases, and emergency relief grants.[8]

A Ziebach County, South Dakota, combination grain and stock farm which was considered successful shows the difficulty of building reserves even under the program most recommended for that area. A study utilizing accounts that spanned a sixteen-year period from 1923 through 1938 revealed a total labor-management income for the farmer and a working son averaging $922 annually over the period, after payment of all farm expenses and allowance of interest at 5 percent on the capital investment. The income had ranged as high as $3,574 in 1928 but fell to a loss of $400 in 1936. It was below the average twelve years, 1923, 1926, and 1929 through 1938. In 1931, 1933, 1934, 1936, and 1938 it was insufficient for farm expenses including charges for depreciation and interest. In the first four of those latter years, the farmer had been delinquent in payment of taxes and had had to borrow operating funds in 1934 and 1936, debts that had been repaid by 1938. He had been forced to reduce his wheat acreage from 217 to 150 acres for want of seed, 50 acres of the tract lying idle in 1938; his young stock had been sold early in the drought; and in 1936 he had cut his dairy herd from eleven to nine cows. Besides these capital reductions, he had withdrawn $434 from his account for depreciation reserve and $1,560 computed as interest on capital investment. Reserves of capital or feed, or both, could have tided him through the occasional poor years before 1931, the analyst concluded, but only with great difficulty could sufficient reserves have been maintained to cover the subsequent period.[9]

The recommended agricultural adjustments of the twenties had imposed particularly heavy capital requirements on farm operators in the area. Diversification entailed expense for more highly bred livestock and a greater variety of farm equipment. Large-scale wheat cultivation, on the other hand, necessitated an investment in expanded land holdings, with increased costs for fencing and taxes, plus a minimum expenditure of approximately $4,000 for a three-plow tractor, combine-harvester, and implements adapted to mechanized operations. Such capital demands accounted for the fact that 67 percent of the owner-operated farms in western North Dakota were mortgaged in 1930, and in seven counties of that area the rate was over 70 percent. Of such farms in western South Dakota and eastern Montana, 56 percent were mortgaged, with two counties in each of those districts reporting rates of over 70 percent.[10]

The new settlers on the margins of established farming districts during the late twenties had had scant opportunity to scratch a toe-hold before the hazard of the drought was upon them. The Gieseker soil studies in Montana made it clear that the lands available for entry were the least desirable—more apt to be deficient in moisture, shallower in depth to calcareous substrata, more broken in terrain, and more susceptible to soil drifting. In many cases they had already been tested and found wanting during the previous period of drought. As early as the winter and spring of 1929–1930, 292 families in recently settled Valley County, Montana, were obtaining aid from the American Red Cross, which distributed aid to about 100 families in Hill County the following winter; and by the end of the summer of 1931, relief had been requested for 40 families, each, in the newly developing districts around Winifred, Roy, and Grass Range in northern and eastern Fergus County. Relief had also been necessary in Petroleum, Golden Valley, and Garfield counties to the east and in five counties of northwestern North Dakota. Between 1929 and 1934 the area used for crops declined by over 100,000 acres in Fergus, Valley, and Wheatland counties of Montana; Morton County, North Dakota; and Corson, Dewey, Lyman, Meade, Perkins, and Tripp counties of South Dakota. All but Morton and Lyman counties had been centers of rapid growth during the late twenties.[11]

Even for those long-established in the region, the readjustment dictated by the experience of the postwar years had imposed serious financial strain. Mortgages and taxing levels assumed under inflated land values still had to be met when products were selling at 25 to 30 percent of the prices prevailing when the obligations were contracted. Farm mortgage debt in Montana had decreased by about one-third between 1920 and 1928, but this decrease was due more to bankruptcies, foreclosures, and other forced sales than to normal transactions. The number of such distress liquidations had been very high in the more recently settled states throughout this period, at a level well above that in neighboring areas and more than triple that in the United States as a whole. As late as 1927 the average number of farm bankruptcies in Montana, rivaled only by the rate in North Dakota, had exceeded the national average by about five fold. Even on the conservatively managed federal land bank loans, delinquency was running at 9.3 percent in Montana and North Dakota in 1930, when the rate nationally was 5.5 percent.[12]

By legislation running from 1918 through 1922, the federal government had authorized special seed grain loans under which approximately $4 million had been extended in the northern Plains. An act of 1928 had released borrowers from repayment on the loans of 1918 and 1919 if the crop had yielded five bushels or less, but collection on the remaining debts had been difficult and the service costs high. By 1932, 74.4 percent of the federal loans in North Dakota had been repaid under the legislation of 1921 and 83.1 percent of those under the act of 1922. South Dakotans, included under the legislation of 1922, had repaid 91.7 percent of their loans. Montanans, however, had repaid only 64.3 percent of the loans of 1921 and 70.6 percent of those for 1922. The unpaid obligations remained a consideration in national land-use planning during the thirties.[13]

Montana and North Dakota had authorized use of county funds also, for feed and seed loans in 1918 and 1919. Applicable through 1922, such loans had totaled almost $7.25 million in Montana and at least $4 million in North Dakota. These debts, funded by issuance of warrants or bonds, had added considerably to the cost of local government. To meet this burden and cope with the difficulties of maintaining services for a dwindling population, property tax rates had risen in North Dakota from $1.11 per $100.00 valuation in 1920 to $1.70 in 1930. Taxation for interest and sinking funds on county bonds in northeastern Montana had risen by 364 percent over the twenty years ending in 1933, and tax levies generally had increased by 169 percent during the period. Under such charges tax delinquency embraced more than 15 percent of the agricultural land in Montana, ranging from 3 to 41 percent in the various counties, even during the relatively prosperous conditions of 1928.[14]

The financial base of farmers in the northern Plains, as M. L. Wilson had warned, was tenuous when the disastrous decade of the thirties opened. After a brief decline in 1930, mortgage foreclosure and forced sale of farms again increased, in Montana over the next two years and in the Dakotas from 1931 through 1933. This period brought the most severe rate of liquidation during the decade, with about 10 percent of the farms in the Dakotas transferred by forced sale during the peak year and 7 percent of those in Montana; but the lower level through the remainder of the period must be attributed to relief measures rather than to improvement of the farm debt situation. Of the owner-operators in Hettinger and Divide counties of North Dakota, 75 to 85 percent reported real estate mortgages in the spring of 1935. While a large number of farmers had left the region by that time, 14 to 16 percent of those who remained were insolvent. Federal land banks and federal land bank commissioners held first mortgages on two-thirds of the mortgaged acreage in the region. The delinquency rate on loans of the land banks rose from 9.3 to 67.4 percent in North Dakota, from 3.9 to 65.9 percent in South Dakota, and from 9.3 to 61.5 percent in Montana between 1930 and 1934. In 1940 it had risen to 72.8 percent in North Dakota, although it had dropped to 40.1 percent in South Dakota and to 34.6 percent in Montana. Delinquency on the land bank commissioner loans was even higher, ranging to between 81 and 86 percent on those extended in North Dakota between 1937 and 1940.[15]

County tax delinquency had also soared during the drought years. Seven counties in Montana, almost all in the northeastern corner of the state, suffered a delinquency rate of 55 percent or more on current levies by the end of 1933. Fergus County, heart of the most well-established dry-farming region, fell just under that level, with a delinquency rate of 54.66 percent, and six other counties of central Montana shared a delinquency record of between 45 and 55 percent. Two years later taxes were unpaid on 78 percent of the land in the Scobey-Plentywood area. As late as 1939 about 72 percent of the 1937 taxes for that district were still unpaid, while an estimated 85 to 90 percent of those on the 1938 levies were delinquent. At the end of the decade the six northeastern counties, together with Powder River and Carter in the southeastern corner, still led the state, with delinquency rates of 30 percent or more.[16]

In North Dakota the fourteen counties comprising the southwestern "Slope" district recorded tax delinquency in 1936 for one or more years on nearly 69 percent of the taxable acreage, ranging to approximately 75 percent in Morton and Slope counties. More than 20 percent of the property taxes in that district had not been paid for five years or more. To the northwest, Ward County had a delinquency rate of 72 percent on rural property taxes for 1937, and 30 percent had been delinquent for five years or more. In western South Dakota, Corson and Ziebach counties and areas extending into southwestern Perkins and northeastern Meade counties listed more than one-half the taxable acreage as delinquent by the end of 1934. Three years later taxes had been unpaid for four years or more on 41 percent of the taxable land of the trans-Missouri region of that state, ranging to nearly 60 percent in Armstrong and Stanley counties and over 44 percent in Corson, Dewey, and Ziebach.[17]

A gradual liquidation was effected in South Dakota and Montana by the end of the decade, but in North Dakota the major movement for redemption on debt and tax obligations was delayed by moratorium until 1941. At that time over 65 percent of the owner-operated farms were mortgaged in fourteen counties of western North Dakota, but the rate was then as high in only three counties of western South Dakota and in no county of eastern Montana. As the North Dakota moratorium on foreclosure actions was lifted, a second wave of forced sales reached high levels in that state through the early forties. Altogether, between 1930 and 1944 the number of such transactions amounted to 66 percent of the farms in North Dakota, 81 percent of those in South Dakota, and 58 percent of those in Montana. Not all involved whole farms, and some constituted multiple sales of a single farm, but the rate far exceeded that of 34 percent for the nation as a whole.[18]

From 1929 to 1939 total cropland was decreased by only 429,623 acres (4 percent) in western North Dakota, by close to 1 million acres (7 percent) in eastern Montana, but by nearly 4 million acres (49 percent) in western South Dakota. That the abandonment of cropping was not greater was again attributable to the fact that in many cases the tracts were added to established enterprises. Average farm size increased between 1929 and 1939 from 551 to 584 acres in western North Dakota, and by 1944, to 714 acres; in western South Dakota, from 710 to 1,145 and to 1,464 acres in the respective years; and in eastern Montana, from 1,010 to 1,284 and to 1,829 acres. Meanwhile, the number of farms declined from 32,968 to 29,896 in western North Dakota between 1929 and 1939 and to 26,631 by 1944. The number was reduced in western South Dakota from 22,484 in 1929 to 16,752 by 1939 and to 14,854 by 1944, and in eastern Montana from 36,324 to 29,797, to 26,631, respectively. Many of the former owners remained on the farms as tenants, many moved into nearby towns or into less arid areas of the three states, and some left the region altogether. The rural farm population in western North and South Dakota declined over 15 and 16 percent, respectively, and that in eastern Montana, about 13 percent during the thirties. Nonresident land ownership and tenancy both increased greatly.[19]

Time has mellowed the reminiscences of conditions as recounted today by "old timers" who survived the experience. They say that they "never went hungry"; cream and butter checks brought some income; older sons and daughters found work away from the farm; extended governmental credit arrangements carried them through the "Depression years." Government investigators from Washington, emphasizing that few of the relief households had such conveniences as electricity, gas, central heating, and running water, often reflected a view of conditions distorted by cultural unfamiliarity; such deprivations were not new to the way of life in the region and not limited to those on relief. A contemporary local study, based on individual records, showed that two-thirds of the dry-land wheat and livestock operators in eastern Montana remained on their land, obtained an average annual gross income of at least $1,000, and, in the view of the analyst, were "fairly successful" throughout the drought period. The author noted that of the third who were "unsuccessful," over 70 percent were attempting to farm on less than 360 acres, a large majority of them using tractors and machinery that had not been scaled for small-farm operations and raising cattle in herds averaging less than twenty head, with little acreage devoted to feed crops and with over one-half the farm unit untilled, in pasture and waste land. They practiced summer fallowing less regularly than did other farmers in the area; only a few raised gardens; and, although three-fifths of them had cattle, they did not even supply the family dairy products. The Montana state relief administrator, placing the number of families requiring long-term aid in the spring of 1935 at 17 percent, agreed that drought and general economic ills accounted for only 36 percent of the cases. Old age had brought about 10 percent of them; 9 percent were part-time farmers who had lost their supplementary employment; the plight of 18 percent resulted from poor management or insufficient land, livestock, and other capital requirements; while only about 13 percent were operating on land deemed too unproductive to furnish a living.[20]

These considerations gave basis for local boosters who in the early years of the period argued that experienced farmers had learned from the difficulties of their homestead days and were "staying with the game and adding to their accumulations every year." Yet the percentage of families receiving some form of relief in the spring wheat region as a whole rose steadily from 7 percent in July 1933 to almost 40 percent by November 1934 and continued at the latter level through the following May. And this proportion was greatly exceeded in many districts. Of the twenty-three counties in western North Dakota, ten had more than 40 percent of the population on relief in March 1935. In Burke County, the percentage was 71; in Divide County, just under 68; and in Mountrail, Sioux, McKenzie, and Williams, above 50.

Between 1933 and 1936, federal aid amounting to $175 or more per capita was distributed in Bottineau County and a double row of counties along the western and southern borders of North Dakota, in the Missouri Valley counties and westward through Corson, Haakon, and Jackson counties in South Dakota, and the southeastern border counties of Montana. Excepting only Ward County in North Dakota and the Black Hills district of South Dakota, the remainder of those two states and the eastern third of Montana formed a zone where relief payments ranged from $119 to $175 per capita, while throughout the central third of Montana they were between $64 and $119 per capita. By the winter of 1937–1938, nine out of ten families in Williams and Divide counties, along the western border of North Dakota, were reportedly living on relief.[21]

Lorena Hickok, a newspaper woman reared just east of the Missouri River in northern South Dakota, who had been sent by Harry Hopkins to check the effectiveness of the relief program as winter approached in 1933, found conditions worst in Bottineau County, in north-central North Dakota, where crops had been lost successively to drought, grasshoppers, hail, and low prices over the past four years. Like other observers, she noted that most of these families had a few years earlier been "considered well-to-do," "land poor," with holdings of "640 acres or so," and livestock amounting to thirty or forty head of cattle, twelve to sixteen horses, and some sheep. She, too, reported that reliance had been placed on "cream checks," but these were dwindling as cows ran dry "for lack of food." She did not find prevalence of hunger, but she did see widespread destitution: "Their houses have gone to ruins. No repairs for years. Their furniture, dishes, cooking utensils—no replacements in years. No bed linen. And quilts and blankets all gone. A year ago their clothing was in rags. This year they hardly have rags." Clothing was shared so that those going into public might be decently covered; but frequently children were kept out of school for lack of shoes, and at home they lacked stockings or shoes: "Their feet were purple with cold."[22]

In Corson County, South Dakota, the relief load had amounted to 52 percent of the families in June 1935, when under revision of the administrative program the relief rolls of the Emergency Relief Administration were cleared of all cases so as to assure greater availability of harvest labor through the summer. Analysts of the Social Service Division of the Works Progress Administration, reviewing those cases the following November, found that their income had dropped from an average of $27.56 per family during the last month on relief to $15.20 for the succeeding month. They had managed to provide a considerable amount of their own food, despite drought and hail damage to gardens; but their cash assets, including income from sales of 1935 crops and livestock, averaged only $123. Only 11 percent of them had found employment in harvesting through the summer, perhaps because many of them had been working their own farms. Five-sixths of them reported one or more immediate cash needs, mostly clothing and bedding and medical and dental care. Property mortgages averaging in excess of $4,000 encumbered the holdings of three-quarters of them; 97 percent of them also owed short-term debts averaging in excess of $1,100 for farm machinery, general expenses, delinquent taxes, interest, rent, and medical bills.[23]

Other surveys showed that relief families, 92 percent of them in the spring-wheat area generally, did raise garden or truck patches, and over three-fourths of them had dairy cows and other cattle. While settlers in the more newly developed areas experienced hardship the earliest, 79 percent of those receiving relief at mid-decade had resided in the area for ten years or more. The average age of these relief recipients was about forty-six; 85 percent of them were between ages twenty-five and sixty-four. Only a few of them had ever received public or private assistance before getting federal relief, and they had received little more than that obtained from the governmental program during the period of its availability. They were, in large part, independent, resilient, and persistently hopeful; but in their hearts lay also bitter anger than their years of bountiful crops had yielded so little profit and pervading fear that all would be lost in the imminent future.[24]

Political tensions aggravated the situation in North Dakota, for there the dissensions that had marked the struggles involving the Nonpartisan League during the early twenties were revived with the rapid development of the National Farmers' Union as an expression of mounting discontent. Organization of the union in the northern Plains dated from around 1913. During the heydey of the Nonpartisan League, it had made little growth, but with the collapse of the latter movement and the growing marketing problems of the late twenties membership in the union had greatly increased. The organization of a central exchange at St. Paul for cooperative sale of petroleum products, a Farmers' Union Terminal Association, also at St. Paul, with which local cooperative grain elevators were affiliated, and a Farmers' Union commission house at South St. Paul, where local cooperative livestock shipping associations found an outlet, had drawn nearly half the farmers of northwestern North Dakota into the organization by the early thirties.[25]

The idea of a farmers' strike, initiated without success by the Farmers' Society of Equity and the Farmers' Union early in the century, had been revived during the price collapse of 1920. Milo Reno, an Iowa farm leader, had favored such action in the latter period. As president of the Iowa Farmers' Union from 1921 to 1930, he won many of the leaders of that body to the proposal. The annual convention of the union in 1931 adopted a resolution calling for "a farmers' buying, selling, and tax-paying strike" unless remedial legislation were adopted. A local unit of a Farmers' Holiday Association was formed in Boone County, Iowa, in February 1932, and mass meetings supporting the idea were held in several other Iowa counties later that spring. On May 3 some 2,000 farmers convened at Des Moines and organized a national association. With Reno as president, they issued a call for a strike to begin July 4. Despite some organizational delay, a strike was officially begun at Sioux City late in August.[26]

Meanwhile, state units were being established throughout the spring-wheat-producing section of the nation. A temporary organization for North Dakota was formed in conjunction with a convention of the North Dakota Farmers' Union (NDFU) at Jamestown July 31. Charles C. Talbott, president of the NDFU and a leader in the Nonpartisan League, chaired the session. Usher L. Burdick, a Williston rancher and lawyer, was elected temporary president. One of the leaders of the Farm Bureau during its formative years and a member of the NDFU from its founding in 1937, Burdick was also a former member of the state legislature and lieutenant governor and had been defeated in the spring of 1932 in a bid for the Republican nomination as a member of Congress. At a two-day conference in Bismarck in mid-September, delegates from thirty of the state's thirty-three counties completed formal organization of the state Farmers' Holiday Association and confirmed Burdick's election to its presidency.[27]

The resolutions of the temporary organization had called upon the "courts and governments" of the state to "invoke special police powers . . . to prevent the forced sale of farm commodities for the satisfaction of liens and indebtedness thereon until the price level for such commodities . . . [should] have reached the cost of production," to withhold from sale their produce except as required to provide funds for production and living expenses, and to "protect one another in the actual possession of necessary home, livestock and machinery as against all other claimants." It was a program calling both for protection of farmers from legal measures to collect debts and for actions to keep crops off the market pending improvement in farm prices. The development in North Dakota drew strongly upon agitation stimulated by a Nelson County farmer, Dell Willis, who at a July 4 picnic in 1932 had put forward a plan urging voluntary withholding of wheat until the price reached a dollar a bushel. Burdick himself had addressed a large gathering near Jamestown in support of the "$1 Wheat" idea the day before his election to leadership in the temporary farmers' Holiday organization. The farmers' strike proposal, therefore, first gave direction to the movement.[28]

Establishment of local units of the Holiday Association spread like wildfire through the state. Units were formed in Bowman and Steele counties on August 7, 1932. On the same day a meeting was called to consider such action in Williams County, but the promoters decided to canvass sentiment among farmers of the area before completing formal organization. That done, a permanent chairman and executive committee were elected on August 19. By the end of the month, 95 percent of the farmers in the county were reported to have signed agreements to withhold their grain. Slope, Mountrail, McKenzie, McLean, Foster, and Ward counties established associations during the latter half of August, and units were formed in Stutsman, Morton, Ransom, and Grant counties during the first week in September. Farmers of Burke, Dunn, Cass, and Traill counties were organized the third week in September;

those in eight other counties by the end of that month; and those in nine more during October. Only one county in the state had not been organized by mid-1933.

All was not harmonious, however, on the issue of whether picketing should be initiated. As confrontations and violence marked the strike effort in Iowa, Burdick emphasized the importance of thorough organization as a preliminary. He believed that picketing would not be necessary if farmers were won to the cause first. Local units, nevertheless, acted independently to enforce the strike called to begin in the state September 20. Pickets were set up at Minot, in Ward County, on September 22, when it was found that livestock hauling had not been halted. From there picketing spread throughout the northwestern counties of McLean, Mountrail, Bottineau, Burke, Williams, and McKenzie and southward into Adams County during late September and early October. It was of short duration, checked by growing opposition amidst journalistic charges that Communists had infiltrated the movement. Picketing continued longest in Ward County, where it persisted until mid-November. Elsewhere it had ceased by October 19, when a heavy snowstorm blocked traffic. Proponents explained that extension of the effort was no longer required since farmers showed little interest in marketing their crops at the prevailing prices. They claimed at the end of 1932 that only about 35 percent as much grain had been shipped in northwestern North Dakota as during the last good crop year, 1929.[29]

Less widespread agitation marked the movement in South Dakota and Montana. As early as July 20, 1932, South Dakota farmers organized a state unit of the Farmers' Holiday Association at Yankton. Barney McVeigh, a Republican who had been speaker of the state house of representatives, was selected president. Emil Loriks, a former supporter of the Nonpartisan League and an early member of the American Farm Bureau, who by the thirties had become active in the Democratic party and the Farmers' Union, was named secretary-treasurer. It has been estimated that the South Dakota Farmers' Holiday Association had about 1,600 members in the spring of 1933, but local units extended west of the Missouri River only into the Rapid City district. Loricks was the dominant force in projection of the movement and in 1933 brought to his role the prestige of his chairmanship of the appropriation committee of the state senate. The South Dakota Farmers' Holiday Association also carried strong support by business and professional groups and by both the Republican governor, Warren Green, and his successor in 1933, Democrat Tom Berry. Such leaders endorsed the idea of withholding crops from market but cautioned against violence in picketing to enforce strike efforts. Roadblocks were occasionally set up at communities bordering Iowa, Nebraska, and Minnesota; but the state organization rejected the national body's call for a strike in May 1933 and while officially supporting the effort of the national movement to close markets that autumn, left compliance to the decision of individual members. Since South Dakota producers had lost much of their crop to drought and grasshoppers that year and benefited more than those of other states in the northern Plains by the extension in October of federal price support to cover corn and hog production, the strike action there had little effect.[30]

Efforts to extend the movement into northern Montana during the fall of 1932 were no more successful. Organizers meeting in conjunction with the Montana Farmers' Union at Great Falls in September issued a call for establishment of a state branch of the Farmers' Holiday Association to take place at Wolf Point in mid-October. It was agreed at that meeting to strike against grain marketing but no action was taken on picketing. Local units were subsequently organized in a few districts, notably in Judith Basin County in March 1933 and in Fergus County the following December. Such efforts were more commonly directed against foreclosure actions than against commercial marketing. For the more radical farmers of northeastern Montana, the revival of the Communist party at this time afforded an alternative channel of protest.[31]

Communist units had been formed in Fargo and half a dozen communities in northwestern North Dakota, in Frederick, South Dakota, and in Sheridan County and adjacent areas of Montana during the early twenties. In the last of these districts, Communists had won control of the local government and maintained a newspaper at Plentywood, the *Producers' News*, under the editorship of Charley "Red Flag" Taylor. The movement had lost much of its influence, however, after Robert M. Lafollette renounced such support during his campaign for the presidency in 1924. Although Taylor, identified as a Farmer-Laborite, held a seat in the Montana senate from 1922 to 1930, his advocacy of the views of Leon Trotsky, seeking world revolutionary struggle, had led to his withdrawal from Communist organizational effort when Trotsky was driven from Russian leadership. Taylor was temporarily stripped of control of the *Producers' News*, and in 1930 he was defeated in a bid for election to the U.S. Senate. A year later, however, he resumed organizational activities, and by October 1931 claimed membership of 1,500 in Sheridan County for the United Farmers League, a Communist-led farmers' organization preaching the evils of capitalism and the necessity of class struggle. With Ella Reeve, "Mother," Bloor of Minot, Ward County, North Dakota, Taylor continued to be active in organizational work for the Communist party throughout the Dakotas and eastern Montana until the fall of 1934. Then, again, divisions within the over-all leadership developed, and by the following year the local movement had once more subsided into infiltration of other farm-protest organizations.[32]

The Communists had initially decried the Farmers' Holiday campaign as a futile effort to resolve the ills of capitalism. The *Producers' News* contended that the proposed wheat strike evaded the real issues—taxes, mortgages, and evictions. Gradually, however, the journal gave cautious support to the movement but urged that farmers redirect the program against the evils of high interest, rents, and taxes. Contingents of "comrades" from Plentywood, Great Falls, Billings, and the mining centers of Butte, Red Lodge, and Roberts in Montana joined "marches" on Washington and the state capitol at Helena in December 1932 to present "demands" for reduction of the salaries of public officials and greater taxation of the wealthy to finance increased relief expenditures. While the Washington meeting carried the endorsement of the Farmers' Union in both Montana and South Dakota and of some thirty-one other farm organizations, it had been organized primarily by the United Farmers League. It presented to congressional leaders resolutions seeking not only increased relief expenditures and improved access to credit but also a moratorium on mortgages, interest, and rents for farmers, through the crisis period; cancellation of mortgages, interest, and debts for those who had through a prolonged time failed to achieve an acceptable standard of living; cancellation of back taxes and a moratorium on current taxes until the crisis ended; and prohibition of evictions. It was a program which, although expressed somewhat more sweepingly, accorded in large measure with the stated goals of the Farmers' Union and the Farmers' Holiday Association, and throughout Montana and the Dakotas the activities of these organizations and those of the United Farmers League became closely intertwined.[33]

The National Farmers' Holiday Association (NFHA) again called for a nationwide strike in May 1933, but the project was abandoned as farmers and their supporters looked to the new Roosevelt administration for relief. As the summer passed, however, and prices rose in reaction to the national industrial recovery legislation, while payments under the various agricultural relief measures lagged, the farmers' discontent revived. The NFHA called for a marketing strike again in November, and in Iowa picketing and violence once more erupted. Milo Reno began to bombard President Roosevelt with demands for enactment of cost-of-production, debt moratoria, and other relief proposals.[34]

In North Dakota William Langer had carried the Nonpartisan League back into political power in 1932, when he was elected to the governorship on the basis of his endorsement of the idea of an embargo on commodity shipments, to be enforced by the state militia. Because of the severity of the drought in 1933, his declaration of such an embargo on wheat shipments in mid-October was recognized even locally as primarily "a spectacular gesture." Langer withdrew the order as related to wheat after two months, but then proposed to extend it to livestock, although it was too late in the year to be effective. The project was dropped in January when the federal district court ruled the actions in violation of federal authority over interstate commerce. The effort had, nevertheless, expressed the farmers' discontent with the administration's program and lent urgency to Lorena Hickok's coinciding pleas for more drastic measures to speed relief.[35]

Meanwhile, the Farmers' Holiday Association directed its efforts toward proposals for debt moratorium. As early as November 1931, the county commissioners of Williams County, North Dakota, had telegraphed the president and Congress, urging that the federal land bank of St. Paul be prohibited from foreclosing mortgages until conditions improved so that the "average farmer" could pay his debts and taxes by installments. Borrowers from the bank in that area organized to reinforce this pressure. Under the terms of legislation establishing the Reconstruction Finance Corporation in January 1932, Congress authorized a five-year deferment of payments on land bank mortgages. By the following winter, several large insurance companies holding farm mortgages had taken similar action. Institutional investors tended to accept some form of moratorium so long as the payment of interest was continued. On the other hand, individual investors and receivers of failed banks, including even such state institutions as the Bank of North Dakota and South Dakota Rural Credit Board, proved less willing. As late as October 1935, farmers of Harding and Perkins counties in South Dakota, attending a convention of the Farmers' Holiday Association at Hettinger, North Dakota, called upon their colleagues to adopt a resolution condemning the threat of foreclosure by the credit board.[36]

In September 1932, the executive committee of the NFHA had directed local units to use all power to stop foreclosures. In November the McKenzie County unit in North Dakota resolved to take action, as receivers of closed banks in the area initiated proceedings for debt liquidation. Several hundred farmers gathered on October 21 to block a sale of chattels on the farm of Nils J. Peterson, near Watford City. An agreement was finally reached by which Peterson retained his chattels and his mortgage was renewed for a year. Forced sales were stopped by Farmers' Holiday Association units at Circle, in Rosebud County, Montana, in January and February 1933, and near Stanford, in Judith Basin County, the following August. Through the similar efforts of units of the United Farmers League (UFL), dispossession was blocked in many areas of the region during the mid-thirties.[37]

The threat of force lay always in the background of such actions, particularly after the Bradley incident in Iowa in the spring of 1933, when a county judge was beaten and threatened with death if he did not sign an agreement to reject writs for execution of foreclosure.[38] Farmers more commonly banded together for "penny auctions," bidding in the property at nominal sums for return to their distressed neighbors. Farmers' Holiday Association councils of defense and UFL councils of action served as committees intervening between mortgagors and mortgagees to work out agreements so that the latter could retain possession of their property or, if they chose to relinquish it, might avoid possible deficiency judgments.

North Dakota voters in the autumn of 1932 had narrowly defeated an initiated proposal for a debt moratorium, but the following February a convention was held at Bismarck to promote the legislation. Attended by Congressman William Lemke, a number of state legislators, and leaders of the Farmers' Union and the Holiday Association, the body adopted resolutions declaring their intention "to pay no existing debts, except for taxes and the necessities of life, unless satisfactory reductions in accordance with prevailing farm prices" were made on such debts. They called upon farmers to organize councils of defense in each county "to prevent foreclosures" and pledged "to retire to our farms and there barricade ourselves to see the battle through until we either receive cost of production or relief from the unfair and unjust conditions."[39]

By executive action, Governor Langer declared a moratorium on debt collection on March 22, and the North Dakota legislature subsequently enacted a succession of laws extending such suspensions through 1943. With less political turmoil, in 1933 South Dakota and two years later Montana provided for redemption of foreclosed real property during periods extended, also, ultimately into 1943. By January 1934, all three states had taken action to delay or suspend action on tax deeds on real property, and by 1939 they had all instituted arrangements by which delinquent taxpayers could pay off their obligations under extended contracts. Other measures made it easier to redeem property sold on tax certificates by permitting, where certificates were held by counties, payment of the original taxes without penalties or interest. Responding further to the demands of agrarian radicals, in 1933 South Dakota, in 1935 Montana, and in 1937 North Dakota also barred deficiency judgments on sales foreclosing mortgages on real property.[40]

Congressman Lemke and Senator Lynn W. Frazier, North Dakotans long active in the Nonpartisan League, carried the demands for relief to Congress. In March 1933 they won passage of amendment of the federal bankruptcy legislation so that farmers could seek agreement with their creditors for adjustment or extension of the time for payment of their debts under supervision of the courts. If an agreement was reached, the farm was appraised at its "then fair and reasonable market value" and the farmer obtained judicial moratorium against further legal proceedings through the term of the agreement, meanwhile remaining on the land subject to payment of rent. During the interim, or at the end of the agreed term, he could redeem the property, free of debt, by payment of the appraised price, less any prior payments on the principal. Amended and renewed from time to time, the measure did not expire until 1949, although it was contested through the federal courts. Meanwhile its sponsors pressured Congress for further ameliorative amendments. After the defeat of one such bill by the Roosevelt administration in the spring of 1936, Lemke renounced his previous support of the New Deal and accepted nomination by the Union party, a coalition of the followers of Father Charles E. Coughlin, Dr. Francis E. Townsend, and Gerald L. K. Smith, to run for the presidency.[41]

Local and national politics were both embroiled in pressures for public assistance. County officials, struggling to maintain normal services with sharply curtailed revenues, found that expenditures for relief nearly doubled between 1930 and 1934. In August 1931 eighteen counties in the drought area of western North Dakota set up a revolving fund for purchase of hay and feed, to be distributed at cost plus transportation charges. Later that year Williams and Mountrail counties, anticipating a congressional appropriation for federal feed and seed loans, issued several hundred thousand dollars worth of bonds on which to contract for stocks while supplies were still available at low cost in the terminal markets. State and county governments expanded construction of public works, Montana authorizing a $6 million bond issue in the spring of 1931 to finance a four-year program of highway development. Local authorities also issued free permits for mining on public lignite holdings, a benefit that in Williams County, North Dakota, provided some 50,000 tons of fuel through the winter of 1931–1932. The credit of local governments was so severely strained through the issuance of bonds and warrants for relief purposes that in 1935 the Montana legislature repealed the authorization to make seed-grain loans in time of drought and to incur debt for poor relief. Bowman County, North Dakota, officially declared bankrupt during the decade, was only one of many counties in the region that approached their constitutional debt limit to meet the relief burden.[42]

There were no community chests operating in the area, and the Salvation Army failed to raise sufficient funds to continue operation in some communities where it had existed through the twenties. Private, volunteer organizations tried to fill the relief need during the early thirties. Churches, civic groups, and fraternal societies collected clothing, distributed food boxes, and engaged in a wide array of projects to raise the necessary funds. A total of some thirty groups joined as a city relief committee at Lewistown, Montana, to survey the conditions, collect and distribute supplies, and try to find jobs for the unemployed through the winter months of 1930–1931. Several county newspapers operated farmers free exchange lists, in which participants most frequently offered to "swap cattle" for anything usable.

Radical groups seized the opportunity to win followers by providing assistance—in some cases extended generally; elsewhere limited to those who were or had been members. After surveying local needs, the Musselshell County unit of the National Farmers' Union requisitioned approximately 100 tons of potatoes, flour, meal, canned goods, apples, dried fruit, sugar, honey and syrup, coffee and tea, salt, bacon and "side meat," lard, soap, and miscellaneous clothing to meet the local need for a year. Similarly, the North Dakota Farmers' Union offered to supply potatoes from a bumper yield in the Red River Valley if county commissioners would send the labor and trucks to harvest them. The Farmers' Union Terminal Association agreed to grind 100,000 barrels of wheat into flour, without commission, if the federal Farm

Board would finance purchase of the grain. Employees of the terminal and the livestock commission of South St. Paul pledged to donate 10 percent of their salary to the drought-stricken counties. Charley Taylor, on behalf of the United Farmers League, arranged a well-publicized exchange of North Dakota coal for South Dakota wheat.[43]

The American Red Cross, however, carried the major burden of philanthropic relief through the spring of 1932. County and local chapters of the organization were called upon for membership fees and emergency quotas, but the bulk of the funding came from the national organization until the summer of 1932, when the newly established Reconstruction Finance Corporation provided a loan. The work had begun as early as the winter of 1930–1931 when the Red Cross spent some $75,000 to supply food, clothing, fuel, and stock feed for 8,000 to 10,000 families in Montana. Relief was initiated in McKenzie, Divide, Burke, and Mountrail counties of western North Dakota and renewed in some sections of Montana as early as July in 1931. It was extended to Williams County, North Dakota, the following month, at which time the organization was assisting 18,000 people in the two northern states. By December, Dunn and Ward counties of North Dakota, and Golden Valley, Musselshell, Petroleum, and Fergus counties of Montana had been added to the list. Wheatland County, the only area in central Montana that had not at that time sought help, called for assistance in February 1932, and by then Billings and several other counties of southwestern North Dakota were receiving food and clothing. South Dakota, notably Perkins County, and Nebraska had also come under the relief program for the winter of 1931–1932. During the year the Red Cross spent over $2.25 million and distributed over $500,000 worth of donated goods for relief in the northwestern states. In Montana $768,473 worth of goods had been provided for 52,155 individuals in twenty-nine counties; in North Dakota, $707,258, for 52,934 individuals, in fifteen counties. Railway corporations, which had hauled into the area donated goods without charge and bulk purchases of the national organization at half rate, were estimated to have contributed $1 million to the effort.[44]

With the coming of spring in 1932 the Red Cross distributed seed packets to each of the 19,634 families on its rolls in the area and closed its operations. By the following September, however, Williams County, North Dakota, was again requesting help. The county commissioners reported the condition of the people "little improved over . . . last year—they have raised but little crops generally, and of poor grade." The Red Cross responded that it, too, lacked the necessary funds. They were able to provide donated beans, mixed vegetables, seed potatoes, and garden seeds, together with cotton cloth and flour milled from stocks of cotton and wheat supplied by the federal Farm Board. With these supplies they attempted to meet the need for general relief in Valley and Musselshell counties of Montana and in Renville, Bottineau, and sections of Mountrail counties of North Dakota during the fall of 1932. But the burden of massive public assistance was now to be passed to programs established by the federal government.[45]

Apart from the federal Farm Board's abortive efforts to maintain prices under the Agricultural Marketing Act of 1929, through which the federal government absorbed the purchase cost of some 95 million bushels of surplus wheat and two large allotments of cotton donated to the Red Cross in 1932, Washington authorities had provided agricultural assistance during the twenties primarily by credit arrangements through the federal land banks and private cooperative associations such as the Livestock Marketing Association and the Agricultural Credit Corporation, operating with governmental support under the Intermediate Credit Act of 1923. Northwestern farm groups in the early drought years accused the administrators of the district institutions centered in St. Paul and Spokane of being more concerned with maintaining liquidity, even reaping a profit, than with rendering assistance to farmers. Like most banking institutions in the region, their officers were frequently identified with livestock interests and regarded such chattels as far better security than land or crops under semiarid conditions. The head of the Spokane branch of the Agricultural Credit Corporation frankly conceded in March 1933 that livestock operators were getting 70 percent of the loans from that office. Even livestock loans were seldom made on less that carload lots; they did not apply to breeding stock; they were limited to a year; and they required 6 to 7 percent interest, backed by first mortgage on livestock or other personal property. Farmers on family-sized holdings, whether or not they were diversified to include a combination of livestock and grain operations, complained vehemently about the lack of adequate credit.[46]

In March 1930 Congress authorized appropriation of $7 million for emergency loans to provide seed, feed for work stock, and tractor fuel in storm, flood, and drought disaster areas of fifteen states, including Montana and North Dakota; but delays in establishing the program held up delivery of the funds until so late in the planting season that few loans were made. By the third week of May only 71 of 145 applicants in Phillips County, Montana, had received checks, and only 19 farmers in Judith Basin County received loans during the year. The following December, Congress by joint resolution authorized appropriation of an additional $45 million for such loans and included provision that they be made applicable for preparation of summer fallow in 1931. The legislation was still further amended in February 1931 to appropriate $20 million also for loans to cover purchase of feed for livestock other than work animals.[47]

A relatively small sum had been lent to farmers of the region in 1930, although a large number received aid in Valley and Daniels counties of Montana. In 1931 some $2 million was lent for seed loans in that state, and $716,000 in North Dakota. Farmers of western South Dakota were added to the program in 1932, with 30,290 loans in that area amounting to $7,094,904 for the year. North Dakotans in 1932 received 38,949 loans, totaling $8,407,737, and Montanans, 18,139 loans, for $4,267,875. Feed loans under the special legislation of 1931 added amounts of $3,274,654 in South Dakota, $3,029,880 in North Dakota, and $1,056,020 in Montana.[48]

The average seed loan was small, ranging around $235; but with the farm price for wheat at about $ .38 a bushel, it required approximately 650 bushels to repay the loan, with interest at 5 to 5.5 percent. The fact that the government in 1931 attempted to institute a stringent collection policy, hiring a hundred additional collectors at the loan offices, contributed greatly to the growing fear and unrest. The repayment rate was poor. Crop mortgages provided the security, and when crops failed, the loans generally became delinquent. Of the few Judith Basin County, Montana, farmers who had borrowed a total of $2,160 in 1930, 43 percent had repaid the debt by the end of the year. Only 8 percent of the Montanans repaid the $2 million debt of 1931, however. A three-year extension on feed and seed loans was finally approved for the northwestern states in 1932. Loans of $100 to $200 were to be spread over two years; larger loans, up to a maximum of $400, were to be paid in thirds over the next three years. When farmers were unable to repay the loans at the end of that crop season, they were permitted to renew them by paying 25 percent of the amount due and extending the crop mortgage through 1933. By the end of August 1933, the South Dakota farmers had repaid 21 percent of their loans for 1932; those of North Dakota and Montana, only 12 percent.[49]

The terms of the loans stirred strong complaint, notably in Renville, Ward, and Williams counties of North Dakota, where political radicalism was mounting. Protesters argued that under the program of 1932 seed loans, limited to $3 an acre, with a maximum of $400, were not adapted to the scale of farming operations in the area. A limit of $1,600 in total loans for the tenants of a single land owner would withhold assistance from "scores," it was contended. Other restrictions were rigid: applicants must raise gardens; loans could not be applied to purchase of machinery or livestock or for payment of taxes or other debts; no loans were to be granted to applicants who had not raised crops in 1931; and borrowers could not apply the funds to farm a greater acreage than had been cultivated in 1930 and 1931. In North Dakota the loans did not cover summer-fallow operations. For a time federal liens on crops barred pasturing cattle on crops too poor to harvest. This prohibition was finally lifted, but only if the borrower executed a mortgage on livestock sufficient to secure the seed loan. The problem was that half the farmers who needed assistance would already have mortgaged their chattels to local credit agencies, which, having sold the mortgages to eastern banks, were powerless to grant waivers. In the view of the editor of the *Williston Herald*, such requirements complicated the whole program of so-called rehabilitation loans to provide feed and seed. Officials of the U.S. Department of Agriculture were "painfully deliberate," if not "outrightly unsympathetic," he contended.

Federal aid was enmeshed in "red tape": "A horde of pussyfooters and special agents have looked us over. They have come into the drought territory incognito and send out garbled reports of actual conditions."[50]

Delays, as waivers on prior indebtedness were investigated on an individual basis, caused the most widespread discontent. At the end of April 1932, when checks had been received for only 1,472 of 2,124 applicants in Williams County, even local bankers and businessmen pressed for action. A representative went to Minneapolis to investigate the problem personally, and the county advanced funds up to one-half the value of the anticipated loans so that farmers could begin planting. Ultimately, the government loans in that one county totaled about $600,000 during the year.[51]

As private charitable effort and the older forms of public emergency action were being overwhelmed by the magnitude of the long-continuing need, Congress established the Reconstruction Finance Corporation. The RFC was set up in January 1932 and originally empowered to make loans re-enforcing the financial structure of such established institutions as commercial and savings banks, credit unions, the federal land banks, joint stock land banks, the federal intermediate credit banks, agricultural and livestock credit companies, and railroads. The Emergency Relief and Construction Act, passed the following July, expanded the authority of the RFC to fund self-liquidating public works, loans to states for direct relief, and establishment of regional agricultural credit corporations to help farmers refinance debts and sustain production.[52]

By January 1933, $1,099,556 had been allocated for relief in thirty counties of Montana, where approximately half the families along the "High Line" and in the southeastern ranch area—notably in Cascade, Valley, Daniels, and Powder River counties—were lacking sufficient food and clothing, and one in three families was "destitute." More than 900 families in Fergus County also received aid. By the following March, North Dakota had been allotted over $500,000—aid for twenty-three counties, about half of them east of the Missouri River. South Dakota, which had suffered heavily from drought in 1932, received amounts totaling nearly $1,804,000 through the following June, much of it for aid in Pennington and Custer counties, but much of it also for relief east of the river.[53]

With the change in presidential administration and the continuous drought through the mid-thirties, the program was greatly expanded. Support for agriculture across the nation underlay much of the New Deal approach to the "Great Depression," but several measures held special importance for farmers of the northern Plains. Among the earliest of the significant changes was a relaxation of terms in credit arrangements. Under the Emergency Mortgage Relief Act of May 1933, interest rates on loans of the federal land banks were reduced from 5 to 6 percent to 4.5 percent on first mortgages negotiated and guaranteed by farm loan associations. In 1935 this rate was lowered to 3.5 percent. The Mortgage Relief Act also funded land bank commissioner loans, so-called rescue loans, based either on second mortgages or on first mortgages which the land banks had rejected. They carried a rate of 5 percent, lowered in 1937 to 4 percent and by 1940 to 3.5 percent. Loans, formerly limited to 50 to 60 percent of the appraised value of the farm property, could now be made for sums up to 75 percent of the value or to a maximum of $5,000. Moreover, borrowers were permitted to re-amortize their old loans under the new terms. On this basis fifteen times as many federal land bank loans, for thirteen times as much credit, were negotiated in North Dakota in 1933 as in 1932. In Montana 898 land bank loans and 5,733 land bank commissioner loans, for a total of $16,572,700, were made in the period from May 1, 1933, to June 30, 1936. Land bank loans, which had represented but 0.3 of a percent of the credit in Haakon County, western South Dakota, as late as 1930, accounted for 23 percent of the first mortgages there in 1935. Land bank commissioner notes covered all the second mortgages in this last district by mid-decade.[54]

Much of this borrowing constituted refinancing. At the urging of the federal Farm Credit Administration, which replaced the federal Farm Board, the governors of the various states appointed community debt adjustment committees in the autumn of 1933 to assess and recommend worthy applicants for refinancing. Lacking legal status at the time, these local committees tried to make adjustments among the various creditors by suggesting extensions of payment period or reductions of principal or interest, to enable borrowers to start over with the aid of partial loans from the federal government. In this manner some $14,500,000 of debt, on 4,319 loans, was reduced in Montana to $10,166,700 between May 1933 and October 1934 helping borrowers to save several hundred thousand dollars in interest. The federal loans were then used to repay over $1 million (10.5 percent) to county governments for delinquent taxes, $750,000 (7.5 percent) to local merchants, $2.2 million (21.8 percent) to commercial banks, $4.5 million (46 percent) to individuals and loan companies, $360,000 (3.6 percent) to insurance companies, and 10.6 percent to miscellaneous other agencies. Under the same program, North Dakota creditors between May 1935 and January 1936 wrote off over $20,660,000, representing about one-third of the affected borrowers' debt. The burden of supplying credit in the region shifted sharply from private lenders to government-supported agencies. Close to 90 percent of the $93 million of the land bank and land bank commissioner loans issued in North Dakota during that period was for refinancing. Little new credit was extended.[55]

Alleviation of the short-term credit situation was also effected in 1933, through establishment of local Production Credit Associations (PCAs), funded initially in large part under the Production Credit Corporation as one of the divisions of the Farm Credit Administration. Borrowers were required to buy stock in the association equal to 5 percent of their loans, and it was anticipated that they would eventually acquire ownership of the lending agencies. The associations could not charge more than 3 percent higher interest than the current bank discount rate, which at the time set a limit of 6 percent on PCA loans. The new short-term credit facilities, which were established throughout the region by January 1934, shortly supplanted the regional Agricultural Credit Associations, which in their brief history had lent only a small fraction of their authorized appropriation because the interest had been set at a minimum of 7 percent, a level designed to avoid competition with private lenders. By the end of 1934, the interest charged by the PCAs was reduced to 5 percent and in 1939, to 4.5 percent. For farmers with a sound credit rating, the cooperatives became the basic source of production credit, with loans running for three months to a year, primarily for planting or harvest expenses. The competitive effect was far-reaching, as commercial banks and other private lenders responded to the pattern of more favorable rates and more flexible maturity dates thus made available.[56]

Other New Deal lending agencies—the Commodity Credit Corporation (CCC), the Home Owners' Loan Corporation, the Federal Housing Administration, the Rural Electrification Administration (REA)—had more limited immediate effect in the region. Crops mortgaged for seed loans could not well be withheld from sale, and poor yields of the drought years afforded little carryover for storage before the crop of 1937. Provision was made for commodity loans to wheat farmers as an accompaniment to the "ever-normal granary" concept in the re-enacted agricultural adjustment legislation of 1938. This afforded both a floor against drastic price declines and a means of financing farmers who held their crop in reserve, but CCC loans amounted to little more than $7 million in any of the states of the region before the end of the decade. Housing loans, even for sorely needed repair and improvement of farm homes were, likewise, not generally applicable in rural areas until 1939. For the state of Montana, as a whole, between 1933 and 1938, the Home Owner's Loan Corporation had extended loans totaling $7,284,979 and the Federal Housing Administration had insured loans worth $5,364,979. Rural electrification in the northern Plains was at this time limited primarily to irrigated communities. REA loans amounted to less than $2.5 million in the entire region before the end of the decade.[57]

The federal Agricultural Adjustment Administration (AAA), on the other hand, provided a major component of support for many farmers. Price relief and reduction of crop surpluses had been the primary considerations in initiating the AAA. Producers who agreed to reduce their crop acreage up to 20 percent were to be paid a subsidy of approximately $ .29 a bushel over and above the regular market return on grain to the extent of 54 percent of their average annual production as calculated over a long-term base period. That proportion of the crop was estimated to cover the amount required for use domestically. Growers were free to sell any excess production from their acreage allotment at open market prices. Owing in part to this program, but even more to the low yields caused by drought and to the government's inflated monetary policy, rising prices increased farm income generally by 60 to 70 percent in 1935 over 1932.[58]

For farmers who suffered heavy crop loss, the subsidy payments were a form of "insurance." Average annual payments ran as low as $110 a farm in the Scobey-Plentywood border area between Montana and North Dakota and $123 in Perkins County, South Dakota, but came to about $200 in Williams County, North Dakota, $235 to $306 for dry-land operations in the Buffalo-Creek district of southeastern Montana, and $317 in Hettinger County, North Dakota. The payments represented about 27 percent of the average cash receipts on farms in the Scobey-Plentywood district and 37 percent of those on farms in Hettinger County in 1934. Between 1933 and the fiscal year ending 1939, AAA payments amounted to $49,368,190 for Montana, $104,624,071 for North Dakota, and $85,986,334 for South Dakota. Analysts reported "a marked tendency" for farmers to expand their base acreage by leasing, even by plowing new land, partly in the belief that larger holdings would provide better opportunity for use of machinery, but primarily because adjustment payments of $300 to $500 would afford subsistence in the event of crop failure.[59]

This "insurance" factor concerned proponents of land-use reform. Little in the AAA agenda of 1934 and 1935 pointed to changes in the regional differentiation of production. There was no enhancement but also no contraction of the cropping specializations, either regionally within the nation or for particular areas within individual states. The percentages of acreage curtailment were applied evenly to the traditional commercial farming programs. The domestic allotment formula represented a compromise that ignored differing regional productive capabilities within the domestic economy.

In the late twenties, M. L. Wilson had believed that dry-land farmers possessed advantages in economy of operation that in the long run would enable them to compete successfully and that American farmers must be able to withstand such competition, internationally as well as regionally, if they were to continue operation. Spring-wheat farmers, however, particularly those of the earlier settled districts, had at the same time complained bitterly of the expanding production in the winter-wheat areas of the central Plains. The pleas of Chairman Alexander Legge of the federal Farm Board for voluntary reduction of wheat acreage in 1931 had won strong endorsement by North Dakota agricultural leaders but sharp rejection by the governor of Kansas. In the midst of such contending divisions, the Christgau proposal had found little basis of support. With the Fairway Farms project in liquidation, Wilson moved to Washington to become head of the wheat section of the AAA.[60]

In that capacity he proved to be responsive to the concerns of dry-land producers. In the drought situation of the thirties, the period used to calculate production history was immediately relevant to farmers of the northern Plains. At first the years 1930–1932 were designated as the base, but local protest quickly brought recognition that "exceptions" must permit a more representative time span for the region. The five years extending back through 1928 and 1929 were accordingly selected, allowing for a difference of over 100,000 acres in the production base of Montana farms. Controversy also arose when acreage in summer fallow was excluded from the record, thus penalizing farmers in those counties where during the late twenties and notably after the onslaught of drought large segments of wheat acreage had been diverted for a year of preparatory cultivation. It was also argued whether the allotments should be established on the basis of county or individual averages.[61]

Wilson's insistence that local committees chosen by the growers themselves should participate in administration of the program relieved the federal authorities of much of the dispute. Permission was granted Fergus County, Montana, for example, to utilize a "Modified County Average," under which applicants reporting "substantial" base-period differences from the county average could apply to their local committee for adjustments. In Hill County, on the other hand, the decision was made to establish the allotments on the basis of individual averages. In the end, the reported acreages tended to be above the estimates projected by the federal Bureau of Crop Estimates. One Montana county was said to have reported an excess of 70 percent, and many counties in Montana and North Dakota claimed acreage 15 to 35 percent greater. It was January 1934 before the differences had been adjusted and individual allotments could be publicly announced. By March, 96 percent of the wheat acreage in Montana had been committed to the program, the highest rate in the nation; and support for the program was high throughout the northern Plains.[62]

Corn and hog reductions were added to the program in the fall of 1933. Since such production was limited in the region, county allotments were not deemed necessary. Individual contracts required 25 percent reductions in the number of hogs marketed and in the size of pig litters; corn acreage was to be cut at least 20 percent. Where applicable, the program proved as popular as that for wheat. Endorsement ran at a rate of over 95 percent in Adams County, North Dakota, in 1934. An estimated 75 percent of the corn production in South Dakota was then under contract as well as 85 to 90 percent of the hog production. Participation increased still greater the following year.[63]

The restrictions of the allotment program were relaxed as the drought intensified. Only a 15 percent reduction below the base acreage of wheat was required for 1934, and it was reduced to 10 percent in contracts for 1935. As the spring of 1935 showed severe injury to the winter wheat crop, virtually all limitations were removed on the spring planting. The wheat contracts for

1936, signed in August of the preceding year, called for only a 5 percent reduction. The corn and hog adjustments for 1935 amounted to only 10 percent of the base average. Farmers who had been urged at first to keep the diverted acreage out of production because surpluses in alternative feed grains would be equally undesirable were called upon to shift the cropping to livestock feeds other than those under AAA regulation.[64]

As early as his presidential campaign speech at Topeka, Kansas, in September 1932, Franklin D. Roosevelt had called for "national planning in agriculture." In the autumn of 1935, he denied that it had been the intention to form the AAA as a "mere emergency operation." He said it was only a step in the process of developing long-term adjustments. Those reforms began to receive emphasis following the Supreme Court decision of January 1936, which invalidated the production control provisions of the program. While growers of winter wheat were assured of payments amounting to $215 a bushel on contracts signed the previous summer, new legislation was drafted to encompass the spring planting. Specific crop allotments were now to be abandoned, but farmers were to negotiate agreements under which they would be paid for shifting acreage from "soil-depleting" crops, generally those covered by the former control program, to "soil-conserving" and "soil-building" crops and for practicing "soil-building" measures on cropland nd pastureland. For 1936 the extent of soil-depleting acreage to be diverted was set at 15 percent. Payments varied according to the yields of the crops to be supplanted and were calculated on the basis of the productivity of individual farms.[65]

Through the rest of the decade, the cropping and tillage practices shifted the focus of the agricultural adjustment program to conservation; but the incentive payments, "helping farmers to maintain these beneficial systems of farming without interruption in poor crop years," remained a vital measure of relief in the drought areas. Exemptions from compliance without penalty in loss of payments were granted when emergencies arose. In July 1936 farmers of drought counties, located principally in the Dakotas and Montana, were authorized to divert acreage for hay or pasture from the 15 percent of conservation holdings. All crops previously designated as soil depleting, with the exception of corn, could then be considered as "neutral" crops when pastured or harvested for hay. Nurse crops seeded with perennial grasses or legumes could be cut, and land designated for summer fallow could be planted for forage such as millet, sudan grass, and sorghums. These "emergency" rulings continued the following year when summer drought or the severity of the winter destroyed soil-conserving crops.[66]

Because of increased production as the drought eased in some areas by 1937, the AAA legislation of 1938 restored the allotment program and established marketing quotas, while retaining the emphasis upon conservation. The measure also offered an innovation in providing for "all-risk" crop insurance, applicable to wheat beginning in 1939. Montana and the Dakotas had operated state hail insurance programs since the early settlement period, at rates in dry-land districts costing 7 to 8 percent of the insured value of the crops. Where participation was voluntary, support for the program was poor. Efforts to provide "all-risk" contracts, initiated by a few private companies prior to 1938, had failed. Compared to the rates for hail insurance alone, the new policies were expensive, approximating between 20 and 25 percent of the insured crop value. Again the response was meager: Montana had 2,683 applicants in 1939; North Dakota, 24,307; South Dakota, 9,457. Support improved somewhat in Montana in 1940, when there were 4,172 policies; but in North Dakota the number fell off sharply. The difficulty of developing a program dependent upon accumulation of reserves to meet a long-term actuarial base was evident. As one local editor explained, after the return of a good crop year farmers had been "buoyed . . . to the place where they feel other good crops will follow in rapid succession." The insurance program was commonly regarded as a measure of relief, not a managerial adjustment.[67]

The work-relief programs of the New Deal were not designed primarily as measures of agricultural assistance. The federal government had assumed responsibility for broad-based, general relief with the establishment of the Federal Emergency Relief Administration (FERA) in May 1933. Initially it had made cash grants to the states for continuance of activities initiated under the loans of the RFC during the closing months of the Hoover administration. Most of those programs had required public service in return for benefits, and that approach continued to prevail as the Civil Works Administration (CWA) was set up within the FERA to provide jobs more quickly during the winter months of 1933–1934. For farmers to leave their land and livestock for employment on public works was, however, difficult and in most cases seasonally limited. Many of the states discontinued participation in the CWA program early in the spring of 1934; but as drought continued in the northern Plains, demands upon the FERA soared to new heights that April. The agency consequently continued to provide relief, both directly and through work projects, for another year. In the spring of 1935, the Roosevelt administration sought to differentiate between what were viewed as the long-term problems of distressed rural areas and the temporary unemployment associated with business depression, so the FERA was supplanted by two new agencies, the Resettlement Administration and the Works Progress Administration (WPA).

For a year or two, administrators of the WPA in the Dakotas barred farm workers from their rolls. Survey of the decline in farm income in one South Dakota district after the program of the FERA was closed showed how vital a part such relief had been. As a result of strong local pressure for resumption of work relief, farmers were gradually restored to public employment. In Montana they had continued to find such work throughout the period. Under the successive programs of the CWA, the FERA, and the WPA, farmers with their teams performed road work, dug ditches, and constructed retaining walls and storage dams along small water courses. They baited roadways, abandoned fields, and range areas to combat grasshoppers. They cleared shacks and fences from public land-purchase areas. They reseeded public holdings of eroded grazing land. Farm women joined in canning and sewing projects. Farm boys found employment in the ranks of the Civilian Conservation Corps, and farmers benefited from the corps' work in building dams, reservoirs, and drainage ditches; constructing bridges, clearing noxious weeds, insects, and rodents from rangelands; and establishing waterways, terraces, and contouring patterns—work that embraced over 64,000 acres of Montana farm and ranchlands between 1935 and 1940. Farm boys and girls were given part-time jobs and student aid through the National Youth Administration. Between 1933 and 1939 the per capita expenditures of these agencies amounted to $198 in Montana, $144 in North Dakota, and $172 in South Dakota, amounting to approximately $106,445,988, $98,041,680, and $119,170,028, respectively. In addition, a variety of federal public works projects—by the Public Works Administration (PWA), the Bureau of Reclamation, the Corps of Engineers, the Bureau of Public Buildings, and the Bureau of Public Roads—also provided off-farm employment, at least doubling the above-noted work-relief expenditures for Montana.[68]

More direct relief was provided through distributions provided by the Federal Surplus Relief Corporation, established in October 1933. The program served the double purpose of disposing of agricultural surpluses acquired under various price-support measures and of making those supplies available to the needy. Both aspects of the effort served farmers of the northern Plains. Cattle and sheep as well as wheat stocks were amassed under the government's purchasing activities, conducted by the Commodities Purchase Section of the AAA, and local relief agencies received large supplies of butter, flour, beef, pork, lard, cheese, eggs, and rice for public assistance grants. Clothing of all kinds, mattresses, blankets, sheeting, rugs, and towels, mostly made in workrooms of the WPA from surplus cotton stores, were also distributed. By 1940 the food-stamp plan had been extended into the region as an additional program for such assistance. Between 1933 and 1939 the federal government had contributed approximately $2 million in such supplies to residents of Montana and over $4 million worth to each of the Dakotas.[69]

The New Deal programs so far discussed were not limited to the northern Plains, nor, apart from reclamation projects, even to the semiarid West, but several measures generated specifically to afford assistance to drought sufferers focused national attention upon regional phases of the relief problem. Seed and feed loans for emergency assistance were continued throughout the decade. The amount of funding available varied somewhat from year to year, ranging at maximums of $150 to $300 per operator in 1933, dependent upon whether payments were authorized for practice of summer fallow, $250 to

$400 in 1934, $400 to $750 in 1935, $200 during much of 1936, and $400 over the remainder of the period. Crop liens continued as the security requirement; interest charges were 5.5 percent, lowered finally to 4 percent in 1941; and the loans were available only to those who could not obtain credit elsewhere.[70]

Such loans were distributed over central and northern Montana and central and northwestern North Dakota eight years of the fifteen between 1921 and 1935. Throughout most of the remainder of eastern Montana and western North Dakota they were granted seven years during that period; in the northwestern quarter of South Dakota, six times; and in the southwestern quarter of that state, five. In many of the dry-land areas of Montana, the proportion of farmers receiving such assistance greatly exceeded one-half during 1931–1935; and as the loans were delayed in the spring of 1936, a northwestern North Dakota editor estimated that approximately one-third of the farmers in the district would be unable to plant any crop.[71]

Between 1930 and 1935 seed and feed loans totaled $42,973,757 in North Dakota, $36,896,506 in South Dakota, and $13,773,090 in Montana. The repayment rate on loans dating from 1921 through 1935 was at mid-decade only 15.5, 19.7, and 23.8 percent, respectively. Thereafter such borrowing decreased markedly, amounting to only $10,488,000 for 1936–1940 in North Dakota; and the repayment rate gradually improved, rising to 25.5 percent in North Dakota, 35.7 percent in South Dakota, and 34.8 percent in Montana for the longer period of 1918–1939. The record, nevertheless, fueled growing demand nationally that reform, not mere relief, be instituted to shape agricultural policy in the region.[72]

Problems caused by drought were not limited to cropping difficulties, and the need for regional relief was not restricted to farmers. Livestock owners were often the first to feel the pressures of drought. When range grasses failed to "green up" through dry spring months and waterholes failed, herds had to be culled for sale on a falling market, and foundation stock had to be fed from sparse hay and feed-grain stores or transported to greener pastures. Costs rose, as livestock prices sagged. Feed-grain loans for maintenance of farm work animals under the earliest of the federal relief measures and extended for range stock in February 1931 were offered again in drought emergency legislation of 1934, 1936, and 1937 and in the rehabilitation programs of the Resettlement Administration and the Farm Security Administration. Large amounts of the sums identified as "seed loans" in the accounts of the federal Bureau of Agricultural Economics were, in fact, loans for production or purchase of livestock feeds—over $7.3 million in Montana and the Dakotas under the amending legislation of February 1931 alone. The Farm Credit Administration reported emergency feed loans for the year ending June 30, 1935, totaling $12,119,681 to North Dakotans, $16,263,348 to South Dakotans, and $1,672,064 to Montanans. Of those sums only 5.1 percent, 15.5 percent, and 26.1 percent, respectively, had been repaid by the end of the decade.[73]

Other programs more specifically directed to assist range rather than farming interests were also provided. As early as mid-May, during the extreme drought of 1934, the AAA set up the Drought Relief Service, which over the next eight months bought almost 8.3 million cattle, at a cost of $111,000,000, to relieve the distress of livestock producers. More than 20 percent of these purchases were made in the Dakotas and approximately 5 percent in Montana. In northwestern North Dakota and southeastern Montana from 60 to 80 percent of the cattle were sold under this program, and in the remainder of western North Dakota and the northern two-thirds of western South Dakota, from 40 to 59 percent. Purchase of sheep was added by September of that year. Conditions were described as worse at the end of June 1936, when 100 railway boxcars of cattle from Custer County, Montana, and 31 cars of sheep from the Deer Lodge Valley, west of the Continental Divide, were shipped eastward to feeding areas or markets. At a drought conference called by the governor of Montana that July, nearly every county in the state reported that it would be necessary to move from 20 to 85 percent of the livestock out of the region. Between 1935 and 1940 the number of cattle in eastern Montana was reduced from 1,090,987 to 700,544 head, and the number of sheep, from 2,810,863 to 2,308,940. The decline was numerically less in the Dakotas but in South Dakota amounted to almost 25 percent of the cattle and over 13 percent of the sheep.[74]

Range rebuilding funds were incorporated as features of the Soil Conservation and Allotment Act of 1936, the Agricultural Adjustment Act of 1938, and the Land Utilization projects under the government's land retirement program. With public funding, waterholes were dug, wells were drilled, grasses were reseeded, and fences were constructed. Most of the development of reservoir and irrigation works was designed to promote production of hay and stock feeds, not cash crops. Indeed, as a program was shaped for agricultural reform in the region, it was not directed toward instituting the mechanized, specialized, commercial cropping that Wilson and his Montana colleagues had envisioned in the late twenties, but toward alleviating the problems of livestock growers—overcrowded, fragmented, and eroded rangelands, scarcity of feed grains, and the lack of accessible water supplies.

The work of the major regional relief organizations after 1935, the Resettlement Administration and its successor, the Farm Security Administration (FSA), is evidence of this emphasis. That reform effort will be discussed below. "Rehabilitation," as an aspect of relief, constituted the more immediate concern of these agencies. They assisted in debt adjustment, in some areas developed group health arrangements, helped introduce better sanitary facilities and housing improvements, sought to improve tenure arrangements, and encouraged cooperative effort for a variety of improvements, from development of water facilities, to purchase of a thresher or tractor, to support of

4-H Club activities. When emergency conditions arose, they also administered feed and seed loans and extended general relief to the needy. More than 1,100 farm families in Williams County, North Dakota, were receiving emergency grants from the Resettlement Administration in March 1936, and 1,500 families there were so aided by the FSA from 1937 to 1939, the heaviest record of public relief assistance in the state. Between 1934 and 1940 outright grants by these agencies totaled almost $22 million in South Dakota, which led the nation; almost $20 million in North Dakota, which ranked second nationally; and about $5.5 million in Montana, fifth in the rankings. "Rehabilitation" loans amounted to $18,218,000, $12,183,000, and $10,668,000 additional, respectively.[75]

Local forces continued to assist in the relief effort. The Red Cross distributed flour, cotton goods, bedding, and clothing. States provided equalization funds to assist needy school districts. Highway construction as well as the projects of the WPA and the PWA were cooperative federal-state undertakings, the last two programs on a ratio of 45 percent federal to 55 percent local contributions. The FERA also worked through the states, nominally on the basis of $1 in federal funds to $3 of local moneys. To meet relief costs, in January 1934 the Montana legislature authorized an appropriation of $750,000 funded from a percentage of the receipts from an income tax. The succeeding legislative assembly enlarged the appropriation by $3 million, financed additionally through allotments of income from an inheritance tax and a wide range of license fees. In 1933 North Dakota initiated a retail sales tax and South Dakota, alcoholic beverage license fees, from which part of the funds were to be directed to welfare activities.[76]

As the drought continued, however, relief programs came increasingly under federal emergency funding. From January 1933 through the end of the fiscal year 1939, federal, state, and local expenditures for various relief programs amounted to $130,604,915 for Montana, $130,258,871 for North Dakota, and $156,603,750 for South Dakota. Of those sums, state and local contributions amounted to only 14 percent for Montana and North Dakota and 12 percent for South Dakota during the first five years. Afterward, the rate of local sponsorship improved slightly, but for WPA-operated projects remained below 20 percent throughout the region during the rest of the period. Over the same span of time, federal loans amounted to $141,835,952 for Montana, $185,055,710 for North Dakota, and $172,206,250 for South Dakota, ranking these states on a per capita basis third, second, and fifth in the nation, respectively, for such assistance. It was a record that identified the region as a problem area of exceptional severity in a land suffering generally from the Great Depression.[77]

During the twenties, farmers had asked for governmental intervention to raise agricultural prices on the basis that business interests had long benefited from the national protective tariff policy. The massive federal funding represented by the drought relief programs of the thirties, on the other hand, began to evoke sharp criticism from rural and urban spokesmen alike outside the ravaged area. Even in the West the editor of the *Spokane Review* complained in the fall of 1935, when national wheat production had again begun to rise, that farmers in North Dakota were receiving an estimated $1,250 apiece, as "extra cash bonus" in "relief and compensation."[78]

Such comment brought a quick retort from a North Dakota editor, who maintained that the region had received "Loans, loans, loans—with but little form of direct relief." The same North Dakota journalist subsequently conceded that relief expenditures had, indeed, been large; but he argued that they represented only ten cents for every dollar lost "during the hard times period," when farm income should have amounted to $945 million under conditions applicable from 1924 through 1928.[79]

Farmers of the area had become defensive. They sensed in the massive credit and relief liabilities a growing vulnerability to extraneous pressures for agricultural readjustment. By the latter thirties, public debate questioned whether the farming system that had generated the income of the late twenties could or should be maintained. The compulsive force of credit and relief administration was being directed to "reform" under a very different agenda.

CHAPTER FOUR

Land-Use Planning for Regional Adjustment: The Resettlement Program

By the mid-thirties, the severity of conditions resulting from drought and declining prices for grain and livestock had brought pleas for assistance that had exhausted local institutional resources in the northern Plains. Political pressures, marked by sporadic violence, had extended the demand for governmental action to the nation's capital. Massive federal funding was being directed to relief of farmers and ranchers of the region as credit and as outright grants. With relief, however, came an administrative impetus to effect lasting ameliorative changes in the local economy.

What those adjustments should be belatedly called forth a large body of social and economic research on the problems of land utilization in the area; but in the immediacy of the crisis, much of that analysis rested upon limited data and faulty assumptions, much of it was completed too late to afford guidance, and much of it was ignored or rejected by the participants concerned in the arrangements. These limitations were all evident in the resettlement program, perhaps the most unsuccessful and unpopular of the efforts, and they reduced the effectiveness of such alternative measures as irrigation, "rehabilitation in place," conservation, and community-based land-use planning. Review of the introduction and regional impact of these programs forms the content of this chapter and the next.

As previously noted, there had been little national concern about planning for utilization of regional land resources prior to the twenties. Major John Wesley Powell's proposals for development of the Plains—as pasturage farms of 2,560 acres, with not over 20 acres cultivated under irrigation, or irrigated farms with approximately 80 acres under ditch—had been rejected in 1891, when the public lands segregated under his direction were reopened to homestead settlement. The subsequent development of irrigation under the Newlands Reclamation Act had not entailed consideration of using the surrounding range areas, a deficiency emphasized in the recommendations of the Bureau of Reclamation during the early thirties.[1] The conception had prevailed that settlers should be free to engage in dry-land agriculture wherever topography and economy of operation favored such enterprise. On that basis, even much of the irrigable land under ditch was not irrigated, and grazing areas were increasingly constricted to hilly, broken, or "badlands" districts.

Land under dry farming could not readily be returned to its earlier state. Environmentally, restoration of native grasses was slow. Economically, it entailed a loss of income both immediately and potentially; capital valuations were degraded even as the costs of investment in a new form of enterprise mounted. Socially, a surrounding community of relationships and services would be disrupted and, in large part, destroyed. Ranchers hoping to recover access to lost stock range and farmers so disheartened and desperate that they were prepared to cut their losses and start over might contemplate such a change with equanimity, but it was not so viewed by a majority of local residents after the initial period of agricultural settlement. The development of land-use planning committed to such an adjustment was extra-regional in perspective, presented as a concern of national interest.

Growing criticism of unrestricted agricultural settlement evolved through the twenties less as a response to the drought conditions that marked 1919 through 1921 than as a reaction to the marketing problem that followed expansion of wheat acreage during World War I. The U.S. Department of Agriculture (USDA) warned in January 1920 that reconstruction of agriculture on a peace basis would be necessary: some wheat and tilled crops should be limited in favor of hay and pasture; mounting land prices and rampant speculation could not continue: "consolidate the gains already made; prepare for the period of competition which is to be expected." A congressional Joint Commission of Agricultural Inquiry focused even more specifically on economic issues in recommending marketing, transportation, and credit reforms as the appropriate response to the postwar farm crisis. In addition to such measures, a National Agricultural Conference called by Secretary of Agriculture Henry C. Wallace in 1922 urged proposals calling for monetary stabilization, improved international trade, better reporting of crop statistics, crop insurance, and expansion of agricultural research and extension facilities. The report advised farmers to adjust their operations to market requirements. A committee of the conference even suggested a 15 percent reduction in wheat acreage. But the problem was still viewed as essentially one growing out of the price decline. When Secretary Wallace introduced his "Agricultural Outlook" reports in 1923, the objective centered on providing farmers *generally* with better information on which to scale their operations. The same generalized approach to a marketing solution was embedded in the McNary-Haugen agitation which dominated so much of the debate on the agricultural situation through the remainder of the decade and continued into the early years of the New Deal.[2]

A Spring Wheat Regional Council, bringing together representatives of the USDA and the state agricultural institutions, was organized late in 1922 and met at St. Paul in January 1923. Prior to the meeting the department representatives organized into two committees, one devoted to marketing problems and the other to production concerns. The latter called for collection of data on physical factors in the region—climate, water supply, topography, soils, and natural vegetation—and on the developmental experience of agricultural enterprise. Drawing upon surveys of settlers' progress recently undertaken by the newly established Bureau of Agricultural Economics in cooperation with the agricultural colleges and experiment stations of Montana and the Dakotas, such studies, cited above, afforded much promise of successful regional development and tended to point to managerial rather than environmental limitations. In final analysis, the St. Paul gathering emphasized marketing problems and concluded that the difficulty that agriculture then confronted was "due far more to the disparity in prices between farm products than to errors in farm organization or practice." Their report commented:

> It is recognized that the types of farming and farm practices in operation at present are the outgrowth of years of experience and accumulation of practical first-hand knowledge on the part of the farmers who occupy the land. In a large majority of cases the combined judgment of farmers of a community is correct as to the most economic organization of the farm business. When there is a well-defined type of agriculture in a region it should be disturbed only when there is reasonable probability of improvement, and then but slowly.[3]

If, however, production were to be reduced, it was logical to question whether that action might not be better applied in some localities than in others. Professor Richard Ely, of the University of Wisconsin, had raised the issue in an address on "A National Policy for Land Utilization" before the National Agricultural Conference of 1922: where should use of land be directed to agriculture and where to other purposes? Lewis Cecil Gray, head of the Division of Land Economics, established in the USDA in 1919 and incorporated three years later in the Bureau of Agricultural Economics (BAE), had been one of Ely's students and pursued Ely's questioning of land-use patterns. As chairman of a Land Utilization Committee appointed by Secretary Wallace in 1921, Gray noted that no attempt at a complete economic classification of the national land area had ever been made but emphasized that for some decades there had been almost no significant expansion of land in forests, grazing, or crops except at the expense of one of those alternative uses. Since at least 1890, he argued, the expansion of crop areas had been into pasture or grazing districts. As a result of the activities of real estate sellers and the desire of settlers to profit through rising land values, that expansion had exceeded the need for cropland. He also contended that the price of wheat would need to be much higher than it was at that time for the crop to be profitable in much of the semiarid region, where, in his view, yields would probably not average over seven bushels to the acre. The results had been bad not only for the developers and the settlers, but still more for lowering the average level of profitability of established farming industry.[4]

It was an argument that eastern opponents of westward expansion had voiced since the founding of the nation, one to which the economics of alternative profitability had given rise successively from the wheat fields of the Connecticut Valley across those of the Genessee country in western New York, those of the Lake Plains of Illinois and Wisconsin, and those of Minnesota and the northern prairies. It had long been leveled against expansion of irrigation under the Newlands legislation, and Gray was a vehement opponent of such development as well as of dry-land settlement. In 1923 he called for revision of public land policy toward systematic direction of agricultural expansion and urged specifically the creation of grazing districts to operate under a permit system on the public lands, contending that "the *let alone* policy of the past few decades has been a source of economic waste and social misery."[5]

In his report for that year, the commissioner of the General Land Office conceded that land in bodies of 640 acres suitable for settlement under the stock-farming homestead legislation of 1916 could no longer be readily found and proposed that the secretary of the interior set apart public lands suitable chiefly for grazing to be leased under regulations insuring preservation of their forage value. He believed it "entirely feasible" to combine small stock-farming homesteads with such leasing units. Renewing the recommendation in succeeding years, he also urged in 1925 termination of further settlement under the existing homestead legislation. The organization of the Mizpah–Pumpkin Creek Grazing Association in Montana in 1928 and enactment of special congressional legislation authorizing the government to enter into cooperative agreements for ten-year grazing leases on public lands in the area formed a model for the ensuing reform effort.[6]

Addressing the Conference of Western Governors at Salt Lake City in August 1929, President Hoover declared that the western states were better able to manage many of their affairs than the federal government was and recommended that, apart from certain mineral rights, the remaining unreserved and unappropriated public lands be ceded to the states in which they lay. He named a Public Lands Commission directed not only to study proposals for such a transfer but also to review the role of the federal government in relation to reservations for forest, mineral, water power, irrigation, and scenic uses. The majority report of the commission, issued in 1930, endorsed the idea of state ownership of the lands and suggested that the states also determine what national forest lands ought to remain reserved. Eastern conservationists were outraged, but so, too, were many westerners. Recent study had shown that the states were ineffective in regulating speculative activities. Small-scale stock operators and farmers were fearful that the larger and wealthier stockmen would block their access to range lands. As members of the Democratic party acquired control of the House of Representatives in December 1931, they refused to consider the proposed legislation.[7]

When the Roosevelt administration came to power in 1933 with conservation as a major policy concern, it was committed to the view that the remaining public lands should remain under national protection. The demands of local stockmen, whose activities were constricted by the limited availability of range, were assuaged by passage of the Taylor Grazing Act in June 1934. Virtually all the unappropriated public domain—over 6 million acres in Montana, over 500,000 acres in South Dakota, and somewhat less than 150,000 acres in North Dakota—was withdrawn from entry and made available for leasing under strict regulation as grazing land. Legislation in all three states provided for organization of cooperative grazing associations empowered to enter into leasing arrangements under which small, isolated tracts might be consolidated and developed as range areas. Under county programs for disposition of property seized for tax delinquency, Butte and Harding counties in South Dakota also developed block leasing plans by which individual operators might acquire preference in leasing such tracts where they were identified as proximate to established ranch holdings.[8]

But the program as envisioned by land-use planners during the thirties did not hold much promise for farming enterprise in the region. Through the previous decade the Division of Land Economics had conducted a broad range of studies projecting national crop requirements—trends in crop and livestock production, both in the United States and abroad; national population growth; changing consumption patterns; the effect of mechanization on feed requirements; land areas available for development; productive capability of lands in production and of those which might be utilized; prevalence of soil erosion; and tenure problems. Working under Gray, O. E. Baker compiled much of the statistical data and calculated their implications.

At first the message had been one of impending scarcity, emphasizing a declining per capita supply of agricultural produce since early in the twentieth century. The report of the secretary of agriculture in 1923 commented that, because of the rapid increase in population, there would "before many years . . . be a home demand for even more of farm products" than were currently being grown. Baker noted that 60 million of the 84 million acres of reported increase in farm land during the previous decade were "unimproved." Most of the land settled under the Enlarged Homestead and Stock-Farming Homestead legislation of the dry-land settlement period had been "permitted by the homesteaders to remain in wild grass." "Transformation of this western range into so-called farm land, in other words, has not involved much increase in intensity of its utilization," he commented. In 1925 he wrote: "It is certain that if the population of the United States continues to increase for more than another century as it has during the past century there is no means by which the present standard of living can be maintained, except by importation of foodstuffs from other lands—which will need their foodstuffs even more than we."[9]

By 1928 a reversal of viewpoint had developed. Remarking upon evidence of three "new facts"—a recent "extraordinary increase in agricultural production," notable changes in consumption patterns, and decreasing birth rates pointing to a "stationary population in about fifty years"—Baker concluded that there would be no need to expand the farm area of the nation for at least a decade; there remained "enough idle crop land in farms to provide for nearly all the probable increase in crop acreage by 1939, and in addition . . . over 100 million acres of plowable pasture in farms, not to speak of the large areas in farms that may be utilized for crops by clearing or draining the land." A few months later he warned: "Unless control and guidance of agricultural settlement is provided for, there is grave danger of recurring periods of agricultural depression such as that which has now continued for eight years—indeed, the condition may be chronic." When Gray and Baker published a graphic consolidation of their data in 1930, under the title *Land Utilization and the Farm Problem*, it focused on joint causal factors for the national agricultural difficulties, "the influence of general overproduction in agriculture and of maladjusted production."[10]

Baker, whose observation of the "dynamic" impact of "modern methods and mechanization" in the Great Plains led him to comment in 1932 that land deemed "submarginal for crops a few years ago" had become "clearly supermarginal," thereafter ceased publicly to discuss policy formation on national land utilization. Gray, instead, rejected as "far too simple" the idea that land planning rested upon "economic land classification for the purpose of drawing an imaginary line between submarginal and supermarginal farm land." Such a conception, he explained, "while useful, assumes a highly commercial type of production and leaves out of account paramount governmental and social considerations." He aimed instead to eliminate rural slums, conserve soil resources, and better group rural population for efficiency and economy in local government. Pointing to recent efforts by the federal Bureau of Chemistry and Soils and the state experiment stations to classify land in individual areas on a basis of its relative productivity as compared with that physically best adapted in the nation for the kinds of crops grown in the area, he found "a surprisingly extensive practice of commercial production on poor land, particularly in the cotton and wheat belts." In April 1934, after the AAA program had been initiated, he raised the question "whether acreage reduction by outright purchase, as distinguished from temporary subsidy, leasing, or the acquisition of easements, should be the method employed to bring production more nearly into line with demand." He opted for "gradual permanent retirement of the lean acres." "For some years at least," he noted, "we shall be interested . . . in whatever reduction may be effected in net productive area, and because of this immediate interest the policy should be pushed as vigorously as possible, and particularly in the areas of poor land employed largely in commercial production."[11]

That farmers and developmental leaders of the northern Plains were, alike, confused by the conflicting pronouncements of the "experts" during the early thirties is understandable. Gray's early concern had clearly been dictated more by desire to reduce the volume of production than by evidence of regional maladjustment in land utilization. In an address before the American Farm Economics Association in December 1930, however, he had delineated the focus of his proposals for "Redirection of Economic Life in Economically Subnormal Areas." He considered reforms "necessary only for areas where the question of where agriculture should stop and other uses begin is undetermined, and for some time to come," he added, "I would confine the undertaking to areas in serious economic distress." As drought intensified in the northern Plains in 1933 and 1934, that region bore the primary thrust of the effort.[12]

Gray projected his views through his roles in a succession of influential positions: head of the Division of Land Economics in the BAE and after 1933 chief of the Land Policy Section in the Planning Division of BAE; executive secretary of a National Conference on Land Utilization at Chicago in November 1931 and a member of the Land-Use Planning Committee that continued the work of that conference into 1934; director of the Land Section of the National Resources Board and a member of the Land-Use Planning Committee that developed its program in 1934; chief of the Land Policy Section of the AAA in 1935; assistant administrator of the Resettlement Administration, 1935–1937; and assistant chief of the BAE, in charge of land utilization, from 1937 until his retirement in 1941. The participation of the USDA in a National Conference on Land Utilization in 1931 was in large measure owing to his efforts.

For several years scholars in the new discipline of agricultural economics had been discussing the problems of farm depression and their relief. Few of them endorsed the voluntarism of the McNary-Haugen proposals. In the late twenties Edwin G. Nourse, of the Brookings Institution, and Rexford G. Tugwell, of Columbia University, had both looked to proposals for "scaling productive operations in conformity with market demand" and in accordance with principles of commercial efficiency. The former had deplored as general difficulties in the way of such readjustment the "incurable belief that every acre that can yield any agricultural product should be put to agricultural use and . . . a widely cherished superstition that any reduction in the quantum of agriculture to be found within the country spells national decay." Tugwell had proposed that planting be limited according to ten-year average yields. Enforcement, he had suggested, could be achieved by denying use of railways and warehouses for produce grown without authorization. In reference to restrictive governmental sanctions, he wrote: "We may offer the farmer assistance on condition of good behavior consisting in growing those crops which seem wise in such amounts as seem desirable, and by such methods as social prudence dictates."[13]

Hoping to channel this interest into support for the work of his division, to attract greater publicity to its recommendations, and to win action in response to them, Gray prevailed upon Secretary of Agriculture Arthur M. Hyde to sponsor a meeting at Chicago bringing together representatives of the federal Department of Agriculture, the Association of Land Grant Colleges, and the major farm organizations, along with influential academic leaders. Gray not only served as executive secretary but presented one of the principal addresses, calling for retirement of "submarginal" lands. The conference adopted recommendations endorsing national land classification, regulation of grazing on the public domain, abandonment of submarginal farms, restrictions upon settlement or re-occupation by farmers in submarginal agricultural areas, regulation of private land colonization projects, limitation of federal reclamation development, and increased public purchase of forest areas. The body also established two continuing committees, the National Land-Use Planning Committee and the National Advisory and Legislative Committee on Land Use, which over the next two years debated such issues as the expansion of agricultural enterprise, current efforts to develop and attract settlers onto irrigation projects, and the "back-to-the-land" movement then luring the urban unemployed into rural areas.[14]

The proceedings garnered widespread publicity and contributed to growing interest in a national program of land use. Both major political parties in the presidential campaign of 1932 espoused policies that called for national regulatory action. The Republican platform urged diversion of land submarginal for crop production to other uses and acquisition of such acreage for watershed protection, grazing, forestry, game reserves, and public parks. Franklin D. Roosevelt, who as governor of New York had led a movement for expansion of forestry reserves in that state, directed particular attention to the problems of land utilization in the eastern seaboard states "and a few others." The American Farm Bureau Federation, the National Grange, and the U.S. Chamber of Commerce, organizations which rarely agreed on governmental program, all adopted resolutions prior to the election endorsing at least the broad principle of a need for a national policy of land use.[15]

M. L. Wilson, addressing the annual meeting of the American Farm Economics Association in December 1932, echoed L. C. Gray's phraseology in acclaiming the "new frontier" of a "rationalized land use" that lay before the country. The Montanan had recognized that his proposal for funding of research to delineate specialized farming areas set up the test of comparative advantage as the basis for regional agricultural adjustment. He still in 1932 wrote optimistically of the developing technological revolution, but as drought combined with low prices to handicap farming in the Plains areas of Montana, he had begun to question the success of the effort there. He explained to the president of the Montana State Farm Bureau as early as March 1931 the need for a policy that would move settlers from the poorer range and dry-farming lands. About the same time he spoke to the Rotary Club of Lewistown of the need to shift attention from farm production to land classification. He reported that the agricultural extension department of the State College had surveyed nineteen counties over the past summer to grade the lands according to yield capacity. Wheat lands were ranked from first to fourth class on the basis of productivity: those yielding twenty or more bushels to the acre, fifteen to twenty bushels, twelve to fifteen bushels, or ten to twelve bushels. Grazing lands, too, were evaluated, on the basis of seasonal carrying capacity for a cow or steer, ranging again through four grades: those requiring less than 15 acres per animal, 15 to 25 acres, 25 to 60 acres, and 60 or more acres. Wilson advised that wheat lands yielding at the level of grade two might be retained in cultivation; those in grade three were questionable; those in grade four should be taken out of production. Much of the Judith Basin, he assured the Rotarians, graded as first class. Over the United States as a whole, however, he estimated that 100 million acres of the total 500 million acres of farmland were "submargin lands—lands which were capable of producing profitably during favorable years but not capable of producing profitably over a period of years." Retaining them in production even during favorable years was undesirable, he commented, because it lowered prices, and in unfavorable years it necessitated seed loans and other assistance.[16]

Wilson, who had worked in the BAE with Gray, Baker, and Howard R. Tolley, had maintained those contacts through the intervening years. He had gone to Washington in February 1930 to discuss the Christgau proposal with Tolley, then assistant chief of the bureau. Wilson had also been active in organizing the National Conference on Land Utilization in 1931 and had prevailed upon a former Montana colleague, Professor E. A. Duddy, of the University of Chicago, to issue the invitations that brought the gathering to Chicago.[17]

How completely Wilson's views had become similar to those of Gray was evident in the Montanan's address before the annual meeting of the American Farm Economics Association in 1932. Wilson called for the federal government to provide leadership in fundamental research on land use, notably in supervising and correlating surveys for a national land inventory; in studying the effect of national and international economic changes and trends in production and consumption in relation to land use, problems of tenancy and ownership, patterns of land settlement, and rural and urban land development; and in supplying technical assistance to states, regions, and local areas on land-use programs, particularly in "problem areas." He urged that the administration of the public domain be transferred to the USDA, that the federal reclamation program be coordinated with planned national land use, and that the role of the Department of the Interior be limited to the engineering phase of reclamation work. He recommended federal acquisition of submarginal lands and suggested that rural zoning by the states might further desired land use.

However, Wilson interjected some moderating ideas that were seldom voiced among the Washington administrators. He believed projects in land-use development should originate and be carried out largely by local people, with the cooperation of all interested agencies, with the federal agencies serving primarily as coordinators in the interest of national policy. He warned that "Sporadic and ill-conceived federal land acquisition . . . [would] do little good," yet he recognized that "the purchase of tillable crop land on a scale sufficient to affect materially the volume of agricultural production would require a sum of money which would be almost staggering." Furthermore, he warned that such a program "would take away the homes and basis of existence of countless farm families." While it might be "socially desirable to shift production gradually away from the submarginal acres, such action must be coupled with planned shifting of the population also." The start must be made with "well matured plans."[18]

With the National Industrial Recovery Act, June 16, 1933, Congress provided $25 million for the development of so-called subsistence homesteads. The program was designed to provide for "redistribution of the overbalance of population in industrial centers" and was originally placed under Wilson's direction in the PWA of the Department of the Interior. Wilson viewed his mandate to include not only the goals of the "back-to-the-land" movement but also to rehabilitate the rural poor. He aimed to avoid conflict with the crop limitations sought by the AAA. Settlers were to produce farm goods for their own livelihood but to work off the farm to provide their necessary cash income without entering into commercial crop markets. In view of Wilson's previous criticism of small-scale, diversified farming and his emphasis upon mechanized, commercial operations, his interest in the subsistence-homestead development was an anomaly. The proposal, nevertheless, provided the format on which the resettlement program was begun.[19]

When the USDA launched the AAA program in the summer of 1933, Secretary of Agriculture Henry A. Wallace and his advisers were concerned that pressure for expanded reclamation projects, to meet both the problem of drought and the need to provide employment on public works, would lead to conflicting policies of land utilization. The long-standing bureaucratic rivalry between the Agriculture and Interior departments, maximized in controversy between Gray and Elwood Mead over land use and farm management on the projects, heightened the need for some administrative coordination. At Wallace's suggestion, President Roosevelt, who had earlier taken a stand against expanded reclamation, agreed in late July to set up a joint committee of the two departments to plan procedures so that for every dollar spent on reclamation another should be devoted to purchase and retirement of submarginal land. On December 28, 1933, Secretary of the Interior Ickes, who was head of the PWA, authorized the Federal Employment Relief Administration to provide $25 million for purchase of submarginal land.[20]

Gray's duties were enlarged to combine his role as head of the research agency, the Division of Land Economics in the BAE, with administration of the land policy section of the planning division of the AAA, which selected the lands to be withdrawn from agricultural production and those onto which displaced farmers were to be resettled. He viewed the funding and establishment of the new section as "the culmination of the work of the old Land Use Planning Committee." Several organizational stages had intervened. In July 1933 a National Planning Board had been established within the PWA, with Frederick A. Delano, the president's uncle and a noted city planner, as chairman. As an adjunct to this board, a new Committee on National Land Problems was formed that autumn, with Wilson, who now held ties bridging the Agriculture and Interior departments, as chairman. The first annual report of the planning board urged that such an organization be given permanent standing as an independent agency. In June 1934, President Roosevelt by executive order established, as successor to the National Planning Board and the Committee on National Land Problems, an interagency National Resources Board, consisting of the secretaries of Interior, Agriculture, War, Labor, and Commerce, the administrator of Federal Emergency Relief, and the members of the National Planning Board. Chaired by Ickes and directed by the Delano board, the new agency received a mandate to prepare by December a comprehensive plan for development and use of the nation's natural resources. Gray became director of the land section, under which Wilson served as chairman of the Land Planning Committee. At the same time, Wilson's return to the Department of Agriculture as assistant secretary to Wallace brought the working relationship of Gray and Wilson even closer.[21]

The report, prepared in a brief five months, presented a broad range of recommendations relating to collection of essential data; land, water, mineral, and public works development; organization of cooperative state and regional planning agencies; and maintenance of continuity in the effort. The body of the program with reference to agricultural development was in the report of the Land Planning Committee. It called for classification of lands "according to problems and probable best uses . . . in order to designate areas unsuited to settlement," repeal of the homestead laws, purchase and retirement of submarginal lands at the rate of 5 million acres a year for an estimated period of fifteen years, "Squaring out or blocking up" of delinquent tax areas, and encouragement of zoning and land purchase by states and counties to promote best uses of the land. Additions were recommended for national and state forests, wildlife refuges, public parks, Indian reservations, and controlled grazing areas. Need for additional reclamation projects was recognized but conditioned by technological and economic criteria. Completion and full use of old irrigating works was stressed as a preliminary to building new ones, and consideration was to be given to construction "of numerous small and seemingly unimportant irrigation operations, which, though essentially local in significance, in the aggregate are of considerable economic importance." Collaboration of the USDA, the Federal Emergency Relief Administration, and the Farm Credit Administration, with state and local agencies and individuals, toward consolidation of some farm holdings and modification in types of farming was urged.[22]

It was anticipated that nationally some 75,345,000 acres would be retired from farming, approximately 20 million acres of cropland, almost 35 million acres of pasture, and the remainder idle land and farmsteads. Problem areas were described as hilly, forested districts; regions of light, sandy soil and serious erosion; and the dryer parts of the Great Plains. In much of the wheat-producing area, farms were found too small to maintain a satisfactory livelihood because productivity per acre was low, and frequent droughts made it necessary to farm extensively and carry reserves of seed and feed. There an estimated 16,000 families were to be displaced and approximately 3 million acres of wheat plus a half a million to a million acres of feed grains taken out of production. Another 20,000 families and 2 million acres of farmland on irrigation projects, where farms were too small or water tables declining, would be similarly affected. An accompanying map of the areas in which it was deemed "desirable to encourage permanent retirement of a substantial part of the arable farming" showed most of the western two-thirds of South Dakota, over one-half of western North Dakota, and large areas of central, northern, and northeastern Montana.[23]

The map had been prepared on the advice of land-planning consultants appointed for each state, but the report conceded that judgments differed on adjacent lands across state boundaries. Studies of soils in Montana from Teton and Cascade counties south and eastward had not at that time been published, and land classification elsewhere in the region was limited to the general categories developed early in the century by the U.S. Geological Survey on the basis of topographic observation—as farming land, farming-grazing land, grazing land, grazing-forage land, non-tillable grazing land, and irrigated or irrigable land. The text explained that the maps were not intended to show the intensity or the magnitude of the recommended land retirement in terms of acreage or population, which might range at levels from 20 to 100 percent of all farms in the broadly sketched retirement zones.[24]

A supplementary report of the Land Planning Committee, prepared a year later, enlarged the areas for crop restriction somewhat in southwestern North Dakota and very considerably in eastern Montana. The first map had included the Judith Basin counties in the retirement area but excluded the

Musselshell Valley and all of southeastern Montana, but the revision exempt-
ed the former district and extended the retirement zone eastward through the
Musselshell country to the North Dakota border. It was a decision that may
have been influenced by the fact that the government was already taking
options on land in Petroleum and Musselshell counties. The later map also
more precisely reflected data available from the Gieseker soil surveys of
Montana, noting the existence of successful dry-farming operations in north-
central, northeastern, and east-central Toole County, the Turner bench of
northeastern Blaine, the Poplar Valley district of Daniels County, and the
benches of southern Chouteau.[25]

Questions remain concerning the coverage of even the second map. Most
of Toole and Blaine counties of Montana were included in the retirement zone,
but they ranked among the least distressed counties in the state. Their wheat
yields and farm income in 1929 had averaged higher than those in even Fergus County; they were areas of slight to very slight drought intensity during
the period from 1930 through 1936; their relief rate through the mid-thirties
was low; and their soils were not severely eroded. In southeastern Montana,
largely excluded from the retirement zone, farm income levels had been lower in 1929 and the rate of federal aid per capita was among the highest in the
state during the mid-thirties. Ranked in a contemporary survey of areas of
intense drought distress, the latter counties were among the worst afflicted.
Similar discrepancies marked the classification in the Dakotas. Williams, Renville, Bottineau, and Hettinger counties were excluded from the retirement
zone in North Dakota but ranked among the most distressed areas, with
among the highest per capita relief rates and major problems of wind erosion.
Tripp, Lyman, and Gregory counties of South Dakota, also omitted as districts
requiring adjustment, had large areas of very severe wind erosion, and all
were within the zone of highest per capita relief expenditure.[26]

The report did not mention the differences in quality that had brought
dry-land wheat producers premium prices through much of the postwar
decade. It suggested that the regional advantages of mechanized operations
had been offset by the introduction of one-plow tractors, small combine-har-
vesters, and leveling adjustments on farm equipment in eastern farming areas. If the back-to-the-land movement continued as a means of relieving industrial unemployment, the authors noted, farming would tend to be more
self-sufficient and less mechanized. "It is important to recognize that mechanization implies a commercial as distinguished from a predominantly self-
sufficient, type of economy," they commented in dismissing the topic. To
them commercial production was clearly not worth consideration.[27]

Neither map was, in fact, used as a guide for land purchase. Sherman E.
Johnson, who was given the assignment to establish the program in the northern Plains in May 1934, already had the work under way in the major areas
of development by the end of the year in spite of the fact that he had had few

guidelines from Washington. In Montana the pioneering-land classification by the Agricultural College, a resettlement project initiated in the Malta area during the late twenties, and the cooperative grazing district in the southeastern district governed the general locale of his project selection. In North Dakota he relied upon recommendations by J. C. Ellickson, who was then engaged in rural rehabilitation under the North Dakota Relief Administration, and M. B. Johnson, a former county agent, who had been recently employed on a cooperative research study of cattle ranching and was described as "well acquainted with the livestock ranchers in the area as well as those who were attempting to grow wheat on land ill-suited for crops." The program in western South Dakota was developed under the guidance and local administration of Oscar Hermstad, county agent at Rapid City, who suggested the area adjacent to the Badlands National Monument as a purchase area. By 1937 they had acquired 3,333,000 acres in the three states, and by 1946, 4,228,000 acres, nearly 2,000,000 acres more than was ultimately to be retired in all the remainder of the Great Plains.[28]

The report by the National Resources Board had been designed to define recommendations entailing legislative support. The Taylor Grazing Act had been enacted just prior to the call for the report. As originally written that legislation had authorized segregation of only 80 million acres of the public domain for organization into grazing districts. The report recommended action that led to amendment extending the withdrawal to 142 million acres and to an executive order covering the remainder of the federal holdings. States were to be asked to provide supplementary planning and zoning legislation. Tenure problems were noted that pointed to further study and ultimately passage of the Bankhead-Jones Act. The attention directed to potential development of small-scale watering projects was to culminate in the Water Facilities Act of 1937. But chief emphasis was given to recommendations for funding of a permanent planning agency and of land purchase, and on neither of these goals was the requisite congressional support forthcoming.

The National Resources Board (NRB) was continued until June 1935, with funding as a relief agency under the National Industrial Recovery Act. Renamed as the National Resources Committee (NRC), but with the same membership, it remained operational under various emergency relief measures through the end of June 1939. Its planning activity remained tenuous, functioning chiefly through the research staffs of other agencies. When President Roosevelt was unable to win the NRC permanent existence under the Reorganization Act of 1939, the commission was absorbed into a National Resources Planning Board and maintained from year to year under legislation supporting independent offices. In June 1943, Congress finally rejected further funding and called for abolition of the board by the end of August.[29]

Financing of the land-purchase program fell far short of the levels recommended by the NRB. Only two-thirds of the initial $25 million was allotted for farmland retirement, the rest apportioned for land purchases in connection with Indian reservations, parks, and wildlife areas. When the program was transferred to the Resettlement Administration in May 1935, further funding of some $48 million was again largely diverted, primarily to the agency's relief activities. In all, only about $27 million was made available for land-purchase projects before 1938. With that sum, acquisitions had to meet the requirements for both "conservation," that is, retirement of so-called submarginal areas from cropping, and resettlement of displaced farmers. In the haste to initiate the time-consuming process of appraising proffered land, obtaining options, checking titles, and clearing liens, far more of the former than the latter category of purchases was negotiated. Consequently, options were carried for years after their negotiation, while settlers to be displaced remained uncertain how long they could remain on their farms and were frequently supported by relief grants. Sustained funding was not assured until passage of the Bankhead-Jones Act in 1937, which authorized a budget not to exceed $10 million for the fiscal year ending in 1938 and $20 million for each of the succeeding two fiscal years. Afterward, funds were provided for completion of several projects designed as resettlement tracts, but by 1940 most of the purchase program had been terminated.[30]

Nationally a total of approximately 11.3 million acres was acquired, of which only about 2.5 million acres were cropland, much of it idle. Slightly more than 6 million acres were in pasture and range, and 2.7 million acres were in forest. The limited amount of cropland suggests that, as Baker had earlier noted, much dry-land development was already operated in conjunction with the running of livestock and that the program of land retirement was directed more toward reorganization and protection of the grazing industry than toward reform of agriculture. It served to consolidate rangelands by reducing the number of small stock-farming enterprises.

The largest projected land-retirement area in the nation, the Milk River project, covering 6,643,837 acres in Valley, Phillips, and Blaine counties of Montana, included some 849,000 acres of unappropriated public domain. The total wheat production from those counties in 1929 had been only 3.2 million bushels, almost a million bushels less than the amount produced in Chouteau County, alone, as the leading wheat producing county in the state. Although designated for land retirement, Chouteau was never included under a scheduled project area. Eastern Fergus County, Montana, in the second largest project, was rough, broken country in which from 73 to 91 percent of the operating units were classified as stock ranches and so-called combination farms. A speaker before the annual meeting of the American Farm Economics Association in December 1935 noted that the land-purchase program no longer appeared as one of production control. On that its effects would not be important, he commented: "It is mainly a long-time program related to the reversal in the trend of frontier development."[31]

Only a small proportion of the projected land purchase was accomplished, but the federal government established immense grazing areas by consolidating the tracts it acquired with remaining holdings of public domain and entering into agreements with cooperating private operators under state-authorized grazing associations. In Montana governmental acquisition of 2,111,000 acres permitted range development of the vast Milk River project area and of 4.1 million acres extending from Petroleum and eastern Fergus counties southward into northern Yellowstone and eastward through Prairie and Fallon counties to link with a tract of approximately 1.2 million acres in the Badlands of western North Dakota.

The government purchased 1,145,000 acres in North Dakota, centered primarily along the Little Missouri River from the 766,000 McKenzie County unit southward through the Slope counties into Bowman. Several smaller projects in western South Dakota led to purchase of 855,445 acres, encompassing 115,817 acres in Lyman, Jones, and Stanley counties, 580,896 acres of Badlands areas extending from Jackson and eastern Pennington counties southward through Fall River, and 155,428 acres in the 500,000-acre Grand River project in western Corson and much of Perkins counties. These lands were cleared of abandoned farm structures; refenced for range use; treated for insect and rodent eradication; supplied with hundreds of check dams, stock ponds, and wells; regulated in grazing use to permit recovery of native grasses; and where necessary, seeded with crested wheat grass to restore denuded areas.[32]

The process of consolidating units and accomplishing the work of range development dragged on for years. Hope that public expenditures for maintenance of schools and roads might be substantially lessened in such areas proved illusory. Stockmen continued to live near their herds for winter care and supplementary feeding, and not all farmers in the designated purchase zones entered into the program. By 1937 only some 300 of 1,200 to 1,400 families in the Milk River project area had contracted option agreements. It was anticipated that another 500 families might be self-sustaining in place if they could gain access to adjacent range; the remaining 400 families clung to holdings under indefinite status. In all, the government had purchased but 970,198 acres for the project by the end of the fiscal year 1939. Along with the unappropriated public domain, this amounted to but 27 percent of the planned retirement zone.[33]

In spite of a campaign by local planning organizations, the extension service, county commissioners, and service clubs during 1938, only about 7 percent of the acreage in the central Montana range area had been transferred to the government by the end of the year, although options were held on another 4 percent of the eastern Fergus County section and 9 percent of the Buffalo-Creek district, recently added to the Musselshell-Petroleum county segment. By 1940 about 55 percent of the acreage for the McKenzie County,

North Dakota, project area and two-thirds of that for the southern extension had been purchased, but less than 30 percent of the land for the Grand River project in South Dakota had been acquired. In North Dakota project areas about one-half of the operating units were eliminated and in those of South Dakota, about 46 percent.

Tracts satisfactory as ranch headquarters were not generally taken over. Unless farms were small and the land was needed for control of access and water, project managers, operating with limited funds, were willing to wait for the owners' decisions. As the rainfall increased, however, fewer settlers gave up their holdings. Government announcements relating to the program persistently reiterated that no pressure for sale was to be applied, yet planning leaders as frequently complained that relief programs fostered an unwillingness to give up unproductive holdings and urged that access to credit and relief assistance depend upon acceptance of the recommended changes. They argued that the cost of relief and of maintaining public services for a sparse population warranted removal in the public interest and even that the right of eminent domain should be exercised to obtain title where land was necessary for development of the project areas. Early in 1936 the district supervisor of the Resettlement Administration in central Montana specifically warned farmers that, because of limited funds, the agency could provide assistance to only those farmers who had sold their land to the government and were compelled to relocate. The record of rural rehabilitation in the Buffalo-Creek Grazing District showed that few loans for "rehabilitation in place" were, in fact, authorized between 1936 and 1938, the explanation being that soil and climate conditions made this a poor area for residential operations. Under a coordinated program the Farm Credit Administration announced a fixed policy not to make seed loans in areas "plainly not suited for crop production." The repeated complaints of delay in extending the loans provided some basis for suspecting deliberate intent to force removal.[34]

Local response to the program was mixed. Government administrators reported "general approval" during the early stages. At a regional planning conference in St. Paul in 1936, representatives of state planning boards, organized in all three states of the northern Plains at the suggestion of the National Planning Board, endorsed range restoration in poor soil areas and assistance for distressed farm families in relocating on nearby irrigated or productive farm lands. Governor Tom Berry, a South Dakota rancher, described by Sherman Johnson as "Perhaps . . . even prejudiced against those who had plowed up the range," was a strong supporter of the land-purchase program. While the political warfare between Governor Langer and the Roosevelt administration may have limited the North Dakota governor's support, his successor, Democrat John Moses, campaigned in 1938 under a call for resumption of governmental land purchase under the Bankhead-Jones Act. The Montana legislative assembly in 1939 adopted a joint resolution requesting continuation of rural resettlement under that legislation.[35]

County officials welcomed the payment of delinquent taxes, which held a priority claim on the proceeds of government land purchases. They also anticipated lowered costs for public services, reduced demands for relief expenditures, and shared receipts from grazing fees as compensation for the loss of tax revenues from publicly owned land. Community government and business leaders were assured that the rising level of regional prosperity under readjusted land use would offset the loss of settlers. Stockmen, who had long opposed governmental regulation, began to call for expansion of the submarginal land purchase program throughout the region and were quick to organize grazing districts under the state enabling statutes. Sherman Johnson, regional director of the land-purchase program in its formative period, reported that farmers too, "were more than willing to sell their lands even at the low price of $1 to $3 per acre, provided they had any assurance at all that they would have other opportunities elsewhere." He found little reluctance to sell because of sentimental attachment to their homes: "This attitude can probably be accounted for by the fact that most settlers in the Great Plains areas have not lived there all their lives," he commented; "also the hardship connected with years of crop failure is not as a rule conducive to developing sentimental attachments to the farm."[36]

Uneasiness and alarm at the direction of the adjustment program were also, however, evident. L. A. Campbell, chairman of the Montana Planning Board, qualified his endorsement by emphasizing that resettlement should be on "nearby" land; he renounced "any policy of wholesale abandonment of the region or of forcing people to move into other states." Such concern swept the region after the first announcements of the scope of the projected adjustments. In January 1934 local newspapers carried reports of Assistant Secretary of Agriculture Rexford G. Tugwell's estimate that it would be necessary to retire as much as 100 million acres from cultivation in the nation. This would not be done piecemeal as under the domestic allotment program, he added, "but by whole farms, tracts and areas." The following July, Elwood Mead seized upon the drought situation in the western Dakotas, western Kansas and Nebraska, and eastern Montana, Wyoming, and California as vindication for the national expenditure on reclamation. Intensive agriculture should never have been developed without irrigation in such areas, and now tens of thousands of people must be moved, he announced.[37]

Frank H. Cooney, governor of Montana, Roy E. Ayers, congressman and editor of the *Lewistown Democrat-News*, and J. B. Scanlon, editor of the *Miles City Star*, while agreeing that the proposal to evacuate dry-land farmers held merit, retorted that with governmental support for construction of reservoirs and flood control dams to provide water for raising stock feed, eastern Montana could support "four or five times the population" it then had. The governors of North and South Dakota protested that they saw no need for removal of families because of "periodical dry spells." Officials in charge of the land-buying program hastily explained that they had no intention of compelling farmers to leave their homes. L. C. Gray termed it "unfortunate" that reports had been circulated of "wholesale removal of the population from the western half of the Dakotas"; only "two small projects . . . , involving less than a million acres" had as yet been developed "in that general territory." H. R. Tolley, head of the planning division of the AAA, announced that the government's program "contemplated the purchase of about 4,000,000 acres of land, . . . only a small percentage of all the poor land in the United States." Mead, too, issued a statement maintaining that he had been misunderstood.[38]

By July 1936, Tugwell was also attempting to moderate the perception. The land problem, he asserted, was merely "one of adjustment rather than of simple retirement. . . . It is all too easy to conclude that all land subject to blowing or easy run-off ought to go back to trees and grass. . . . It is not only impractical, but unnecessary." Still, he added, "Adjustments must be made in many cases, and individuals must cooperate in making them."[39]

Local Democratic editors in that election year responded by renouncing the "Evacuation Bugaboo," with explanations that the federal government had "no thought of forcibly moving anybody . . . from places which they desire to hold and occupy." Such an intimation was "wholly unwarranted by the actual facts." Republican editors agreed that the assurances seemed to take care of farmers, but still queried whether the governmental planners were considering the mercantile and business enterprises that had grown up around farming communities: "If we leave it all to the experts in distant Washington the results may be disastrous, indeed, to many." Disturbed by the large-scale, independent migration of farmers out of drought areas that summer, the Resettlement Administration issued a leaflet advising them to remain on their farms unless they had adequate resources and knew where better land and jobs could be found. The agency then redirected its major efforts from resettlement to "rehabilitation in place." The following December, Tugwell, who had been directing the resettlement program, resigned.[40]

In an effort to calm the fears of local community leaders, the U.S. Department of Agriculture and the state experiment stations began numerous studies to assess the costs of the program to counties and school districts. While payments on delinquent taxes and purchases of lands that states and counties had taken over in debt foreclosure brought welcome immediate revenues, they were of short-term impact. Unless service costs could be reduced to correspond with the declining tax base as lands were acquired by the federal government, major losses of operating income threatened. As the North Dakota land utilization project was expanded into the Slope area, the Adams

County Lions Club was assured that the program would not encompass more than one-third of the property in a district. A year later, however, when federal holdings in Billings County amounted to 38 percent, petitions were circulated and sufficient signatures attained to force a vote on dissolution of the county government. Only 45 percent of the property in that county remained in private ownership, and part of that was not returning taxes. In McKenzie County, 40 percent of the land was publicly owned and an additional 19 percent was more than five years tax delinquent. In one Montana county, the program projected federal holdings of over one-half of the area. At the end of the decade, both the North Dakota and the Montana legislatures called upon federal authorities to extend grants in lieu of taxes upon property that had been taken over.[41]

Local supporters had initially anticipated that as much as 50 percent of the revenues from grazing fees might be distributed to the local governments. That amount was authorized on land held under the Taylor Grazing Act; but under the Bankhead-Jones legislation, by which Congress belatedly gave its sanction to the land-purchase program, the payments were set at 25 percent. In either case, the revenues remained slight during the protracted period of range restoration. Studies projected that they would remain limited for ten to twenty years before the grasses could sustain profitable operations. As late as 1947 Fergus County, Montana, was, in fact, receiving only a little over $1,000 annually as its share of the fees on the area purchase by the government for retirement as grazing land.[42]

The low appraisals offered on government land purchases affected assessments on the remaining property. In 1937 the State Board of Equalization called for a reduction in valuation amounting to 30 percent in Musselshell and other counties of eastern Montana under the argument that the prices at which the government was purchasing lands fixed a cash value that could not legally be exceeded in assessments. County officials protested, and the reduction was lessened to 15 percent, but study showed that the federal purchase appraisals on agricultural tracts in the Milk River project area did average 30 percent lower than their assessed value. When the government bought county holdings of delinquent tax land, the prices offered were still lower. Petroleum, Musselshell, and Golden Valley counties of Montana accepted options on their holdings at a mere $1.20 an acre.[43]

Settlers fared only a little better in surrendering their lands. The program in 1938 projected an average purchase cost of $3.00 an acre. Milk River project lands through June 30, 1939, were acquired for an average of only $2.66 an acre. The tracts in the Badlands districts of South Dakota were bought at an average of approximately $3.20; those in Perkins and Corson counties of that state, at from $3.96 to $4.42; and only those in the older-settled Jones, Stanley, and Lyman counties ranged as high as $6.20. Yet the Land Planning Committee's report to the National Resources Board in 1935 had mapped land values in many of the retirement areas at from $10.00 to $19.00 an acre, and census reports for that year had listed them as averaging $5.48 to $7.71 an acre in the Milk River counties of Montana, $6.99 to $8.96 an acre in Billings and McKenzie counties of North Dakota, and $6.10 to $6.16 an acre in Corson and Perkins counties of South Dakota.[44]

Settlers who sold out to the government could barely hope to clear their debts, with little surplus to carry into new ventures. The delay while options were initiated, appraisals made, titles searched, funds allotted, and payments finally drawn extended for years. Even after the congressional authorization of the program in 1937, few options were officially accepted before June of the following year. Meanwhile, debts and taxes mounted as additional liens to be met before the title could be transferred. The organizational arrangements, so different from the "well-matured plans" that Wilson had seen as necessary in advocating removal in 1933, bred frustration, discouragement, and resentment among the displaced.

That removal was not as popular as administrators claimed is evident in their repeated allusions to the reluctance of settlers to sell their tracts. Scarcity of funds did not alone explain the limited extent of land purchase and the continuing necessity for provision of schools and public services in the purchase areas. Tugwell and his successors who administered the resettlement program under the Farm Security Administration into the mid-forties attributed the dissatisfaction to hostility of community leaders, town businessmen, and their close associates, the agricultural extension agents, large-scale farmers, and the American Farm Bureau. Such a reaction did mark the effort in the South, and it was reflected in the later congressional hearings on the Farm Security Administration, but it was not generally evident in the northern Plains. There the leadership, at least initially, encouraged and endorsed the program. Protests, instead, came from groups of settlers who, like the residents of Musselshell County, Montana, and Perkins County, South Dakota, opposed the actions of their county commissioners in selling delinquent tax lands to the government. In both instances, large numbers attended public meetings to challenge the action. Petitions were circulated in Perkins County on which, reportedly, all but one freeholder in one township and all but two in another registered their opposition. The effort of creditors, and particularly of the South Dakota Rural Credit Board, to foreclose mortgages based on property values far higher than the government appraisals sparked much anger, reflected in the resurgence of the Farmers' Holiday movement in western South Dakota in 1935, but also in hostility to the land-purchase program.[45]

Settlers who recalled high grain yields in favorable years—those who like the North Dakota editor that counted the cost of relief as but a small fraction of the return on such a crop—were not prepared to give up the struggle. One Perkins County resident expressed his defiance in doggerel:

What is this new plan that of late
they figure for the Sunshine State?
Submarginal land is here they say
and should be taken right away
out of production of all grain
because we do not get the rain.
The "Brain Trust" figured out a plan
to move us people where we can
all raise each year a bumper crop
where water flows without a stop,
a little irrigated farm
where drought and 'hoppers do no harm.
The time we've worked and sweat out here
to build up homes from year to year
has all been wasted so they tell,
and helped make surplus stocks as well.
Our homes and farms will have to go,
our towns and stores won't have a show,
and grass will grow
upon the street
where friends and neighbors used to meet.

We've sunk our bottom dollars here
and toiled and sweat for many a year,
we've built our churches, schools and towns,
they must be classing us as clowns.
If they would pay us what we've spent
in cash and blood and sweat and rent,
we might take notice of their plan,
although I'm not a "Brain Trust" fan.
Far better leave us folks alone,
we'll struggle through and save our home.
We'll have a crop once in a while
and all our folks will pack a smile.
Submarginal our land forsooth,
it maybe is half the truth,
but we have made our homes out here,
and if things are some out of gear,
we're no worse off than other states,
and we have food upon our plates,
though debt and credit may be twins,
the man that bucks them always wins.
No irrigated farm for me,
I'll stay right where I am "by gee."[46]

Administrators of the Resettlement Administration and the Farm Security Administration, reporting on the resettlement aspect of the program, emphasized the small proportion of displaced farmers who remained to be relocated. More significantly, the record indicated that the vast majority struggled to make removal arrangements independently. Of the 24,128 families initially residing in land-purchase areas, nationally, three-fourths had relocated by 1942 without government aid, and only 9 percent of those relocated had been resettled on farms provided by the government. Approximately 901 families in the Milk River project area of Montana held tracts subject to purchase. By early 1940, 403 of these families had moved outside the area independently; another 40 had bought irrigated farms in the locale with their own resources; an unspecified number had moved out of agriculture for employment in nearby towns; 156 had been resettled on scattered irrigated units built for that purpose by the government; and 79 remained yet to be placed. At that time it was decided that 44 of the remaining displaced families would be able to relocate themselves; 9 were considered "problem cases for other agencies to handle," and only 26 would require placement.[47]

This project area was not a new vision when first activated in the spring of 1934. Irrigation in the Milk River Valley dated from the nineteenth century, and a federal reclamation project had been in operation by 1917. H. L. Lantz, an extension agent of Phillips County, had worked with area business leaders and land owners around Malta during the mid-twenties to develop agricultural readjustment featuring utilization of the irrigated area as a basis for resettlement of dry-land farmers from the surrounding benchlands, but until 1934, the project had lacked funding for land purchase. M. L. Wilson brought it to the attention of the Roosevelt administration, and it provided a model for the subsequent resettlement program. The first resettlement land was put into production in 1936, but by early 1940 only 11,757 acres were being farmed under the program.[48]

It had been a long-delayed process and an expensive one. Through June 30, 1942, the capital investment for Milk River resettlement had been over $2 million; the cost of maintenance and taxes had amounted to $220,539; management costs had totaled $102,716; and the operating income had come to $115,400. Notwithstanding the prosperity of the prewar boom, operating income during the fiscal year ending on that date had totaled $30,579, and operating expenses, including management costs, $53,064. Most farmers had lost money in the enterprise, several of them as much as $6,000, apiece.[49]

Displaced farmers from the central Montana grazing area were resettled principally on the Fairfield Bench in the Greenfields Division of another old reclamation undertaking, the Sun River project. The Greenfields Division had been in operation by 1919, an irrigation district had been organized in 1926, and a contract of responsibility for construction costs had been accepted by the following year. From 1935 through 1936 options were taken by government agencies to buy most of the land already developed. Nearly 100 families were resettled there in 1937. During the first two years the average acreage income was only $11.31, annually. By 1942, 129 units, with a total of 12,722 acres, were in operation. The total capital investment had been $1,159,572; cost of maintenance and taxes had amounted to $158,992; management cost, $62,429; and the total operating income had been $91,448. During the fiscal year 1945 the operating expenses and management costs had come to $42,858, and the income had been but $13,160. About half the settlers were delinquent in repayment of their loans, and the total scheduled payments on loans exceeded the gross farm income. It was one of nine resettlement projects surveyed at that time by the Bureau of Agricultural Economics and the Farm Security Administration, projects deemed to be "relatively successful," the less successful having been excluded from the sample; but the record of the nine was described by the analyst as "shockingly bad . . . , examples of what should be avoided."[50]

Those who were resettled on government projects were described in the same survey as "discouraged, often cynical, and sometimes antagonistic toward society." The reaction is understandable, but it boded ill for the success of the resettlement effort. Other problems also marred the program. Studies indicated that most of the settlers had a record of debt and tax delinquency in their previous operations. They now undertook farming under the difficulties of irrigation, with which they were inexperienced. Presumably, government supervisors were to afford guidance. Farm and home plans were the basis on which the system operated, but few supervisors were trained in development and use of these tools. They, themselves, frequently looked upon them merely as routine imposed by "Washington," and settlers complained that they were useless "red tape." The latter questioned the qualifications of the college-trained "experts" assigned to provide leadership and resented the control over loans and continuance on the projects. On the Fairfield Bench friction developed to the point that a Resettlers Adjustment Association was organized, which carried its protest to the county commissioners, the governor, and the state's representatives and senators in Congress.[51]

In the anxiety to get settlers in purchase areas to enter into the removal program, agents appear to have exaggerated the benefits to be attained. Settlers had no advance knowledge of the size of the units to be offered to them. Farms on the Milk River project ranged from 40 to 160 acres in size; those on the Fairfield Bench provided 80 acres. They would not yield more than a subsistence income under the type of irrigated operations practicable in the region. Sugar-beet marketing was dependent on the availability of long-delayed sugar refineries in the vicinity, and profitable dairying was contingent upon access to sizeable urban centers. The principal use for irrigated produce was as livestock feed; but pasture areas were not included in the allotments for settlers in the project areas, and there was much complaint that they were not granted range rights when surrounding ranchers organized cooperative grazing districts.[52]

A third Montana project, at Kinsey Flats, along the Yellowstone River in Custer County, was not developed until 1938 through 1941 and reflected some lessons learned from earlier experience. Designed to serve settlers displaced by the extension of the grazing projects in eastern Montana, it offered units for eighty families and covered 19,076.42 acres, of which but 6,100 were irrigable. It not only provided somewhat larger units but also incorporated access to adjacent range acreage. While the older resettlement ventures had emphasized an irrigated, small-farm economy, with diversified cropping that included dairying, hogs, sheep, poultry, small grains, potatoes, peas, and alfalfa, with sugar beets as a cash crop, the later ones centered upon range livestock and grain. Approximately 14,000 acres of the Kinsey Flats tract were developed for cooperative grazing use. Administration of this project also differed in that seventy-four of the families were organized as a community project, Kinsey Flats, Incorporated, which leased 9,800 acres from the government and sublet individual farms to its members. The government's capital investment up to June 30, 1943, had been $737,625; taxes, $4,214.97; and management cost, $44,154.54; but total income had yet been only $33.56. During the fiscal year ending in 1942, management cost had been $7,554, and $5,006 had been paid in lieu of taxes; no income had been reported. No effort was made to sell individual units to the families prior to 1945. By then the corporation had lost control of the adjacent grazing area, and operations were necessarily shifted from range cattle to more intensive cropping.[53]

Three other projects were also developed late; and while not designated specifically as part of the resettlement program were intended, under congressional directive, to provide preference in entry "to families who have no other means of earning a livelihood, or who have been compelled to abandon, through no fault of their own, other farms in the United States." The Buffalo Rapids project, an extension of the Kinsey Flats development along the lower Yellowstone in Dawson County, Montana, had been begun during the mid-thirties by the Bureau of Reclamation to provide work relief and develop irrigation. In 1940 the undertaking was expanded to the opposite side of the river and identified as a project in the Water Conservation and Utilization program (WCU). Initially, the project was to provide 15,000 acres of irrigable land to be settled by local dry-land farmers whose holdings were then to be used cooperatively as grazing lands in conjunction with the irrigated tracts. Part of the irrigated area was also to serve as feed base for small ranchers and other dry-land operators remaining in place who lacked an adequate winter feed supply. The project would have involved readjustment of 434 farm families. Seventy of them would have been rehabilitated in place and 219 by resettlement. The remaining 145, however, would have had to move or be moved elsewhere.[54]

The first part of the project, ultimately encompassing 26,123 acres, was irrigated in 1940, and federal funding had been approved for the expansion, amounting to 21,172 acres, when the war delayed operations. Two units were irrigated by mid-decade, but a third was not completed until 1950. Project lands were sold between 1946 and 1950, fully developed for irrigation, but those in the expansion lacked housing and other supplementary buildings. Again the area of dry land that could be connected with irrigated operations was very limited.

The Lewis and Clark project and the Buford-Trenton project, both near Williston, North Dakota, were also developed as small-scale irrigation undertakings designed to be integrated with dry-land and range operations. The first consisted of 4,800 acres brought under irrigation in 1940 by the state Water Conservation Commission, funded by the North Dakota Rural Rehabilitation Corporation. The other, containing about 10,500 acres, had been built early in the century by the federal Bureau of Reclamation, but because of its expense and lack of use had been discontinued in the mid-twenties. Under the WCU program, the U.S. Department of Agriculture acquired a total of 14,800 irrigable acres for subdivision into farm units. A block of 65,000 adjoining acres was also sought for organization of a grazing district.[55]

The Buford-Trenton project was designed to provide readjustment for 265 farm families—35 by rehabilitation in place, 115 by resettlement on the site, and 115 by removal elsewhere. Nearly 40 percent of the grazing area was owned in 1941 by the federal or state governments or the Bank of North Dakota, and the federal government then took options on another 22 percent in lands subject to county tax foreclosure; but ranchers unwilling to sell held the remainder. While further development lagged through the war years, the options expired and much of the land was acquired independently by individuals seeking to expand personal holdings. In limited production by 1943, the project lands were not fully operational until 1951.

In planning for these projects it had been recognized that provision must be made for established dry-land farmers continuing within the designated grazing area. Those over age fifty were not deemed suitable for resettlement under a new type of operation, and some, it was foreseen, would not care to make the change. A temporary arrangement contemplated allocation of grazing privileges for use of the restricted rangeland and irrigated tracts on which to raise winter feed. These perquisites were, however, to be withdrawn if the operator left the unit. There was no recognition that dry-land production represented a desirable segment in the land-use structure.

About ten displaced families from western North Dakota were resettled among the occupants of 140 new homesteads on 28,043 acres in Cass and Traill counties of the Red River Valley in 1936. The tracts were occupied between 1937 and 1939 and proved generally successful under diversified farming in this eastern district. Capital expenditures, however, had amounted to approximately $1.4 million by June 30, 1943. Operating costs had come to $104,706, and total income, $194,279.[56]

Resettlement was accomplished in western South Dakota primarily under the Belle Fourche irrigation project. In all, 288 operators were displaced by government purchases in the Badlands areas, 126 by those in Perkins and Corson counties, 48 by those in the south-central counties, and 13 at old Fort Sully from land committed to Indian use. In 1936, 74 units were provided on a tract of 4,540 acres purchased for resettlement purposes near Sioux Falls, and subsequently 30 families were placed on scattered farms, totaling 5,988 acres, in eastern South Dakota. Two other farms, covering 2,031 acres near Spearfish, in Lawrence County, were purchased for resettlement but never redeveloped. Capital expenses for these projects had amounted to $673,435 and operating costs, $109,706 by 1943. Total income was $71,025. Other families received assistance through loans and grants by the Farm Security Administration, but contemporary governmental analysts conceded that the resettlement effort was incomplete at the close of the period. WCU projects providing 12,000 acres of supplemental irrigated acreage at Rapid Valley, in Pennington County, and 16,210 acres of new development at Angostura, in Fall River County, were intended to provide preference in settlement for displaced farm families, but were among the works interrupted by the war effort.[57]

The cost everywhere had been heavy. Proponents of the program emphasized that the benefits were also great. The standard of living was much improved on the resettlement projects over that which the displaced settlers had had. In the new communities, they had better schools, churches, and recreational facilities. Houses of four to six rooms were provided, with walls finished on the inside and painted on the outside. Good water was piped indoors, and in connection with the earlier subsistence homestead projects complete bathrooms were supplied. Such expenditures, however, had increased the debts for which the settlers contracted. Many considered the standard of living designed by Washington social workers far more elevated then they could afford. One elderly woman on the Fairfield Bench was so outraged when the FSA provided an electric refrigerator, purchased by the agency as part of a bulk order, that she hauled it off the project and gave it to her son.[58]

Congressional pressure was begun to induce the FSA to transfer ownership of the units to the settlers as early as 1937, far short of the extended loan periods which land-use planners had regarded as protection against introduction of land speculation and abandonment of recommended utilization. By June 30, 1943, over half the farms had been sold into private holding. Congressional hearings extending through 1943 and 1944 focused primarily on the operation of the program in the South, but evoked sufficient general dissatisfaction that beginning in 1944 Congress attached a requirement in the appropriation acts for the Department of Agriculture specifying that the rural projects be sold to private owners as quickly as possible. County FSA committees were directed to appraise the farms on the basis of their long-term earning capacity, and loan contracts were offered providing for forty-year terms at 3 percent interest under variable payment arrangements. In 1943 farms that had been developed at an average unit cost of $8,228 on the Milk River project were evaluated at an average price of $7,871; those on the Fairfield Bench, developed at an average cost of $8,498, sold at $7,236; and those in the Red River Valley, developed at $7,763, sold at $6,345. The costs did not include debts incurred for management, taxes, and other operating expenses, then heavily delinquent. No sales had been made at that time on the Milk River Valley or Sioux Falls units.[59]

While Title III of the Bankhead-Jones Act of 1937 had finally given congressional sanction to the land-purchase program, that action had been dictated largely by recognition that settlers' operations had been disrupted while option proceedings were held in abeyance for lack of adequate funding. Two years earlier Senator Burton K. Wheeler of Montana had argued for such a commitment as, specifically, an issue of "breaking faith with the farmers if the government refused to take up the options." Nationally, the area projected for purchase had already been curtailed drastically from the original estimates of necessary readjustment, and few new purchase areas were approved for development under the Bankhead-Jones legislation. After the summer of 1939 funding for the program declined greatly. Meanwhile, the focus of operations under the Resettlement Administration and its successor the Farm Security Administration had shifted to "rehabilitation in place." At a press conference on July 7, 1936, reacting in that election year to the controversy centering on the programmatic emphasis identified with the Resettlement Administration, President Roosevelt had protested that "nobody ever had any idea in their [sic] sane senses of depopulating the country." With drought again severe that summer on the Plains, he had appointed yet another committee to survey the region and expand the agenda for long-range reform.[60]

CHAPTER FIVE

Alternative Planning

The Great Plains Drought Area Committee, appointed by President Roosevelt on July 22, 1936, was called upon to outline "the essentials of a long-time program looking toward betterment of economic conditions" in the region. The chairman was Morris L. Cooke, then administrator of the Rural Electrification Administration. Other members included Hugh H. Bennett, chief of the Soil Conservation Service, several drainage and irrigation engineers, and Tugwell and Wallace, the only ones who had been associated with the previous planning effort. The recent publication of the *Drainage Policy and Projects* study under the National Resources Committee had added new groundwork for the approach, an emphasis marked by the appointment of so large a proportion of hydrographic experts. Cooke, himself, had experience as chairman of a Mississippi Valley Committee, which in considering the combined problems of dust storms on the Plains and downstream flooding in 1934 had stressed the importance of broad-based erosion control measures extending back to the upper valleys. Cooke had also been a guiding force in preparation of a report known as the *Little Waters Study,* prepared for the Soil Conservation Service, Resettlement Administration, and Rural Electrification Administration in November 1935. Bennett's presence further assured support for an emphasis upon erosion control. Given little over a month to prepare their report, the Drought Area Committee held a series of meetings with local leaders in the affected areas to learn more background information.[1]

Though the committee's recommendations reiterated much that had been urged in the earlier studies of the National Resources Board—and were accordingly criticized by Frederic A. Delano of the latter board, as unnecessary duplication, they brought to the fore alternative programs that were to shape major readjustments over the next several years. In September, President Roosevelt appointed a second Great Plains Committee, again chaired by Cooke, to extend the preliminary survey. The ensuing publication, entitled *The Future of the Great Plains*, issued the following December, was primarily an educational effort that had been recommended in the August report, but it has won recognition as a comprehensive summary of federal interagency views "on the maladjustments and desirable changes in adjustments to the resources and risks" of the region at the culmination of the mid-thirties drought experience.[2]

While the Cooke committees renewed the proposal for retirement of submarginal land from commercial production, they announced that the "fundamental purpose of any worthwhile program must be not to depopulate the region but to make it permanently habitable." Their recommendations were to take form in the redirection of the Resettlement and Farm Security programs into "rehabilitation in place," in enactment of the Bankhead-Jones Farm Tenancy Act of 1937, in the crop insurance provision of the AAA legislation of 1938, in the formulation of the Mt. Weather agreement for better coordination of federal with state and local planning activity, in the establishment of a "continuing territorial agency" for consolidated interagency approach to regional problems, and, programmatically, in new attention to the importance of fuller utilization of the available water resources. This last objective encompassed not only expanded irrigation development, primarily through small dam construction on the tributary streams, but also renewed emphasis upon tillage procedures to promote fuller utilization of natural precipitation. Both were avenues of development that the early land-use planning studies under L. C. Gray's direction, when not in outright opposition, had largely ignored.[3]

The record of the older, and larger, federal reclamation projects had not improved greatly, despite the adjustments instituted during the twenties. Use of irrigable acreage increased considerably with the onslaught of drought—from 64 to 87 percent on the Huntley project, on the Yellowstone River; from 19.4 to 34.4 percent on the Milk River project; from 23.4 to 71.8 percent on the Sun River project—all in Montana; and from 35.7 to 68.9 percent on the Belle Fourche project in South Dakota. But the repayment record continued to be poor. In 1932 Congress had authorized a moratorium on regular construction charges for 1931 and a 50 percent reduction on those due for 1932. Moratoria had again been declared for those due in 1934 and 1935, and an extension of time was granted for those due in 1938.[4]

Water shortages had been severe on most of the projects, notably on the Milk River and Belle Fourche. The promotional effort to attract new settlers to the latter project had consequently been checked; the project's record-high population of 2,807 in 1935 fell sharply to 2,190 by 1937, and thereafter continued downward to 1,540 by 1946. Heavy soils had also posed serious farming difficulties at Huntley and along the Milk River, and the small size of the units, as little as 40 acres on the Huntley project, had proved so uneconomic that the number of operators, originally set at about 680 for the installation, was readjusted in 1936 to 309, on larger, mostly 85-acre, farms. P. L. Slagsvold, of the Montana Agricultural Experiment Station, who surveyed the projects in that state during the mid-thirties, found that many had proved disappointing. The net income was low, whether compared with irrigated farms in other areas or "in some cases even as compared with that from dry farms."[5]

With crop surpluses mounting nationally, this was not a record to encourage massive expenditures for constructing new reclamation works. The National Resources Board had estimated in 1934 that 34,329 units in existing irrigation developments were available for settlement, but only 334 were reported in Montana and 31 in South Dakota. Moreover, much of the land described as irrigable on those projects was unirrigated.

State planning bodies, organized at the urging of the National Planning Board, nevertheless looked upon irrigation development as the primary agricultural adjustment for the drought. The Montana Planning Board was established by merely renaming an existing state Water Conservation Board to assume the additional planning function. The conception was little different in North Dakota, where the Planning Board, when discontinued in 1939, was supplanted by a Water Conservation Commission. Such bodies in all three states opposed all proposals for federally administered, multipurpose valley development of the Columbia or Missouri river basins because of their concern to preserve upstream flow for local irrigation. WPA projects for most counties were focused on dam and ditch construction for use of the smaller streams, and pressure was persistent for PWA development of larger irrigating facilities.[6]

In that time of drought, reclamation appeared to provide not only a means of extending work to the unemployed but, for many of its proponents, the only potential for agricultural expansion. Railroad companies, chambers of commerce, immigration bureaus, and similar promotional agencies viewed the intensive development on irrigated holdings as profitable to community business interests, regardless of the low net income to individual farmers. The possibility that such works might increase the production of hay and feed grains attracted the support of the politically powerful livestock industry. And Washington planners, who shared the belief that the region never should have been opened to agricultural settlement, agreed that if resettlement were not to entail depopulation, the program must encompass expanded irrigation.

President Roosevelt's initial reluctance to authorize such construction gave way to a decision that dry-land retirement should be balanced by expansion of irrigated acreage, a criterion interpreted under the view that ten acres of dry land were the equivalent of one under irrigation. The National Resources Board's sharp criticism of the high cost of irrigation was moderated in the report of its Water Resources Committee, calling for extended development of drainage basins; and the report of the Great Plains Committee, while still skeptical of the utility of large-scale projects, recommended supplemental irrigation where conditions were favorable and facilities could be constructed inexpensively. Two years later the report of the permanent Northern Great Plains Committee, established upon the recommendation of the latter body, noted that, locally, irrigation projects of considerable size were possible and "should replace the cash-grain and small-scale stock rearing" in the many areas where such development had failed. One of two main sections of recommendations for regional rehabilitation, as further developed by this organization in 1939, centered upon the "fullest practicable use" of resources through irrigation.[7]

Even under the impetus of such support, the area irrigated was increased during the decade only from 1,594,912 to 1,711,409 acres in Montana and from 9,392 to 21,615 acres in North Dakota, while it was decreased from 67,107 to 60,198 acres in South Dakota. Relatively little large-scale development was undertaken in the region. The largest project was the Fort Peck Reservoir and related damming on feeder streams of the Missouri River, begun in 1933 and largely completed by 1938. Built by the Army Corps of Engineers primarily for flood control and maintenance of navigation on the lower river, the safeguarding of water supplies for irrigation in the upper basin and provision of hydroelectric power "consistent with the requirements for navigation" were supplementary concerns. Approximately 84,000 acres were projected as newly irrigable. Provision of electric power, not authorized until 1942, permitted additional irrigation on a small scale by pumping operations at numerous points in northeastern Montana and northwestern North Dakota. Hope for extension of irrigation onto 1.3 million acres by gravity diversion from below Fort Peck into the Souris River district and eastward across North Dakota to Devil's Lake, projected by W. G. Sloan, of the U.S. Bureau of Reclamation in 1941, formed the basis for the Missouri Basin Development program of the postwar period, but failed even then to bring large-scale irrigation to the western Dakotas.[8]

The need to supplement the erratic supply of water for existing projects led to construction of the Fresno and Chain-of-Lakes dam and reservoir on the Milk River northwest of Havre, the Ruby River, or Upper Musselshell, project, and the Dead Man's Basin project on the lower Musselshell—all in Montana. Indian lands received some of the largest developments—the bulk of the expanded irrigable acreage from the Fort Peck dam, some 30,000 to 50,000 acres in Rosebud and Custer counties from the Tongue River, a portion of the benefit from the Fresno storage unit, and additional acreage in the Flathead country of northwestern Montana. Efforts of the Montana Water Conservation Board to win approval for expansion of the acreage under the Milk River project onto the Saco Divide were rejected until a pumping unit was authorized to serve a small area as an aspect of the war effort in 1942. Hopes for construction of facilities to reclaim an area estimated variously from 190,000 to 250,000 acres in Hill and Chouteau counties by reservoir development on the Marias River were also dashed when the Bureau of Reclamation estimated that construction costs would amount to more than double the authorized acre limit.[9]

The largest facility under construction in North Dakota during the thirties was the projected 7,000-acre Lewis and Clark installation along the Missouri River near Williston. With the addition of the revived Buford-Trenton proposal, the scheduled development in the early forties amounted to some 21,000 acres; but, as previously noted, the completed works encompassed considerably less acreage, and the latter project was delayed more than a decade by wartime scarcities. Hopes for a giant dam on the Heart River, west of Mandan, were rejected by the Army Corps of Engineers in 1941 and also reduced considerably when later materialized. In western South Dakota little more than repair work was undertaken at Belle Fourche because of the pro-longed severe depression and consequent heavy loss of settlers. A dam on Flat Creek, about 10 miles south of Lemmon, approved as a project of the Civil Works Administration in 1934, initiated development of the Grand River Valley; but more extensive facilities on that river were also rejected during the decade. The Rapid Valley project, to provide supplementary water for 12,000 acres near Rapid City, and a dam at Angostura, to bring new irrigation for some 16,000 acres along the upper Cheyenne, were both approved as projects of the Bureau of Reclamation by 1941. The former, however, was suspended at that time for further planning, and work on both was interrupted by the war.[10]

More important for readjustment of agriculture in the region were myriad smaller check-dams, lakes, and reservoirs constructed on tributary streams feeding into the larger rivers. They marked a return to the flood irri-gation of hay production that ranchers had used at the end of the nineteenth century. Agricultural extension agents in stock-raising districts, notably W. H. Lamphere, of Custer County, Montana, had been urging such development during the early thirties, and at that time the federal Bureau of Reclamation had emphasized the need for integration of grazing areas with irrigation projects so that the Montana installations would be profitable. By the autumn of 1936, leaders of the Montana Agricultural College were calling for such a broad, integrated approach, combining stock raising with supportive agricul-ture. The shift in emphasis has been noted in the altered development of the later resettlement projects.[11]

Irrigation for commercial cropping was still not economic on most large-scale projects in the region. It was estimated that the repayment period for the government expenditures on federal projects, nationally, would average about 50 years beyond 1934, even if there were no additional moratoria. Pay-ment for the Chain-of-Lakes storage project was calculated to require 250 years. Crop income on the Sun River project averaged only $12.30 an acre in 1940; for Belle Fourche, $23.36; for the Milk River project, $26.80; for the Low-er Yellowstone, $39.45. With such low returns, construction costs had to be held to a minimum and the development had to support complementary en-terprise. Assuming that relief expenditures were a necessary public service, not to be tallied in the reckoning of project cost, local authorities had drafted programs for a vast array of small-scale water conservation measures even before enactment of special legislation called for such work. The WPA, Rural Resettlement Administration, and Soil Conservation Service (SCS) reported over six hundred water-conservation projects already built or under construction in eastern Montana by the end of 1936.[12]

In August 1937, the recommendations of the Great Plains Committee led Congress to authorize both the Interior and the Agriculture departments to develop facilities for water storage and utilization, but neither measure provided the necessary appropriation. The Farm Security Administration, however, initiated the program that year by an allotment of $5 million from relief funding for construction of facilities to aid in resettlement and rehabilitation of needy farmers. For the next several years, the appropriation bills for the Department of Agriculture carried the sum of $500,000, from which an amount not to exceed $50,000 a project was to fund construction and maintenance of reservoirs, ponds, diversion dams, pumping facilities, windmills, and similar installations for small-scale irrigation. With assistance by the state extension service, Montana farmers and ranchers constructed over 13,000 such projects between 1933 and 1942, affording improved or new water supplies for some 220,000 acres, as well as stock-water facilities.[13]

In 1939, Congress enacted the Wheeler-Case bill, authorizing the secretary of the interior to undertake construction of Water Conservation and Utilization projects in the Great Plains and the arid and semiarid areas of the nation, and in the appropriation bill for the Interior Department provided $5 million for the purpose. Moreover, that sum was to be approximately doubled by the value of labor and materials supplied by the WPA, the Civilian Conservation Corps, and other federal relief agencies, for which reimbursement was left to the discretion of the president. Under amendments in 1940, the expenditures for WCU projects were limited to $1 million per project; the work was to be undertaken in consultation with the secretary of agriculture; and the supplementary contributions by other federal agencies were specifically stipulated as non-reimbursable. Settlers on the projects under either the Water Facilities or the WCU programs were required to enter into contracts requiring repayment of appropriated construction expenditures over forty years, beginning only after the elapse of a ten-year development period dated from the initial delivery of water.[14]

With the infusion of such funding, the Bureau of Reclamation and the state water conservation boards, in cooperation with the WPA and other federal agencies, began construction not only of the Buffalo Rapids, Lewis and Clark, Buford-Trenton, Rapid Valley, and Angostura projects, designed to meet resettlement needs, but also of a large number of smaller, 5,000- to 10,000-acre irrigating works throughout the region. Among them were Ackley Lake in Judith Basin County, Willow Creek in Gallatin, Bainville in Roosevelt, Poplar River in Daniels and Roosevelt, the Little Missouri in Carter, all in

Montana, and Cartwright, off the Yellowstone River in McKenzie County, Zahl and Beaver Creek in Williams, the Upper Cannonball in Slope, and Cedar Creek in Grant and Sioux in North Dakota.

The report of the Great Plains Committee has been criticized for failing to recognize the future impact of irrigation by pumping. Since underground aquifers were very deep throughout most of the northern Plains, and electricity was not yet widely available in rural areas, pumping away from streams was prohibitively expensive for crop watering. The annual report of the Montana Agricultural Experiment Station in 1933 had questioned whether the kind of crops that could be grown in that state would ever warrant deepwell pumping for irrigation. Meeting with leaders from the northern Plains at a Bismarck conference, members of the Great Plains Committee had been urged to direct special investigation to the possibilities for pumping river water for irrigation of adjacent benchlands. Such pumping became a feature in much of the small-water development, notably at the Lower Yellowstone, Buffalo Rapids, Lewis and Clark, and Buford-Trenton projects. County extension agents were also active in assisting individual farmers, with small homemade pumps and power from tractors, to lift water for no more than a dozen feet sufficient to irrigate a family garden and furnish feed for livestock. The Montana Extension Service reported that 140 such projects, to irrigate a total of 5,600 acres, had been completed in that state in 1937 and an additional 66, for less than 2,500 acres, in 1938. "Coal, beaver dams, kerosene and windmills," the *Williston Daily Herald* noted at that time, were "all helping the progress of irrigation in the upper Missouri territory."[15]

Although constructed by the Reclamation Service, the WCU projects were developed under the direction of the Farm Security Administration and, like the small-water projects generally, were designed to promote integration of forage production with grazing utilization of adjacent range. In many cases they provided storage for mere flood run-off and tended to run dry through the summer, but they permitted hay cropping and some cultivation of farm gardens. When, late in 1936, the Resettlement Administration and subsequently its successor, the FSA, shifted the focus of their reform from resettlement to "rehabilitation in place," they looked to such sources also as the basis for maintenance of a diversified, small-farm economy, combining livestock enterprise with production of reserves of feed and seed and cultivation of a major portion of the family food supply.

The rehabilitation program had been initiated by the FERA in April 1934 so that farmers might become independent of relief. It was anticipated that with credit assistance to readjust outstanding debts and supervisory guidance in farm and home planning, farm families might at least supply their own livelihood. It was intended, however, by the federal social workers to be an "acceptable level of living," one that included such modern amenities as pressure cookers and refrigerators, indoor household water and plumbing, adequate health care, life insurance, regular school attendance, even in some areas maintenance of high-school dormitories, and participation in community activities. Where conditions made these standards impossible, relief grants or loans were made available to finance them.

The provision for supervisory guidance as an educational instrument that would enable relief clients to rise above the need for further assistance was acclaimed by proponents as a major innovation, but critics questioned its effectiveness. Farmers protested that it was personally intrusive, even involving the household operations of their wives. In 1939 those receiving grants in Adams County, North Dakota, for example, were required not only to maintain adequate gardens, to include adequate feed for support of livestock, and to store and can as much of the produce as possible, but also to keep detailed records of their income and expenditures, to maintain clean homes and farmsteads, and to attend scheduled group discussions on "practical farm and home practices." Analysts within the system complained that because many of the supervisors were ill-trained, overworked, and unsympathetic with the program, it "tapered off largely into one of advancing loans to needy owners, tenants, sharecroppers, and a few former agricultural laborers, to all of whom the usual avenues to credit were then closed."[16]

Numerous contemporary studies indicated that relief families held far fewer livestock and less acreage than the regional average, but the rehabilitation loan program during the thirties failed to provide sorely needed funding to replace breeding stock lost during the drought years or for purchase of additional land and machinery. Of the WCU projects, particularly, the chief of the Reclamation Service warned in 1940 that the FSA plan to rehabilitate 1,100 families on each 100,000 irrigated acres established a level of operation that had so far provided an "existence, precarious at best, . . . for families settled only one-third as thickly on typical areas . . . selected for development." Capital outlays for expanded operations were, however, regarded as the basis for a commercial level of productivity discountenanced under the relief effort and left to be served through normal channels of credit. As late as 1939 the North Dakota legislature found that no provision existed under the various federal agencies for credit to re-establish foundation herds. The FSA at that time undertook to promote a livestock-breeding program through cooperative purchase of sires; but not until 1942, as part of the "Food for Freedom" effort, were rehabilitation loans authorized so that relief clients were able to make improvements in farm housing and buildings, purchase or rent equipment, or acquire supplemental land needed for adequate farm holdings. By means of loan restrictions, coordinated FSA, SCS, Farm Credit Administration (FCA), and other USDA programs specifically undertook to curtail cash-crop wheat production. Rehabilitation loans were hailed as "the staunch defender of the family-type farm."[17]

Without capital investment for expanded operations, the "self-sustaining" economy yielded little surplus for liquidation of debt and assumption of new obligations elevating the standard of living. Despite the stipulated supervisory guidance toward better management practices, the repayment record of rehabilitation borrowers in the mid-thirties was notably poor. Their resources on entry into the program had been depleted, and their starting loans were meager. After nine years over 25 percent of the loans made in 1935 and nearly 17 percent of those for 1936 were canceled as uncollectible; 25 percent were still being carried for collection but were no longer considered rehabilitative; another 25 percent remained under supervisory management, and only 22 percent of the 1935 group and 28 percent of those for 1936 had been repaid.[18]

The high production costs of irrigated farming and the meager returns on irrigated or diversified operations fostered persistent hope that increased rainfall would permit a return to dry-land wheat cropping. Even the integrated economy of livestock grazing with irrigation depended upon supplemental cultivation of feed grains—oats and barley in northern Montana, corn at lower altitudes in Montana and the Dakotas, barley and rye in North Dakota, and oats, barley, and sorghum in South Dakota. These were crops that could be dry farmed more economically than irrigated; and if soil moisture was adequate for such production, it was also likely to be sufficient for the far more profitable cropping of wheat. Premiums for high-protein wheat climbed during the scarcity of the drought years. Wheat produced in north-central Montana, from the Judith Basin to the Canadian border, carried an average protein rating of 15 percent in 1937, with a premium of eight cents a bushel. In some districts it ranged as high as 19 percent, at an increased return of two cents for each additional half-percentage point. Both the Greater North Dakota Association and Montanans, Incorporated, protesting vigorously when production controls reduced the regional wheat allotments the following spring, argued that such wheat was never in surplus; the premiums testified to the demand.[19]

The annual report of the Montana Agricultural Experiment Station for 1933 had echoed the themes voiced by L. C. Gray at the Chicago Conference on Land Utilization: Montana was primarily "a live stock state, and all planning for the future use of its agricultural lands must recognize this fact." Yet the same report had noted that certain areas of the state had "no other alternative than to produce wheat." Responding to the argument that the nation had a wheat surplus, Director Linfield reiterated: "*but there is no surplus of the high quality wheat that Montana produces.*" A series of bulletins issued by the experiment station for the guidance of local planning groups during the latter years of the decade continued to point to the profitability of wheat production for the older districts of dry-land cultivation—the northeastern and east-central counties as well as those from the Judith Basin north and west to the Canadian border. In remarks at Williston, North Dakota, during the summer of 1937, M. L. Wilson himself questioned the growing commitment to irrigation and reasserted his own interest in the dry-land wheat farmer, "the man who has learned to operate on a large scale to get volume production though the yield per acre may be small." Neither he nor the USDA would try to make irrigation farmers out of dry-land operators, he announced. Analyzing the cropping pattern at the end of the decade, the federal marketing statistician for Montana noted that larger acreages in alfalfa seed, sugar beets, barley, corn, and sweet clover indicated a greater diversification, aided by the development of irrigation; but wheat, he added, from an acreage standpoint, remained the most important crop in the state.[20]

Water-conserving technology, promoted by the Soil Conservation Service and incorporated in the revised Agricultural Adjustment program, held greater importance than resettlement, irrigation, or diversification among the recommendations of the Great Plains Committee and other planning agencies of the late thirties. A Soil Erosion Service had been established in the Department of the Interior in 1933, initially as a vehicle for unemployment relief; but as dust storms carried western soils eastward during the next two years, pressure had mounted for a national policy commitment to soil conservation. In legislating the Soil Conservation Act of 1935, Congress reassigned the program to the USDA and renamed the agency the Soil Conservation Service. Already during the summer of 1934 the Montana Extension Service and the state planning consultant had begun a series of meetings directing the attention of county planning boards to the use of strip cropping and other methods of controlling soil blowing, the need for regrassing abandoned farm lands, the value of the state's soil survey, and the importance of planned land utilization. Much has been written of the purported rivalry between local extension workers and the agents of the SCS relating to direction of the educational effort, but in the northern Plains, county agents at that time regularly led expeditions of farmers to survey the results on the nearest SCS demonstration plats.[21]

By 1937 there were two such projects in Montana, each encompassing 32,000 acres under demonstration tests—one in the Power-Dutton area of Teton County; the other between Froid and McCabe in Roosevelt County. Much of the acreage was operated by cooperators who signed an agreement to work under SCS direction for five years. Emphasis was placed on range development, through dam construction and seeding with crested wheat grass; but farming areas featured use of strip cropping, shelterbelts of trees, and contour cultivation—that is, following the contours of the land along a level line, crosswise of any slope, regardless of direction.[22]

In the eastern borderland areas of the Plains, a 49,000-acre tract was developed near Bottineau, North Dakota, and 49,280 acres in Gregory and Tripp counties, South Dakota. Here more attention was given to farm management, again with strip cropping and shelterbelts planted on the north and west sides of each farm, but with a diversified six-year pattern of crop rotation. The schedule provided for corn, wheat, oats in which sweet clover was also seeded, a second year of the biennial clover, fallow, and again wheat. Since the terrain near Bottineau was very nearly level, an S-shaped system of strips was substituted for contouring, covering entire farms except for areas in pasture or trees. Crop strips were uniformly 100 feet wide; and between each strip was a ribbon of grass 20 feet in width, usually a mixture of brome, western wheat, slender wheat, and crested wheat grasses. The S shape was intended to prevent wind from any direction blowing far on cultivated sections before encountering a grass buffer strip. In place of every fourth grass strip, a shrub and tree buffer was planted. Occasionally alfalfa could be planted in place of grass and used for winter feeding of milk cows, which each cooperator was required to keep. As further protection against wind damage, use of pulverizing cultivators was discouraged; most of the work was to be done with a lister and duckfoot cultivator so as to leave the surface soil rough. Where weeds were a problem, a one-way cultivator might be employed, but the stubble was to be left as a trashy cover.

There was little new in these proposals. Apart from contouring and rotational provision for green manuring, the latter practice having long since been rejected as unsuitable for dry-land cultivation, this was a technology that had been recommended by local agricultural experts at least a decade earlier. Even tree planting had been urged by early proponents of dry farming. In his *Soil Culture Manual* of 1902 Hardy Webster Campbell, who first popularized dry-farming methodology, had cited not only the beauty, comfort, and profit in tree production but its value in breaking the force of hot winds. Thomas Shaw, noting the virtual absence of such windbreaks on the benchlands of Montana, had also recommended tree planting. The Judith Basin county agent, reporting the existence of only 2 plantings in that county prior to 1926, had stressed the development, and the dry-land experimental farm at Mandan, North Dakota, which had supplied the early demonstration plots, had added 15 more in Judith Basin County by 1931. Beginning in 1928 the State University of Montana at Missoula began selling trees at cost for shelterbelt development and by 1935 had provided nearly 1.5 million trees over the state. At that time more than 250 shelterbelts had been established in Fergus County and over 209, some 80,000 trees, in Cascade. Although only 27 plantings had been set out in Williams County, North Dakota, by 1935, the Mandan farm and a state forestry nursery at Bottineau had provided over 60 for adjacent McKenzie County by 1931.[23]

The most significant difference in the program of the SCS was the introduction of contouring. As late as 1935 the chief soil specialist of the Soil Erosion Service in South Dakota had emphasized the older form of stripping at right angles to the strongest winds. The following year, however, the SCS announced plans for cooperating Montana ranchers to contour furrows on rangeland and contour regrassing of abandoned farm tracts. The first contour lines in Fergus County, Montana, were run by the SCS on a demonstration ranch near Moore in the winter of 1937–1938, and the first in Williams County, North Dakota, by a cooperating farmer in the Little Muddy Soil Conservation District (SCD) during the summer of 1939. By the fall of 1940 some 2,325 acres in McKenzie County, North Dakota, had been contoured, again by an SCD cooperator.[24]

The new practice was not popular with farmers. They objected to delays caused by establishing contour lines, complained of difficulty in maneuvering machinery along curving lines, and deplored the irregular appearance of their fields. Moreover, until widespread acceptance of the damming lister, introduced in the northern Plains in 1938, blowing continued to occur where contours ran in the direction of the wind. Experiment indicated that the practice was most useful for moisture conservation on row crops or fallow but had far less value for spring-sown small grains, which normally were less subject to water erosion. Only the incorporation of regulatory action and incentive payments generated acceptance of the innovation.[25]

In August 1936, President Roosevelt summarized the multifaceted functions of the SCS: (1) erosion prevention on publicly owned lands, including the newly acquired range-purchase areas; (2) operation of some 150 demonstration projects scattered throughout the nation; (3) direction of Civilian Conservation Corps camps assigned to provide the necessary construction forces for these operations; and, finally, just being proposed, (4) extension of technical guidance to associations of private cooperators undertaking programs of soil conservation. The effort to develop plans for five-year management systems with individuals was not to be discontinued; but, convinced that the problems of erosion could not be resolved on isolated farms, the SCS was initiating a form of organization that would provide for "relatively uniform application of erosion-control practices to all lands within a watershed or other naturally bounded area." With congressional authorization, the secretary of agriculture had been permitted to require state enactment of the necessary enabling legislation as a condition for receipt of federal expenditures for erosion control. In alluding to operations under this mandate, the president had spoken somewhat prematurely. It was late in February 1937 before he presented to the governors the format for a standard soil conservation districts proposal to be submitted for action by the appropriate legislative bodies.[26]

Within the next three weeks all three states of the northern Plains authorized establishment of such districts, but each imposed more restrictive requirements than the president had recommended. The Montana law was found to have omitted necessary constitutional safeguards and could not be applied until it was revised in 1939. The model law had required only a majority vote of the occupiers of land in the proposed organizational boundaries for setting up or discontinuing a district and in referenda on acceptance of its regulations. North Dakota accepted these criteria except in demanding a two-thirds vote on the regulations. South Dakota and Montana required, as well, a two-thirds vote for organization of a district. South Dakota furthermore limited votes on organization and discontinuance to land *owners*. The effects were to delay organizational development through the remainder of the decade. By 1940 only three districts had been established in each of the western Dakotas, and none had been set up in Montana. The Soil Conservation Service, which rented caterpillar tractors and other heavy equipment for constructing water holes and small-water dams for stock, finally in 1941 informed local planning groups that they would have to form districts if they wanted to continue such use.[27]

Farmers in many areas at first opposed the regulatory restrictions of the commitment. Supervisors were authorized to "develop comprehensive plans for conservation of soil resources . . . within the district, including specification of engineering operations, methods of cultivation, the growing of vegetation, cropping programs, tillage practices and changes in use of land," even its retirement from cultivation. They could require *all* occupiers of land in the district "to enter into and perform such agreements . . . as to the permanent use . . . of [their] lands as would tend to prevent or control erosion." While in practice the regulatory authority depended primarily upon persuasion, supervisors could carry infractions before the public courts and call upon the state legal officers for the requisite support. Soil conservation districts became governmental subdivisions, "a public body corporate and politic, exercising public powers," lacking only taxing and bonding authority. Districts could not be discontinued in less than five years; referenda on such action could not be held oftener than every three years, and those for supplementation or repeal of land-use regulations, not more frequently than once in six months. According to some analysts, soil conservation districts presented a more effective force than zoning as a means of controlling land-use development, because soil districts regulated established land users while zoning, it was thought, could restrict only new operations.[28]

More immediately, the programmatic influence of the SCS took form in the modification of the AAA after the Supreme Court decision nullifying the production control and taxing provisions of the original measure. Under the Soil Conservation and Domestic Allotment Act of 1936 farmers were offered payments for shifting acreage from "soil-depleting" to "soil-conserving" (hay) crops or to "soil-building" (alfalfa or sweet-clover) crops and for conducting "soil-building" practices on cropland or pasture. As first applied, the program paid a farmer at per-acre rates varying according to his average yield over the period 1923–1932 for reducing the acreage in soil-depleting crops, generally those in surplus—wheat, oats, barley, flax, potatoes, corn to be used for grain or forage, grain sorghum, and safflower in the northern

Plains—by minimums of 15 percent. He was also to be paid for seeding the diverted acreage in grasses such as alfalfa, sweet clover, and crested or slender wheat grass and for use of such farming procedures as strip cropping, green manuring, terracing, and weed eradication. In designated counties summer-fallow land could be counted as soil-depleting acreage if strip farming was practiced; the required minimum soil-conserving acreage was then reduced from 15 to 10 percent. A stockmen's program provided payments also for such operations as deferred grazing, range reseeding, developing springs, digging wells, dam construction, water spreading, fencing, and destruction of sagebrush. Benefits for shelterbelt planting were added for Montana operators in 1937.[29]

When production and surpluses continued to increase, Congress further revised the program in the Agricultural Adjustment Act of 1938. While retaining the conservation features of the previous legislation, the new regulations reduced the soil-depletion acreage allotments and incorporated penalties for exceeding them. Fergus County, Montana, for example, received an allotment of 281,721 acres for 1938, a 20 percent reduction, and but 210,191 acres for 1939, a further 25 percent reduction. A provision, restricted to farmers of the ten Great Plains states, permitted that land subject to serious erosion be taken out of production and returned permanently to grass, as "restoration land." For such withdrawals, farmers were to be recompensed fifty cents an acre annually for three years. Where serious blowing would occur without measures of control, operators under the arrangement were expected to provide the necessary protection. If they failed, a charge of $1 an acre was to be levied against the AAA payments otherwise earned on their farms.[30]

The restoration effort called for development of "permanent" vegetative cover on 875,000 acres in Montana, 1,025,000 acres in North Dakota, and 550,000 acres in South Dakota, all land that had been under plow in the past five years. The Montana state AAA committee explained that "restoration" was merely new terminology for such cropland as that on which payments had been made previously for permitting return to native vegetation; land so listed in 1936 and 1937 was to be included. This included much idle farmland that had been left untilled as a temporary response to the unfavorable conditions. Contemporary analysts of the Buffalo Creek Grazing District recognized that the program would have "only a slight effect on adjustment because much of this land . . . [was] already being allowed to revert to grass." They contended that it would serve principally large stock operators who had acquired some abandoned farm land or some "sod" [sic] and corporate or nonresident land owners who had not been able to get satisfactory leases for their holdings. Judging by payments made for the practice in 1939, the largest amount in the decade, somewhat less than 492,000 acres were "restored" in Montana, 688,000 acres in North Dakota, and 938,000 in South Dakota. What "permanent" meant seems to have been questionable even at the time. In

1940, still within the first three-year compensation period, the payments covered only 130,000, 188,000, and 236,000 acres in the three states, respectively. Writing at the end of the decade, L. C. Gray renewed his call for continuation of the federal land-purchase program because the new approach could not meet the problem of assuring continued nonuse when the land was "subject to new stimuli of favorable prices and rainfall."[31]

Strip farming was the most successful measure promoted under the AAA program in the region. In 1930 a mere 2,520 acres were reported under the practice for Montana, but by 1940 dry-land farmers of that state were much more commonly using it, for a total of nearly 2.25 million acres. Hill County led with about 300,000 acres under such cultivation, but Sheridan, Chouteau, and Roosevelt each had over 250,000 acres. In the southeastern counties, at lower altitudes and where livestock figured more prominently, corn was commonly substituted for fallow in alternate strips with wheat. Stripping was conducted in 1940 on about 1.75 million acres in North Dakota and on some 886,000 acres in South Dakota. Contour stripping remained infrequent, with but 2,268 acres in Montana, apart from operations on the SCS demonstration projects, and none reported for the Dakotas.[32]

The conservation program generally won strongest adherence in the Dakotas. Over 94 percent of the cropland in both North and South Dakota was covered by applications for such payments in 1940, compared to 84 percent in Montana. Ward and McLean counties in western North Dakota were among that state's leading participants; and Dunn, Billings, and Mercer counties were particularly active in the range phase of the program. Montana led in reseeding pasture and rangeland, with over 2.5 million acres of deferred grazing, compared to a little over 1.25 million in South Dakota and less than 110,000 in North Dakota. Montana also led in new seeding of alfalfa, with some 122,000 acres; South Dakota listed only about 73,000, and North Dakota, only about 54,000. But new seedings of other legumes and grasses totaled over 1.6 million acres in North Dakota, with less than half that amount in South Dakota, and only about a third of it in Montana. In South Dakota individual efforts to develop water supplies through construction of springs, reservoirs, and dams as an aspect of the AAA program more than doubled those in Montana and were almost seven times greater than those in North Dakota. Work on water-spreading systems was a more common practice in North Dakota than in Montana, and none was reported for South Dakota. Terracing held little place anywhere in the region.

State and local committees at least nominally held a voice in formulating the schedules of approved conservation practices. Under federal mandate, statutes providing for state participation in the AAA program called for the drafting annually of a state agricultural plan to be submitted to the USDA. In North Dakota the state Agricultural Extension Service was assigned that duty directly; in South Dakota the extension service shared the function with the state Agricultural Advisory Executive Committee; and in Montana it fell to a state Agricultural Conservation Board of seven members, six of whom were appointed by the governor from nominees proposed as representatives of a pyramid structure of district, county, and community conservation committees. The seventh member of the Montana board was chosen from one of two nominees by the State Agricultural College, but the position was routinely filled by the state extension director in his official capacity. Under all the arrangements, the AAA structure of community elections to select delegates to county committees, who in turn elected representatives to district committees, and the latter in like manner to state committees, provided advisory bodies for consultation in the drafting of the state plans, in educating constituents on the terms of the regulations, and in verification of their performance under them. M. L. Wilson's insistence upon the importance of local participation in administration remained an important key to enforcement. While representatives of community and county committees commonly mirrored the views of the more vocal local leaders—extension agents, journalists, and interested businessmen—the system gave greater expression to small community constituencies than most planning agencies operating in that period.[33]

Local participation in planning for agricultural development, as noted previously, was not new in the region during the thirties. Continuing the work begun in the late twenties, the Montana Extension Service had held a series of so-called land utilization meetings in the spring of 1931. In Fergus County, for example, M. L. Wilson had addressed representatives of the twenty-two organized community clubs on the importance of land classification as a basis for "intelligent readjustment" in place of "the long and painful pressure of economic requirements." A committee of that body in cooperation with the railway agricultural development agents and the extension service had then invited farmers and stockmen to draft plans for local land use. Among the principal recommendations of these meetings was the need for more flood-water irrigation and production of livestock feed.[34]

When in December 1933 the National Planning Board urged the organization of state planning agencies and the PWA allotted funds for assignment of consultants to assist them in project development, the regional governors responded quickly. In Montana a special session of the legislative assembly was called, which in January 1934 established a Water Conservation Board and provided authorization for state cooperation with federal agencies in constructing self-liquidating works for water conservation. In the Dakotas the governors in February named temporary boards for planning activities. Again there was strong support by powerful promotional interests for the organizational effort, but again it reached out to embrace local participation.[35]

Montanans, Incorporated, called a state planning conference to meet in March 1934, at which some 300 delegates agreed to divide the state into twelve planning districts to formulate a coordinated and systematic program for development of public works. At the district planning sessions spokesmen for the state Water Conservation Board, the State Agricultural College, and the Federal Emergency Relief Administration joined county agents, school and highway officials, state political leaders, and community businessmen in an attempt to "give the various county and community representatives an understanding of the planning program and the organization requirements," to "make possible the machinery . . . [to] give community expression in presenting and working out the most permanent and long-time projects." District meetings, in turn, led to county meetings as assemblies of representatives from community boards, each of which was directed to draw up plans "for some constructive project."[36]

By 1935 state planning boards had received legislative authorization throughout the northern Plains, and in all the states they were specifically authorized to advise and consult with municipal, county, district, and other local planning bodies. Every county in western North Dakota, all but Teton, Rosebud, and Dawson counties in eastern Montana, and all but Shannon and Todd counties (Indian areas) in western South Dakota were reported as so organized by 1938.[37]

This development was directed toward the need for designing programs of public works; and as a response to drought, it turned first to provision of various forms of irrigation construction. But, as reported by the government-appointed land consultants, the discussions extended to the range of proposed land-use recommendations. In June 1935 the Montana consultant summarized the state recommendations as: "restoration of submarginal lands to range use, the supplying of supplemental water to lands . . . under irrigation, the development of irrigated areas to balance feed production with grass resources, permanent rehabilitation of stranded rural families, encouraging the development of small mining industries and the employment of relief families on useful projects . . . essential to this program." The number of relief families in each district to be "Rehabilitated in Place," "Reabsorbed," "Resettled," established on "Subsistence Homesteads," and otherwise adjusted had already been determined. A series of maps defined the land according to four grades, as well as grazing, timber, and forest areas; the irrigated tracts, those partially irrigated, and those proposed for irrigation; the organized and proposed grazing districts; the areas of mineral deposit; the existing and proposed rural electrification; and the transportation facilities.[38]

It appeared that remarkable progress had been made the first year. By the second year report maps were available to indicate areas of overgrazing and erosion, but only a preliminary map had been drawn of the "ultimate preferred use of the principle [sic] areas of the state." At the same time the state planning consultant cited difficulties in getting a staff organized and reported that he was attempting to coordinate his work with research studies under way at the state college. Meanwhile, there had been a reduction of the public educational meetings, "until," he explained, "the present studies are brought to the point where they can be presented for consideration." It seems probable that the progress of the first year had been less the result of community discussion than the work of the consultant.[39]

Without the groundwork of early land classification like that on which the Montana mapping had been based, the state planning boards in the Dakotas emphasized their need for preliminary fact-finding. The North Dakota board reported that because of the financial distress of the state, "and in some measure also . . . political upheavals within it," infrequent meetings had been held. The Greater North Dakota Association had, in fact, paid the traveling expenses for members in attendance. Both the Dakota boards nevertheless contended that the publicity attending the planning effort constituted a major initial accomplishment.[40]

Their reports indicate that the recommendations of the National Planning Board carried the support of the state boards, probably under strong pressure from the appointed land consultants, well before the fact-finding data were available. All of them supported and recommended extension of the federal land-purchase program in their states. The discomfort of the North Dakota board in the role was apparent, however, when the chairman called a statewide planning conference to meet in January 1935 with the observation: "We do not intend to continue serving as the court of last resort when it comes to declaring certain areas of the state as submarginal or to determining the best means of providing water conservation for other areas." A year later the governor of Montana also emphasized the goal: "to place the responsibility of planning for the future where it belongs—in the hands of our own people." Maps were being prepared for the use of the counties, he explained: "But with all this we know that the best program for land and water use development can be worked out by home folks, who know the counties' history, the soil, topography, water resources and general conditions. It is a proposition of Montanans doing their own planning and maintaining the state's identity."[41]

During the latter half of the thirties the planning effort shifted increasingly to a local focus, both in program analysis and in participation. The history of this development in Fergus County, Montana, was perhaps exceptional, for it evolved upon the background of County Agent Dan Noble's leadership during the preceding decade; yet even there the work until 1935 had centered in agricultural discussions at community clubs, without much formal organization for specific planning. That year the county farm-week economic conference, an annual event since the twenties, set up a county agricultural planning committee, apparently in response to the organization of the state planning board. The county body, in turn, named four topical subcommittees to consider problems of land use and water, livestock, crops, and farm and home.[42]

The following year a two-day farm economic planning conference, drawing fifty to seventy-five participants from all sections of the county, renewed the appointment of a Land Use and Water Committee to draw up recommendations for consolidation with those of similar bodies in thirty-nine other counties as the basis for state planning recommendations. After "open forum discussion" of the committee report, the conference called for permanent retirement from cultivation of 5 percent of the cropland in the western part of the county; reseeding to soil-building crops on an additional 15 percent of the land in the county; use of strip cropping, "where practical," to control soil blowing; use of sweet clover as a conservation crop; reduction in land valuations for range purposes; and organization of grazing districts. The conference agreed that, under existing moisture conditions, the ranges were overstocked but contended that the livestock holdings were not too large to be carried under more favorable circumstances. While they estimated that probably 75 percent of the land in the county was overgrazed, they concluded that grazing districts, with range-water development, would correct the problem. They provided no endorsement for the large-scale federal land-retirement program already under way.[43]

The same four topical committees were appointed at the farm-week conference in 1937; and after holding a series of twelve community meetings, attended by a reported 475 residents, and sending out a circular letter to ascertain general sentiment on a proposal for a permanent planning program, the leaders called an assembly in November to organize formally what was denominated the first Fergus County Agricultural Planning Committee. Each chairman selected one person, who with the original four selected seven others, representing "different areas and in most cases different interests," to form the county body. The following year the first community meetings for agricultural planning in the eastern part of the county were held at Roy, Winifred, and Grass Range. By 1939 twenty-eight community committees were organized. The county structure was completed in 1940 as bylaws were framed and officers, selected by and from the group of twenty-eight community committees, were chosen to form an executive board, with representatives of the federal program agencies, the state extension service, and the county commissioners serving ex officio. Nine topical subcommittees were operating in the county planning effort under this organization at the end of the decade.[44]

Development of county planning bodies was gathering momentum over the rest of Montana and in the Dakotas during the same period. An Adams County Agricultural Planning Board was organized at Hettinger, North Dakota, in March 1936; and a four-county regional meeting, subsequently identified as the Upper Missouri Basin Regional Planning Board, brought together representatives of Williams, McKenzie, Divide, and Mountrail counties at Williston, North Dakota, the following October. A Ward County planning committee was set up in 1937 and formed the basis for expanded organizational activities into the early forties. By the summer of 1939, land-use planning committees were active in Bowman, Burke, McKenzie, Ward, Divide, Mountrail, and Sioux counties of North Dakota. Material on such programs was being incorporated in the public school curricula of several of the counties the following year, and Burke County had published a ninety-two-page booklet on its recommendations. In South Dakota land-use mapping projects begun in Perkins and Tripp counties in 1938 were nearing completion at the end of the decade, and Corson, Grant, and Meade counties had undertaken similar work.[45]

It is notable that so much of this activity was under way before the revised organizational development under the Mt. Weather agreements in the spring of 1939. The work had, however, added to administrative problems which made structural revision necessary. Under the New Deal programs, increasing discord had developed, not only among the federal administrative agencies, but also among local, state, and federal cooperating bodies. Planning boards of all the states complained that the federal agencies operated without coordination and frequently at cross-purposes. North Dakota political leaders were particularly critical, charging that heavy relief expenditures created enormous local debt but effected little long-range reform. Many of the agricultural extension agents and the local business and commercial groups upon whose support they had long depended resented the fact that educational functions normally performed under state and local control must now be shared with federal operational supervisors—AAA, FCA, FSA, SCS, irrigation district, Taylor grazing district, Forest Service—whose channels for administrative decisions extended back to Washington.[46]

The framework of a revised planning effort had been shaped in part, at least, at a five-state conference in Bismarck, North Dakota, in April 1938. It was the first meeting of a Northern Great Plains Advisory Board, or Committee, established at the request of President Roosevelt a month earlier, in fulfillment of the call of the Great Plains Committee for creation of a continuing territorial agency to promote governmental readjustments that would better integrate the efforts of the federal government with those of the states, counties, and local communities. The aim was not to displace existing agencies but to afford continuing study of the regional problems and "by consultation, education, persuasion, and guidance" to coordinate the work of all toward a common end. Under a regional definition expanded to include Nebraska and Wyoming, the body brought together representatives of twelve federal bureaus and twenty state agencies, who adopted an idea presented by the North Dakota State Planning Board calling for development of a "Master Plan," with the operations of all state, regional, and federal agencies coordinated to its requirements.[47]

The problem centered in the division of responsibility. At the annual meeting of the Association of Land Grant Colleges and Universities at Houston, Texas, in the fall of 1937, it had been agreed that the association and the USDA should each appoint a committee to attempt to work out a mutually acceptable arrangement. The agreement developed at Mt. Weather, Virginia, in July 1938, and subsequently endorsed in individual agreements between the department and the state agricultural colleges, centered the structuring of planning activities in the USDA, where an agricultural program board, chaired by a land-use coordinator and including the heads of the action agencies, the chief of the BAE, and the department's directors, advised the secretary. That autumn Secretary Wallace designated the BAE as the department's chief planning agency and reorganized it to encompass a division of state and local planning, with a representative in each state, and an interbureau coordinating committee in Washington to correlate all program plans from the field.[48]

Within the states, advisory committees were to assist county committees in an effort to formulate agricultural plans for their constituencies and to review and approve all plans prior to forwarding them to the BAE. The state committees were to be chaired by the director of the state extension service, with the state representative of the BAE as secretary. The membership was to include, besides these officials, the chairman of the state AAA committee, the state coordinator of the SCS, the state director of the FSA, a representative of the Forest Service, representatives of any other agencies administering land-use programs in the area, "and a number of farm people, at least one from each type-of-farming area within the State," preferably also members of county land-use planning committees.

The initiating bodies in the planning effort were to be the county committees, consisting of the county agent, at least one member of the local AAA committee, the rural rehabilitation supervisor of the FSA, representatives of other state and federal agencies administering local land-use programs, and a "substantial majority" of farmers. The chairman was to be a farmer, and the county agent, if the group so desired, the secretary. Plans proposed for local development were thus to be drawn up in conjunction with the administrative officials responsible for translating them into action. Identified as "Cooperative Land Use Planning," the structure was hailed as "A New Development in Democracy," "Democratic Planning in Agriculture."

It is questionable whether the planning committees were really representative of average farmers. A contemporary study of the organization in about fifty counties dispersed through twenty-two states indicated that local committee members were elected in only five states, including Montana and North Dakota, and the elections were conducted in open meetings, without written ballots or regular electoral procedures. Elsewhere the members were selected by the county agents, who, anxious to file a good report, tended to draw upon men and women with whom they had established working relationships characterizing them as "the leading" farmers and homemakers. In sixteen of the states studied, women were serving on county committees, although they usually constituted only a small percentage of the whole membership. Representation of special interest groups—farmers' organizations, stockmen's associations, commercial clubs, bankers' associations and other business organizations—was seldom stipulated but often dominated the selection process. Tenants and nonresidents had small voice in the proceedings. Residence distant from the community center also inhibited participation, particularly if, as was customary, meetings were scheduled for winter months, when farm operations were less demanding. As evidenced in central Montana, activity year after year was dominated by a relatively small number of the same individuals, who occasionally transferred their leadership role into political prominence.[49]

The Ward County, North Dakota, committee, publicized as a model under the new plan, evolved from a nucleus already organized as an older planning committee, similar to that described above for Fergus County, Montana. Additional members to a total of twenty-four were selected by the county agent for the initial meeting. This group then divided the county into twelve communities, of three or four townships each, delineated on the basis of types of land, topography, local trading and meeting centers, and accessibility of meeting places. Farmers were then invited by mail and radio to public assemblies at which seven or eight committeemen were elected, including at least one for each township in the community. The community committees each elected their own chairman, who served on a county body of about thirty, including local representatives of the governmental action agencies. During the spring of 1939, when the planning proposals were being framed, twelve community and three county meetings were held, with participation reported as including 600 of the 2,700 farm families in the county.[50]

State committees, formed from the more prominent leaders at the county level, were still less representative. A formal election process to establish farmer representation was not instituted in South Dakota until the fall of 1940, and the new delegates were not seated until the following January. The North Dakota Advisory Council in the fall of 1939 had called for nomination of additional farmers as prospective members, but the process was still not elective in 1942, when the director of the extension service informed county agents that the farmer members had been "selected" and remarked: "It is our hope that some method can soon be developed providing for the democratic election of farmers to the State Committee by the counties themselves." In Montana the farmer representatives were appointed by the director of the state extension service without explanation of the selection procedure.[51]

While the USDA planning directive stipulated merely that "a number of farm people, at least one from each type-of-farming area within the State," should be included on the state committees, it was anticipated that there would be as many as there were state and federal administrative personnel.

In practice, however, the meetings in the northern Plains included a dozen or more agency representatives with rarely half as many laymen. The BAE regional planning office urged limitation of the number of official participants and expansion of the popular membership, but to little avail.[52]

At the local level, planning work was designed to progress through three stages: preparatory planning—organizing the committee and assembling data; intensive planning—studying the problems, agreeing on the facts, and deciding on the objectives for improvement in the various land-use areas of the county; and unified program planning—deciding specifically what should be done and by whom, that is, developing what was known as a "unified action" program. In view of the scope of the resettlement effort and the planning activity already under way, the lack of requisite data reported by these committees for the preparatory phase is remarkable. A BAE survey of the progress in rural land classification through December 1935 showed heavy reliance on crop-history records of the AAA, which dated only from 1928 and embraced but two years not marked by widespread drought. By 1936 the Gieseker reconnaissance soil surveys extended to less than a third of Montana's counties. When the state Land Use Advisory Committee reviewed proposals for unified action planning, local committees cited need for land-use mapping in Fergus, Richland, and Wibaux counties and complained that they lacked even basic soil surveys for Flathead, Fallon, and Richland. Data on location and size of individual land holdings, ownership of water rights, range conditions, ownership characteristics, gross income, individual crop yields, bankruptcy and foreclosure history, population movement, the availability of facilities for transportation, education, health, and recreation—a total of twenty-eight factors cited as requirements by the Resettlement Administration in surveying part of Fergus County—needed to be assembled and mapped. Working with statistical compilations gathered by WPA labor, the Montana Agricultural Experiment Station published between 1936 and 1941 a lengthy series of studies embracing such information, culminating in 1940 in a detailed graphic supplement on necessary farm adjustments for the state. Elsewhere in the region progress was slower.[53]

North Dakota reports indicate that no mapping was under way in most counties as late as April 1939. Some work had been undertaken by a AAA committee in 1936, purely statistical at first. Later the information was used to develop recommended adjustments in optional land use. Describing the process, the state extension agent in land use informed the advisory council that the work had been done primarily by an "ex officio and selective" committee, who had met two or three days during the year and "relied on experience and opinion in part, in a hurried inventory of land use possibilities." The first detailed survey was prepared for the Ward County Land-Use Planning Committee and published in February 1941. Conducted with WPA labor by the North Dakota Extension Service, it utilized AAA statistical data and aerial surveys and SCS mapping of soil types and topography to identify individual farms within that county on the basis of soils, topography, type and size of farm, and tenure. On that basis five type-of-farming areas were outlined and two meetings were held with a farm planning group in one of those areas to develop recommended budgets for 320-acre and 480-acre cash-grain farms and a 480-acre diversified operation. Although recognizing that additional data might be needed under varying conditions, the analysts offered the study as a model for the remainder of the county and elsewhere in the state.[54]

The South Dakota Agricultural Experiment Station had collected data, not yet analyzed in 1935, relating to evaluation of county-owned land, but found that "rather poor records" complicated its task. The report of its study, published in 1938, noted that in the absence of soil surveys, soil maps, and land classification, tax assessments were based upon uniform, aggregate valuations. When county planners in Hand and Perkins counties in the summer of 1940 requested soil surveys, the state Land Use Program Committee debated their value at length before informing the counties that they knew of no assistance that could then be given but that "everything . . . [would] be done to secure help as soon as possible." Six months later the committee approved in principle a subcommittee proposal for legislation establishing a soil and land classification and evaluation system, but refused to sponsor the measure.[55]

The goal as the new planning program began was to develop a "unified action" agenda for one county in each state, a concentration of effort which it was hoped would produce demonstrable results as encouragement for similar work elsewhere. A plan was put into operation in Teton County, Montana, in 1940, and proposals for a similar project in Roosevelt County were sent to Washington the following year. By March 1940, Daniels and Judith Basin counties had also drafted reports, the latter of which had been submitted to the state committee but returned for further work. Pondera and Richland county reports were in "preliminary stages"; Chouteau County land-use maps were completed and those for Fergus County almost so. Mapping "in many other counties" was under way. A year later reports for Valley and Fergus counties were being drafted. In North Dakota, Sargent County, in the eastern half of the state, had undertaken such planning before the Mt. Weather agreement, and afterward, Ward County was authorized to develop a program, which was the only one in operation by mid-1942. In South Dakota, Tripp County was selected in 1941 to initiate such a plan for a west-river county. Fourteen other counties of the state had begun land-use mapping by March 1940; two counties east of the Missouri had completed preliminary reports by the following January; but no "unified action" program had been fully approved before general termination of the effort.[56]

Extension agents wrote enthusiastically of the local response: "farmer participation is all that can be expected and more." Near the end of the first year, the North Dakota extension agent in land use wrote: "We have seen enough of county planning already in the short time it has been operative to feel safe in saying that no one can doubt the ability of farm people to plan wisely and well." Yet not all agricultural educators favored the concept of "democratic planning." At a symposium of the Farm Economics Association in 1939, Professor John D. Black questioned whether "a group of leading farmers gotten together in any county . . . [could] be trusted to know what is best for the county." He believed that they would "make poor suggestions in very many cases" and suggested that the success of a few venturesome farmers with new systems of land use or cultural practice might better form the basis for recommendations than the prevailing experience.[57]

The committees did question and reject some of the recently introduced programs. As the work began in North Dakota, one farmer committeeman from Golden Valley County expressed "sympathy for the 'back to grass movement'" but urged that it not "be used to dodge a multitude of agricultural issues." Planning activities were inhibited by lingering bitterness from the resettlement program in Fallon, Valley, and Flathead counties of Montana; and in South Dakota a state committeeman grumbled: "I do not believe we can make human beings into 'checkers' and 'set them here and there.'" An important aspect of committee activity, as originally conceived, was evaluation of local proposals for water facilities, but the Montana committee found that few farmers wanted to cooperate in an effort to set up demonstration projects. A member of the Montana state committee noted in 1941 that available federal funds were not being used because operators were unwilling to commit themselves to the expense and "red tape" of such development. North Dakota farmers also complained of the cost; and, while the South Dakota committee reported considerable enthusiasm for the work in January 1940, it noted by the following September that thirteen counties, mostly west of the Missouri and designated as top-priority authorizations, had failed to express any interest in the matter.[58]

Nearly all state committees stressed the need to maintain family-sized farms and questioned whether the federal programs sufficiently protected that institution. During lengthy discussion of this topic, a subcommittee of the Montana state body noted that many of the governmental efforts—specifically those of the resettlement and rehabilitation programs of the FSA, the work of the SCS, the tenant purchase program, reclamation development, the lending policy of the FCA, and the educational focus of the extension services—encouraged family-sized enterprise; but the full committee nevertheless urged the county and community planning groups to observe what changes were occurring in farm size and particularly what effect was being produced by organization of cooperative grazing districts and operation of the Land Utilization projects. Water facilities proposals for both the Heart River and the Little Muddy watersheds were sharply criticized in the North

Dakota State Advisory Council for requiring large reductions in the number of operating units. When the unified action plan of Hand County, South Dakota, called for enlarging farm holdings, the state committeeman representing the FSA, Emil Loriks, formerly president of the South Dakota Farmers' Union, protested that extension of such a proposal over the entire state "would displace perhaps 12,000 to 15,000 farm families"; and that, he predicted, "would probably be only a preliminary step in the process and in another year or two with further advances and improvement in our power and machine farming, we would no doubt be clamoring for still larger units and ultimately fewer operators in order to show larger profits."[59]

Committeemen were, however, unsure what constituted a family-sized farm. A representative from Brown County, South Dakota, chairman of a AAA committee, raised that question in 1940, commenting: "I have about come to the conclusion that a one-half section will not make a living for a family, but I have neighbors who are making a living on a one-fourth section." The issue was a major point of controversy in the membership appeal between the National Farmers Union and the American Farm Bureau Federation and represented not so much a reasoned management decision as an emotional identification with organizational social distinctions. In the transitional farming areas of the western Dakotas and eastern Montana, where vacant land was scarce, the commitment to small farms voiced for some supporters a protest against land planners who advocated resettlement as a measure to facilitate better adjustment of operational scale. For other advocates it signified the practicability of extending familiar agricultural patterns into the dry-farming effort. Experience, however, was steadily eroding the force of such connotations; it was not merely the demand for profits but adjustment to the requirements of competitive survival that dictated the search for economies of scale.[60]

The county planning program disintegrated as the nation was caught up in the war effort. Perhaps such a result was inevitable as climatic conditions and farm prices improved. Also, discord between local leaders and Washington planners, between federal agencies themselves, and between rival farm organizations supporting particular programs and agencies was in full play by 1941. In January of that year an interbureau coordinating committee of the BAE drafted and sent to the state advisory committees an eighteen-page memorandum of "Suggestions for a Unified State Agricultural Program to Meet the Impacts of War." Discussions in the state committees were wide-ranging, highly generalized, and more reflective of social philosophies than of specific plans. The state proposals looked to long-term arrangements— soil and land classification; disposition of state and county land holdings; production geared to a domestic, primarily self-sufficient economy; marketing reform; and development of employment opportunities through decentralization of war industries and stimulation of recreational or handicraft enterprises. Fear of overexpansion of agriculture with a repetition of the postwar depression of the early twenties was a dominant concern. Sherman Johnson, who was then working in the farm management division of the BAE, expressed amazement at how little realization existed in these planning sessions of the imminent effects of the European war upon the domestic economy. His suggestion to a Montana extension conference in September that farm machinery purchases should be made while such items were still available fell on deaf ears. The administrators of the AAA protested, with some justification, that the planning committees were not dealing with practical problems.[61]

That summer, when the BAE proposed expanded production goals for 1942, officials of the AAA, concerned about surpluses, stated that the goals were excessive and requested the authority to frame such decisions. While this planning remained a function of the BAE, the power of the agency was being greatly reduced. In July 1941, Secretary of Agriculture Wickard, who had long worked in the AAA and had little regard for the planning activity, set up a new system of local boards, chaired by AAA officials, to serve as the administrative link for coordinating USDA defense programs.

Reflecting upon the organizational problems of the planning work some months later, the Montana Land Grant College–BAE Committee, which served as an executive committee for the Montana State Agricultural Advisory Committee, stressed the need for over-all plans in counties as a guide to action programs and noted the need for leadership in formulating agricultural programs in the state. They cited a lack of participation by educators, researchers, and technicians and difficulty in getting participation of federal action agencies in assisting on projects. As a remedy, they proposed that plans be made offering responsibilities in leadership to the federal agencies, "such as they have not had in the past."[62]

A memorandum written at that time by the SCS to the BAE in response to a request for assistance to a North Dakota county planning committee reinforced these observations and further indicated the problems of developing interagency cooperation. "As an action agency," the official wrote, "our experience has indicated that *we can get more conservation on the land per dollar expended in assisting soil conservation districts than in any other manner known to date.*" While the service believed "*the objectives of the local and State land use planning program to be fundamental*" and would "endeavor to render all practicable assistance" to the work, the extent to which the SCS could make such commitments was "circumscribed by the congressional and Departmental directional guides." He recognized that in many instances those commitments had "not been sufficient to carry out the recommendations initially deemed desirable by local land use planning committees."[63]

The state and local planning program was terminated in 1942, when Congress yielded to pressure from the American Farm Bureau Federation, and with scant opposition from Secretary Wickard refused to provide further funding. Two years earlier the giant farm organization had mounted an opposition to the federal agricultural planning activity generally, whether by the AAA, the SCS, the FSA, or the BAE, as a threat to the local leadership of the county extension agents. Responding to charges of agency duplication of effort, Congress in 1941 reduced the budget of the BAE for the following year by half a million dollars. While this was a setback, forcing the bureau to curtail financial assistance to the state advisory committees, it did not halt the program. The South Dakota Land Use Program Committee appears not to have met after March 1941, but the committees in Montana and North Dakota continued until the summer of 1942.[64]

In budgetary legislation of 1942, the House of Representatives, still at the urging of the Farm Bureau Federation, proposed a further reduction amounting to approximately $1 million from the funding for the BAE. The cuts would have fallen on the economic research as well as the program development of the agency. Secretary Wickard defended the "fact-finding" but not the planning activity. After conference on the House and Senate bills, the reduction in appropriation was compromised, but the measure carried a stipulation that no part of the funds could be used for state and county planning.

While the executive committee of the Montana State Advisory Committee a year earlier had pointed to the need for greater leadership from the federal agencies, only one state extension director, E. J. Haslerud, of the North Dakota Agricultural Advisory Council, defended the program before Congress. His announcement to the state body in June 1942 that funding would probably not be available for their activities during the coming year caused consternation. All present were reported to have expressed a desire for the work to continue, the farmer members even offering "to attend meetings at their own expense so long as their personal prosperity permitted it." At that time nearly 900 communities over the state were organized under the program. Nearly 3,500 committeemen had served on local committees during the past year. Summarizing the local effort some years later, Haslerud noted that more than 4,000 community gatherings and nearly 300 county meetings had been held, involving 27,000 farm families, nearly 40,000 different individuals.[65]

Historians, commonly viewing the county planning effort as an aspect of the administrative power struggle, have minimized its accomplishments. Professor John D. Black in his contemporary criticism had, with prescience, questioned whether under the organizational structure the unified programs could "be converted into action with sufficient vigor to insure their continued vitality." The county planning structure had clearly not been efficient; yet as the above survey indicates, the program had stimulated community participation which had heretofore been lacking. The work of the local committees included organization of weed-control districts; grasshopper and rodent control; promotion of irrigation and stock-water development; soil, grass, and water surveys; preparation of land-use maps and determination of leasing values for county-held lands; study of population changes and community patterns in relation to rural electrification, schools, roads, health, and recreational facilities; assistance in national forest development and fire protection; and review of the activities of the extension service, SCS, FSA, and FCA. Much of this involvement was merely informational, as administrative officials reported on their programs, but discussion sessions afforded opportunity for popular participation through suggestions and recommendations. As one farmer committeeman informed the South Dakota Land Program Committee in 1940: "We have been getting a great deal of information together about the county. I do not know that we are just sure what we are going to do with all of the information. It is all so new. The people, however, are for it and they go right after it. They don't all agree, which at times makes it harder to know what to do, but they are surely becoming better informed about the county in many ways."[66]

During the decade a variety of planning agencies had pursued a relatively limited range of proposals for readjusting the regional economy. Initially the land-use planning problem had been viewed as one of a wheat surplus which, as drought swept the northern Plains, planners hoped to counteract by conversion of drought-stricken wheat-producing districts to rangelands for livestock. Dislocated resident farmers would be resettled on small, irrigated farms intended to provide a diversified, self-sufficient economy but not to compete in commercial markets. The Federal Emergency Relief Administration, the Resettlement Administration, and the Farm Security Administration all retained the latter management goal in guiding general relief efforts. State planning boards and the Great Plains Drought Committee, ultimately with congressional support, expanded the focus in promoting irrigation development, but again primarily as a support to an economy based on livestock. In its demonstration projects, the Soil Conservation Service had also stressed livestock rangeland development for much of the region but in climatic transitional zones had proposed tillage modifications to benefit small, diversified dry-farming operations. The county unified-action programs had followed much the same conceptual format, stock-water and range development, with attention to diversified cropping under dry-farming procedures in eastern border areas.

Yet late in the decade it was apparent that cash-crop grain farming was not only still prevalent but was the most profitable type of operation where terrain and soil conditions were suitable. The planning recommendations of the Montana Agricultural Experiment Station (MAES) and even of the federal Bureau of Agricultural Economics incorporated large areas of such land utilization in the localized studies that were being published at the end of the decade, although in most cases addition of some livestock and an accompanying cultivation of feed crops were seen as desirable supplements. The larger-scale operations, 480 to 1,280 acres, were always found to be most successful. Some planners were even contending that rotational, diversified cropping and the addition of livestock did not stabilize farm income.[67]

Writing in 1939 on the "Type of Farming Modifications Needed in the Great Plains," Elmer Starch, who had moved from his post at the Montana Agricultural College to serve as regional director for the Resettlement Administration and subsequently the Farm Security Administration, emphasized the goal of "flexibility," to be achieved by a combination of enterprises that might be used according to the variability in weather conditions, taking advantage of good grass *and* good farming circumstances as they arose. He recognized, however, that diversification within the bounds of a given farm unit of one or two sections of upland was not practicable in the region. Instead, he recommended wheat as the best crop for benchlands, since it was better adapted to an arid climate than any other cultivated crop. Net income, he warned, might be reduced if the farmer in that situation incorporated alfalfa and oats in the rotation. Indeed, he continued, with such cropping, nothing would be gained in stability since it had "been the experience of farm operators that wheat is the last crop to fail, with the possible exception of native grass."[68]

Starch proposed that diversification could be achieved by development of a pattern under which the farm headquarters was located within an irrigated area, where feed crops might be grown, while the same operator raised wheat on not too distant benchlands and ran cattle on a cooperatively managed grazing reservation. This was the integrated approach that the MAES advocated during the latter years of the decade and which the later resettlement projects were designed to permit. If individual farmers were not able to achieve such an operational balance, the hope was that at least the regional economy might benefit under internally correlated marketing arrangements.

Efforts of planners to diversify the economy and, specifically, to increase livestock production had not been successful from 1929 to 1939. The acreage in harvested cropland, it is true, was decreased by 30 and 36 percent, respectively, in eastern Montana and western North Dakota and by 50 percent in western South Dakota, and a reduction in the acreage of harvested wheat almost mirrored the pattern, with declines of 32, 38, and 52 percent, respectively; but large increases in the acreages of idle land or cultivated summer fallow—46, 224, and 339 percent, respectively—accounted for much of the difference. In the two northern states, the reduction in harvested wheat acreage amounted to little more than the limitations imposed under the AAA adjustment program. In western South Dakota it mirrored an accumulation of sizeable reductions in the acreage of every crop but sorghum, and the increase in sorghum only slightly offset a 72 percent decline in corn acreage. Increases in the acreages of oats, corn, and sorghum in eastern Montana and of corn and sorghum in western North Dakota provided some offset to major declines in the acreages of barley, rye, and alfalfa, but the acreages in feed grains and hay crops were generally contracted throughout the region. While the acreage in plowable pasture was increased somewhat in the western Dakotas, it actually declined slightly in eastern Montana. Meanwhile, the number of cattle was reduced by approximately one-quarter in eastern Montana and western North Dakota and by over one-third in western South Dakota. The number of sheep was increased by about 24 percent in western South Dakota but decreased by about 22 percent in eastern Montana and changed only slightly in western North Dakota, where the total remained small.[69]

The average size of farms was increasing—by 27 percent in eastern Montana, 6 percent in western North Dakota, and 61 percent in western South Dakota—as the number of farms was reduced by 18, 9, and 25 percent, respectively. Yet Montana planning studies indicated that about one-third of the dry-land farms ranged between 260 and 590 acres, while under conservative long-run yield and price estimates at least a section would be required to maintain a satisfactory living ($2,000 in gross income annually). Unless the larger holdings were broken up for redistribution, almost one-half the farms in the northeastern counties would fail to meet that criterion. The situation was somewhat better in north-central Montana, where but 17 percent of the farms could not provide such an income, yet that estimate would require a reduction of 331 families in the district. Similarly, a study relating to north-western South Dakota concluded that 41 percent of the farms were 720 acres or less and most would need to be "displaced" if a satisfactory income (gross income of $1,462) were to be achieved. To get units of 480 acres in western North Dakota would have required displacement of 290 farms (18 percent) in Divide County and 171 farms (14 percent) in Hettinger County. Small wonder that the issue of appropriate farm size provoked controversy in the county planning effort, as it had in the resettlement program! The dilemma of too small land units, whether as farm or ranch and whether under irrigated or dry-land cultivation, barred profitable operations over much of the region.[70]

The subsidized reform of the AAA approach had proved effective in readjusting individual cropping and management practices, but the program was still criticized for its failure to reallocate regional land-use patterns. The generalization of cropping and tillage regulations, the compensation for compliance on an acreage basis without reference to productivity, and the pressure for maintenance of cropping histories tended to forestall adjustments that might otherwise have developed as an economic response to local conditions. Whether the investment in irrigated communities and range development might ultimately effect the desired changes in land use or, as an alternative, support the flexibility required for dry-land development were questions yet unanswered.

The degree of public support for land-use planning was to be crucial to maintenance of the adjustments under way. While for many Montanans the presence of former local leaders—M. L. Wilson, Elmer Starch, and Chester C. Davis, to name only the most notable—among the administrative bureaucracy, had conveyed a sense of ameliorative intercession, regional drought sufferers still complained of delays, ignorance, even arrogance in their relationships with federal administrators. Tugwell's earlier published proposal that farmer assistance be conditioned upon "good behavior" and reforms be enforced by denying use of privately owned marketing facilities revealed attitudes that kindled outrage. The contemporary Russell Lord has provided an accurate account of the popular perception of agency officials:

> eager-faced, immature technicians and academicians . . . zealous novitiates in governmental action [of whom] few, very few, had ever been elected to anything; and had never even tried to be. . . . But most of these unelected fledgling New Dealers were operating beyond strictly technical lines; they were up to the eyes in action programs and in active policymaking. In this their lack of a definite mandate and their lack of definite political experience proved decidedly a weakness. They were bound to arouse a mounting resentment and rebellion on the part of Congress, particularly; and they did.[71]

Recognition of this reaction underscored the importance of the movement to incorporate local planning, but it came too late to save the institutional framework of that program, as well as the Resettlement Administration, the Farm Security Administration, the National Resources Board, the land-planning division of the Bureau of Agricultural Economics, and finally the BAE itself. Too late to forestall in the drought-stricken Plains a developing sense of regional alienation. Too late, also, to offset the generalization of conceptions of regional definition and identity.

Local planning for dry-land areas in succeeding years evolved under the more circumscribed framework of the soil conservation districts, the agricultural conservation and production committees (successors to the AAA bodies), and the county agricultural extension programs. The legislation establishing the old state planning boards had been repealed in the Dakotas with the development of the Mt. Weather program. The Montana State Planning Board had not been funded from 1937 to 1939; and although continued in 1939, the board limited its activities almost solely to programs of water development. More broadly based regional planning took form intermittently through the Great Plains Agricultural Advisory Council, providing informational exchange among technical and extension personnel of the agricultural colleges and the USDA, but without an organizational role for farmer representation.[72]

While the council, which was developed in accordance with the recommendations of the Great Plains Drought Area Committee, has continued to meet for over half a century to study and discuss regional problems, its agenda since 1946 presupposes a regional unity of climate, agricultural production, farm technology, and market outlets that is of limited applicability.[73] As dry-land agriculture entered a prolonged period when drought was less a concern than farm-price relationships, the adjustments entailed a realignment of focus to national and international trading considerations for which the conceptualization of unitary regionalism accorded little cognizance to local and individual differences. Such generalization became one of the costliest legacies of the drought crisis of the thirties—and for agriculture in the northern Plains, potentially the most damaging.

PART THREE

Recovery Years
Realignment of Priorities

CHAPTER SIX

Prosperity Returns

Under local initiative, agricultural adjustments in the northern Plains during the twenties had assigned priority to the problem of drought and, in that approach, had centered upon the technology of dry-land cultivation. Efforts to win recognition for protein measurement in grading of wheat and M. L. Wilson's emphasis upon measures to maximize economy of production had reflected growing concerns related to marketing problems, but they also called for adjustments directed to the advancement of dry-land operations. By the end of the decade, the principal reforms of methodology and management for such agriculture had been identified.

The severity of the drought during the thirties, sharply challenging these developments, had led to the interjection of national programs based upon very different conceptions of the problems. Their primary concern over the past decade had been the general marketing situation, for which the drought represented a remedy. Payments to regional farmers to relocate out of commercial competition appeared an approach to reduce crop surpluses. When farmers failed to respond, a program of "rehabilitation in place" was instituted, re-enforcing the previously developed methodological approaches but rejecting the management recommendations of regional specialization. Credit was limited for capital expansion, and crop and cropping quotas restricted production.

Recurrent drought has always been a concern for dry-land farmers in the northern Plains; but in the half century since the thirties, there have been only two periods of prolonged drought, and neither of those was regionally generalized for longer than a year. Drought has been scattered spatially and temporally. Such episodes require a variety of measures for risk management, perhaps even short-term relief, but they need not constitute a deterrent to long-term profitability of operations. With the improvements in crop varieties and methodological developments available, the adjustments to drought became more dependent upon educational guidance and expanded opportunity for managerial flexibility than upon further technological advances.

For some fifteen years the demands for food during the war and postwar reconstruction also forestalled the marketing crisis. The role of the United States as the primary creditor nation and sponsor in the rehabilitation effort opened trading channels long closed. Prosperity prevailed both regionally and nationally. But the agricultural innovations of the postwar years—use of higher yielding plant varieties; of fertilizers, fungicides, and insecticides; of larger and more powerful machinery; of expanded crop acreage—increased yields exponentially. While vastly raising production costs, they regenerated marketing problems.

Productivity became the focus of governmental attack, as planting allotments and marketing quotas were reinstituted in the mid-fifties. New programs were designed to restore the northern Plains to grass—not because of regional drought or low yields, not because of noncompetitive crop production costs, and not because grass afforded a more profitable use of the land. Although drought had prevailed in Colorado from 1952 through 1956, it was experienced in Montana only during the first and last years of the series, and above average precipitation marked the intervening three. Production values on wheat in Montana reached heights from 1950 through 1956 that were not exceeded before 1963 and thereafter not again repeated before 1972. Low rainfall was recorded in the Dakotas only in 1952, and diminished yields there were owing to plant disease, not climatic conditions.[1] In enactment of the Great Plains Conservation Program of 1957 under a broad generalization of regional characteristics that linked Colorado, New Mexico, Oklahoma, and Texas with Montana and the Dakotas extended eastward to the 98th meridian, government planners negated the localized marketing specialization of the Christgau proposal. They also assigned new status to conservation as an adjustment priority.

Popular reaction again halted the program for retirement of regional cropland. Production in Montana reached amounts above the peaks of the fifties from 1974 through 1983, and production values held above those of the fifties every year but one from 1973 to date. Under aggressive marketing, the surplus problem was alleviated as exports soared beginning in 1968. Despite some unevenness, they remained at levels not heretofore attained through the next two decades.[2]

Again, five years of drought in the lower Plains, from 1974 through 1978, revived the call for cropland retirement as a conservation measure, and again it was extended largely toward the dry-farming areas to the north. Under federal planning effected in legislation of 1985, nearly 13 million acres of cropland were designated for withdrawal in Montana and the Dakotas, including over 8 million of the 15.7 million acres then cultivated in eastern Montana. Under long-range projections, only a little over one-third of the cropland in the Dakotas, Nebraska, and Kansas and about one-fourth that in the mountain states from Montana and Idaho south to the Mexican border are scheduled to remain in production.[3]

During the recovery years, adjustment priorities for dry-land agriculture were nominally realigned from drought to marketing demand and finally to conservation, but in national land-use planning, the marketing concern has held primacy since the early twenties. As discussed in the following chapters, the problem of continuing crop surpluses has shaped the current conservation programs. Sustained by misconceptions of regional generalization and an array of social, cultural, and political predilections, the governmental agenda of production control was directed primarily to the wheat-producing Plains areas, at an estimated cost of over $20 billion dollars, approximately $5.4 billion for the northern states, under the expectation that cropland retirement will in time obviate the mechanisms of price-supported marketing. Yet after three successive years when average annual precipitation was deficient, culminating in 1985 with wheat production the lowest since 1937, bids were negotiated in Montana for retirement of only 28 percent of the eligible acreage during the first seven signup periods under the new legislation. Payments averaging approximately $400 an acre over ten years induced relatively few dry-land farmers in that state to give up use of their land.[4] Since the economic recovery of the early forties, prosperity has not been continuous in the northern Plains, but it has elevated expectations.

Rainfall returned in average amounts over most of the region by 1940. Central Montana and western South Dakota continued their poor record a year longer but then joined southwestern North Dakota with above-average levels of precipitation until 1949. South-central and north-central Montana reported deficient moisture again from 1943 through 1946, and northwestern North Dakota had relatively light rainfall in 1945, but peak amounts in the spring growing months alleviated these conditions. In all, the decade presented one of the longest sustained periods favorable for cropping since agricultural settlement of the region.[5]

Heavy yields reflected the improved climate. Eastern Montana, which had produced an average of 8.9 bushels of wheat to the harvested acre in 1929, 10.5 bushels in 1934, and 12.6 bushels in 1939, yielded an average of 18.45 bushels in 1944. Western North Dakota yields averaged 9.7 bushels in 1929, 3 bushels in 1934, and 9.2 bushels in 1939, but 16.75 bushels in 1944; and western South Dakota yields averaged 10.7, 5, 7.5, and 10.4 bushels, respectively. Individual counties registered much higher averages in 1944—Teton and Big Horn in Montana, almost 23 bushels; Chouteau and Pondera, over 21; Cascade, Fergus, McCone, and Sheridan, over 20. No western county in the Dakotas did as well; but in North Dakota, Williams County yielded almost 20; Golden Valley and Mountrail, over 18; and Burke, Dunn, McHenry, McKenzie, McLean, Oliver, Stark, and Ward, over 17, contributing to a western regional average higher than that for the eastern half of the state. Local journalists again reported yields of 30, 40, and 50 bushels to the acre in districts that had been recommended for retirement five years earlier. In 1944, when the federal government suspended the national crop-insurance program because of excessive losses, wheat growers in the northern Plains had accumulated a credit balance equivalent to several million bushels that served to cover deficiencies elsewhere.[6]

Wartime limitations on the availability of farm machinery and labor posed difficulties for agriculture. In November 1942 manufacturing of farm machinery was reduced to 20 percent and repair parts to 13 percent of the amount produced in 1940. When as late as mid-October 1942 some 60 percent of the threshing was yet to be done in North Dakota, the governor protested that the situation was "desperate." The Montana Legislative Assembly that winter petitioned government leaders to reconsider the quotas and to modify military deferment policy for agricultural assistance. Only slight relaxation was authorized a year later when the quotas were scaled to production in 1940 or 1941, whichever were greater. After a decade in which capital expenditures had been limited by low income, Fergus County, Montana, with a total of 1,356 farms, received quotas of only seventeen grain drills, forty-two combines, and fifty-five tractors in 1943 and thirty-four grain drills, thirty-two combines, and fifty-seven tractors in 1944. The Farm Security Administration tried to ease the problem by promoting cooperative use and acquisition of the available machines but was discouraged by congressional criticism that the action was "communistic." Trucks, tractors, and harvesting implements were still "virtually impossible to buy" as late as the spring of 1946, but increased hand labor served to meet the harvest demand. Housewives, urban residents, school children, migrants from Mexico and the South, German and Italian war prisoners, Japanese internees, and army trainees were called on to help in the fields. Governor Moses of North Dakota at the end of 1942 lauded 48,000 students who had spent nearly 716,000 working days to harvest the state's "greatest crop."[7]

Acreage quotas also limited the cropping of wheat until the spring of 1943, but production had reached new levels in all three states beginning in 1941. Through the end of the decade, the trend continued above the level of 1940, when the amount was considerably greater than the average of the thirties. Acreages in flax, barley, oats, and alfalfa were also increased throughout the region during the war years. At the same time the number of cattle on the ranges approached the record levels of the pre-drought period in Montana and western North Dakota, while greatly exceeding them in western South Dakota. The number of sheep was reduced slightly in the northern states but expanded by nearly one-third in South Dakota. Hog production, too, increased, even in the Plains area, as part of the "Food for Freedom" program.[8]

Apart from a brief period in the spring of 1940, when the German blitzkrieg disrupted foreign commerce, farm prices soared. By the autumn of 1939 federal administrators announced that the increase in commodity prices had virtually eliminated any potential government loss on loans advanced to farmers on stored wheat. The seasonal average price received by farmers for wheat, nationally, increased from $ .68 a bushel in 1940 to $ .945 in 1941, $1.10 in 1942, $1.36 in 1943, $1.41 in 1944, and $1.50 in 1945. Beef cattle prices for regional farmers rose from $6.30 per hundred-weight in 1939 to $6.80 in 1940, $8.10 in 1941, $9.50 in 1942, $10.00 in 1943, and, after sagging to $8.90 in 1944, to $10.10 in 1945. The ratio of prices received to prices paid regionally rose from 69 percent in 1939 to 77 percent in 1940, 85 percent in 1941 and 1942, and 95 to 97 percent from 1943 through 1945.[9]

Gross income from wheat, alone, rose in Montana from $31.5 million in 1940 to $98.7 million in 1944, dropping only slightly in 1945 before continuing upward. The rise was constant in North Dakota, from $62.9 million in 1940 to $370.3 million in 1947, and in South Dakota, from $17.9 million to $128.1 million during that period. By 1950 annual net farm income in Montana amounted to $228.5 million; in North Dakota, nearly $262 million; and in South Dakota, $245.8 million. A single year's net income was well in excess of the total payments for relief in the region during the thirties; in North Dakota, more than double that amount.[10]

This was a decade when the debts of the drought years were rapidly liquidated. Farm mortgage debt in Montana was reduced from $66.1 million in 1940 to $31.5 million (52 percent) in 1946, the best record of debt reduction in the nation. North Dakotans' debt declined from $141.2 million to $75.2 million (47 percent) and South Dakotans', from $127.7 million to $89.8 million (30 percent). Delinquency rates on farm mortgages held by the federal land banks, which had stood at 72.8 percent in North Dakota, 40.1 percent in South Dakota, and 34.6 percent in Montana in 1940, fell to 28 percent, 11.6 percent, and 22.7 percent for those states, respectively, by 1946. The FSA reported in June 1946 that nearly 69 percent of its rural rehabilitation loans to North Dakota had been repaid in full, 90 percent of the amount when due, and nearly $2.5 million additionally paid as interest. In South Dakota principal repayments on such loans amounted to 63 percent and in Montana, 62 percent. In Fergus County, Montana, nearly 70 percent of the crop and feed loans for the extended period 1918 through 1937 had been repaid by 1942, and the rate of repayment on current loans was close to 90 percent. Public welfare officials of North Dakota noted that many of the former relief clients were repaying even the money "given" to them during the lean years.[11]

President Roosevelt had called on farmers in the summer of 1942 to help fight inflation by diverting a large share of their expanding income to debt retirement. The scarcity of farm equipment and other consumer goods in wartime necessarily limited such expenditures, but forced liquidation accounted for much of the debt reduction. Forced sales occurred at the rate of 201 per thousand farms in North Dakota, 152 in South Dakota, and 94 in Montana during the six years from 1940 through 1945, while nationally the rate was only 54.[12] In that pressure lay the opportunity for much needed expansion of operations by those farmers with the resources to continue in the area, but also the threat of speculative development. Efforts to assure continuity of the reform programs of the thirties became the primary concern of agricultural planners with the outbreak of World War II.

Promotional interests had been chastened but never quashed by the drought experience. The Greater North Dakota Association had responded to the more favorable conditions of 1935 by releasing a new publicity pamphlet. The state of Montana at that time had also initiated a major campaign, designed to attract tourists if not settlers; and Montanans, Incorporated, had revived the effort in 1937. As drought shifted to the southern Plains at mid-decade the editor of the *Montana Standard*, while renouncing any desire "to capitalize [on] the sufferings of others," had proclaimed that the nation looked to Montana and the Dakotas "to produce in abundant substitute for states whose acreages have been destroyed and whose crops will not be harvested. . . . Throughout the vast stretches of this state plans are being laid for an active farm campaign that is greater than any Montanans have known in years. . . . The years of disappointment and dissatisfaction are past."[13]

With the soaring agricultural production of 1940, local editors acclaimed the development as far better than the "false prosperity found in manufacturing war supplies." The drought "cycle" had run its course, "with a substantial period of good moisture years promised ahead." Readers were once again assured that dry-land farmers had "found that they can grow good crops even in years in which the rain and snow fall are far under normal, providing the moisture comes during the growing season," or because they had learned to diversify, or because of "greatly improved tillage methods." Montana farmers were "definitely establishing this state as one of the most successful wheat growing regions of this or any other country."[14]

In 1941 the Greater North Dakota Association produced a new film on the agricultural development of that state, and in 1944 they carried the "success story" to representatives of advertising agencies and business executives in gatherings at Chicago and New York. The South Dakota legislature in 1943 authorized the governor to appoint a commission, with an appropriation of $15,000, to assemble and disseminate information on the uses and development of the state's natural resources. As the war ended Montanans, Incorporated, again became active and in 1946 reorganized as the Montana Chamber of Commerce. "Like a verse of 'America the Beautiful,' " ran one of its advertisements in the *Des Moines Register*, "Fergus County seems to sing of bounty and fertility."[15]

Real estate advertisements appeared in newspapers again, and creditors sold their holdings of foreclosed property. A North Dakota statute, enacted in 1933, requiring corporations to dispose of their rural real estate within ten years, may have accelerated the development in that state; but sales by insurance companies, commercial banks, federal and state credit agencies, schools and charitable institutions, and other such entities were heavy throughout the region during the early forties. They represented 58 to 74 percent of the land sales in South Dakota and about one-third of those in central Montana at that time. The Bank of North Dakota, which had reported with satisfaction sale of 266 farms in 1939, sold 1,107 in 1941, 1,180 in 1942, and 1,767 over the next two years. The South Dakota State Rural Credit System, state boards of school and university lands, and the federal land banks were also pushing heavily for sales in the period. The federal land bank at Spokane marked the changed situation as it reported Montana farm transactions valued at nearly $900,000 in 1939, a 58 percent increase during the first six months of that year over the record of the same period in 1938. The St. Paul land bank recorded sale of 2,526 North Dakota farms in 1941 and another 1,619 in 1942.[16]

County governments provided the strongest impulse for such sales. Half the North Dakota counties had held auctions of delinquent tax lands as early as 1938, and in January 1940, the governor called for an end to further postponement of such action. One-fifth of the property in Ward County changed hands the following year, with the county as land owner in over one-third of the transactions. In Montana 55 percent of the land sales in Phillips County between 1940 and 1946 were by the county, as were more than 40 percent of those in Rosebud, Daniels, and Fallon counties. By July 1944 82 percent of the land acquired by tax deed in Fergus County had been placed back on the tax roles, and much of the remainder had been put into production under lease. Only 5 percent was being withheld as "nonproductive" in nature.[17]

The extent of this liquidation of public holdings marked the failure of planners' hopes to retain such tracts under governmental control. Selling agencies normally announced that "only agriculturally suitable land" was being sold, that efforts were being made to enable resident farmers to expand their enterprises, that sales were being made to tenants who sought to move up the "agricultural ladder," that this was no "wildcat land boom"; but supervisors of the newly formed grazing districts and representatives of county planning bodies repeatedly said that "temporarily favorable conditions" would bring resumption of cropping on "unsuitable lands." The North Dakota legislature in 1941 alluded to the current purchase of agricultural lands under tax-deed proceedings by individuals who could not "make efficient use of such lands," by some who were putting them to uses that might "result in serious or total damage of such lands and lands adjoining," and, in some instances, by those who had no expectation of completing the purchase but intended to use the tract as long as possible under the initial downpayment.[18]

In 1933 the Montana Legislative Assembly had stipulated as state policy: "To promote the conservation, protection and development of forage plants, and for the beneficial utilization thereof for grazing by livestock . . . ; to put into crop production only such lands as are properly fitted therefor; . . . to extend preference in sales and leases of lands to resident farmers, stockmen, and taxpayers; to gradually restore to private ownership the immense areas . . . which have passed into county ownership because of tax delinquencies." The other states of the region wrote similar statements of policy for administration of tax-deed lands at the end of the decade. They all expressed a commitment to conservation; all called for a distinction between agricultural and grazing utilization; all permitted ten-year grazing leases to organized grazing associations, but only South Dakota stipulated that these leases could not be interrupted by sale; all provided extended redemption periods to former owners, but all sought to end public ownership of the tracts; and all assigned to county commissioners the authority to maintain, and in large measure to define, the criteria. Only the Montana statute specifically assigned to the county agricultural planning committees an advisory role in this process, but by 1940 debate on the problems of county tax sale crowded the agenda of the state and local planning organizations.[19]

The discussions emphasized the need for the land surveys and classification in which the committees were engaged, but made evident that few of them were prepared to meet the demand. As late as September 1941, the Montana State Advisory Committee adopted a resolution calling for delay of sales and leases for cultivation of state or county holdings "until such time as a soil survey and classification of each individual tract involved may be had, and, if necessary, until such time as adequate legislation may be passed, to assure the future use of these lands in accordance with sound land use policies and planning." Meanwhile, debate on the ramifications of the problem continued in all the states: the desirability for increased farm size which fueled demand for the land, the need for county revenues which dictated sale of the tracts under competitive bidding, the possibilities for governmental and institutional consolidation that might reduce such costs, rates and length of tenure if counties leased the lands while retaining ownership, and finally the extent of production requirements as war spread through Europe.[20]

The pressure of events left little time for extended inquiries. In March 1941 *after* the legislature had recently enacted regulations for administering tax-deed lands, the North Dakota Agricultural Advisory Council came to a "consensus" that it would be "advisable" for a subcommittee to work with the County Commissioners Association on legislation affecting that subject, but the principles to be initiated remained poorly defined. They were to emerge only gradually, in relation to specific situations. In December the council accepted a report by its committee on the FSA that was highly critical of that agency's failure to fund tenant purchases of land in Burke County and called for extension of such a program on a state-wide basis. At the same time, the council endorsed "for policy purposes" the report of a county committee urging limitation of financing by the Farm Credit Administration to a total of 640 acres and its restriction to resident operators.[21]

Although county planning committees in Montana were early assigned an advisory role in handling tax-delinquent property, the state Land Use Advisory Committee delayed formulation of a statement of principles on land transfers until its final meeting in June 1942. The Technical Committee on Tenure proposed that "*When and if* the problem of unstable lease arrangements . . . [was] great enough to hinder needed production for war purposes, and *when and if* the problem of land transfers through sales might inflate land values and hamper maximum production," a range of measures might be instituted, including establishment of land-transfer clearing centers in the counties; reversion of tax-delinquent lands to the state rather than to counties; unified administration of state and county holdings by a state land commission or, where such lands were encompassed in areas of federal jurisdiction, by the appropriate federal agencies, with revenues above the costs of administration reverting to the state; and zoning to prevent "further settlement in areas temporarily attractive because of high war prices." After the discussion, the proposals were yet to be referred to the county planning bodies for comment on local support.[22]

Only South Dakota enacted a rural zoning law to deter threatened speculative development. In Corson County, where about one-half of a proposed restricted area was owned by the Standing Rock Indians, much of the remainder of the tract had been reverting to public agencies through mortgage foreclosure or tax deed proceedings. Arguing that if they were to build up satisfactory operating units, competition of cash-wheat producers for strategic lands must be prevented, ranchers with the support of the Corson County Land Use Planning Committee in 1939 prevailed upon the local representatives in the state legislature to introduce a proposal authorizing zoning restrictions. The measure was not enacted then, but in response to an appeal by its supporters, the South Dakota Agricultural Experiment Station in cooperation with the Federal Bureau of Agricultural Economics instituted a survey of its potential value.[23]

The study noted the cost of maintaining schools, bridges, and other public services for the sparse population, but concluded that zoning was not a practical means of achieving the desired end. The analysts commented that the proposal ignored the necessary complementary relationship between grass and croplands and the ease of transition between the two uses. The Corson County Planning Committee itself had recognized that any zoning restrictions would have to permit cultivation of feed crops and that restrictions would have to be applied to existing operating units, a relatively drastic and complex use of police power in the view of critics. Even restriction upon settlement posed problems, for wheat farming in the region was no longer dependent upon residence on the land, while ranching entailed year-round residence at headquarters sites and maintenance of access roads.[24]

Nevertheless, the report concluded that zoning offered some promise in combination with attempts to increase public ownership through programs of land exchange, credit, and relief, with guidance of settlement into communities where public services could be furnished economically. The influence of the cooperating federal sponsors was apparent. The representative of the federal Indian service forcefully argued for the measure before the state Land Use Planning Committee, and the Corson County Planning Committee renewed its agitation for the legislation. At the legislative session of 1941, a zoning proposal was finally approved, but no action appears to have been taken under it.[25]

Couched in generalized application for the promotion of "health, safety, morals or the general welfare of the community," the measure very broadly authorized the board of commissioners of any county to regulate and restrict the use or occupancy of land for trade, industry, residence, recreation, agriculture, grazing, irrigation, water conservation, forestry, or other purposes upon assent of the majority of those voting at a regular primary or general election. Once established, however, changes in the regulations, restrictions, or boundaries of such zones could be effected only with the approval of at least 80 percent of the owners of the land included and with the unanimous vote of the board of county commissioners. The legislation thus required broad-based political agreement upon policy objectives at a time when they were rapidly disintegrating under expansive pressures. The measure had been designed less to initiate land-use reform than to protect the interests of an entrenched monopoly. The coupling of so loosely defined an authorization with such restrictive constraints upon modifications in administration made it popularly unacceptable. Instead, the more focused programs and the more limited constituent bases of soil conservation and grazing districts provided the principal structural framework for land-use control.

While there was local reluctance to impose restraints upon agricultural expansion in the region during the war years, there was wary concern among agricultural planners that the experience of the World War I period and its aftermath should not be repeated. Beginning as early as September 1939 repeated warnings were sounded. In the spring of 1941, the National Resources Planning Board urged efforts to block "possible disarranging effects of war and preparations for war." County planning committees formulating unified action programs were to emphasize "that long-time objectives . . . must not be lost sight of as steps . . . [were] taken to meet the immediate situation" of war and defense.[26]

Payments for conformity with soil-depleting allotments under the AAA were reduced 10 percent for the 1939 production and eliminated by 1944, but additional funding was diverted to enhance the soil-conserving allotments under the program. In Montana county Agricultural Conservation Program (ACP/AAA) committeemen were also directed to monitor whether plowing of sod posed an erosion hazard to the community. With elimination of constitutional and administrative problems in establishment of soil conservation districts, organization and expansion of these institutional arrangements were also accelerated. Land in such districts was increased in Montana from about 1.25 million acres at the end of 1940 to 24 million acres five years later. An additional 7 million acres were covered by grazing districts in the state. In 1945 there were nearly 14.5 million acres in soil conservation districts in North

Dakota and 14.75 million acres in South Dakota. Meanwhile, extension workers urged farmers to continue strip cropping and to extend shelterbelt tree planting. A writer in the BAE's *Land Policy Review* in the spring of 1943 explained that "the simpler practices—contouring, strip cropping, wind stripping, care of pastures and woods, mulching, use of green manure crops and rotted stable manures, rotations, liming and fertilizing" were measures that would "do most toward quickly bringing about the increased crop yields . . . so keenly needed" for the war effort.[27]

Following the bumper crop of 1938, officials of the USDA undertook to counteract accumulating carry-overs of wheat. An increase of about 9 percent over-all in the acreage allotment for 1940 was not intended to permit larger individual shares, but rather to accommodate the "many new farms . . . being included in the program." The allotment for 1941 was reduced by 5 to 7 percent, and that for 1942, by 12 percent. The authorized corn allotment was also curtailed by about 12 percent the latter year in response to accumulating stocks. Marketing quotas, instituted on the basis of producer referenda beginning in 1941, further reinforced the restraints upon wheat production over the next year and a half. At the same time goals called for increased production of meat animals, dairy products, poultry, and some truck crops, and price supports were increased in the spring of 1941 to promote these emphases. The Perkins County, South Dakota, extension agent aptly proclaimed the program one of "diversion" rather than "increased production." It would, he announced, "allow a man to cut down when the emergency is past with the least painful effects."[28]

Congress, however, responded to pressures for more immediate benefits. Under the legislation of 1941, the loan rates for corn and wheat, among other products, were set at a minimum of 85 percent of parity, and in the fall of 1942 they were raised to 90 percent provided the president should not determine that retention of the lower percentage was necessary to prevent increased costs of feed for livestock and poultry. Such a determination was applied in respect to wheat and corn until 1945, when under specific congressional action the higher rate was restored. Under the Steagall amendment, which extended the authorization for the Commodity Credit Corporation, farmers also received assurance that the high loan rates would remain applicable for two years after the termination of hostilities.[29]

Secretary of Agriculture Claude R. Wickard himself succumbed to demands for expanded wheat production in the spring of 1943, when carry-over stocks had been sharply reduced by the need to use the grain as feed to meet the requirements of the "dangerously" expanded livestock herds. In February marketing quotas on wheat were suspended, and farmers were permitted to increase plantings as long as they also met at least 90 percent of their goals for acreage devoted to war needs, such as potatoes, flax, soybeans, grain sorghums, dry beans, and peas. Production of wheat in Montana that year reached a level not attained since 1928, and that in North Dakota was the largest to date. For 1944 North Dakota farmers were asked to increase their wheat acreage by 32 percent; Montanans, by 8 percent.[30]

Such adjustments did not signify recognition by agricultural planners that wheat cropping represented the most profitable use of regional resources. It was anticipated that the grain would be used as feed to supplement insufficient reserves for the expanded livestock industry. In the spring of 1943 Congress authorized sale of 150 million bushels of wheat for that purpose from the stocks of the CCC, at a loss to the government estimated at $ .40 a bushel. Some 60 percent of the total U.S. wheat production for 1943 was used for feed. The reserve was reduced from over 622 million bushels on July 1, 1942, to less than 317 million bushels on the same date in 1943, and carryovers were meager the next two years. Yet goals for number of livestock in the region were increased annually.[31]

Regional complaints forced adjustments in other phases of the cropping program designed for the war effort. At local planning meetings, farmers protested that they could not profitably raise feed grains and that the area was not well adapted for dried beans and peas, soybeans, corn, potatoes, and flax. Reinstitution of crop insurance and its application to flax in 1945 cushioned the continued request for flax to replace the loss of imports, but the acreage in some areas had to be reduced to avoid repeated cultivation on the same land. Montana's goals were lowered for oats, peas, and Irish potatoes in 1945, and for the following year those for flax, corn, beans, and peas were eliminated, while those for rye and potatoes were greatly reduced.[32]

At the same time, the goal for Montana's wheat acreage was lowered 3 percent for 1945 and remained stationary at that level for 1946. In the autumn of 1944, convinced that the war was nearly over, the War Food Administration reiterated fears of developing surpluses. Local leaders shared that view and were concerned about expanded wheat cropping if there should be recurring drought. The editor of the *Williston Daily Herald*, of North Dakota, wrote of the beef industry as "the mainstay of our country" and, citing the "tremendous surplus of wheat," dismissed the argument presented to him in January 1943 that the acreage needed to provide a 300- to 400-pound increase in a beef animal could yield 1,800 pounds of food in wheat for the war effort. The editor of the Lewistown, Montana, *Democrat-News* in this same midwar period alluded to the problem of postwar food needs as a burden for which the nation would "never be paid." "The drain on American resources to help mend a broken world may be most worthy," he continued, "but there is no escaping the fact that it will also be very damaging to the productivity of our lands unless food growing programs are well supervised." A press release from the Montana Agricultural Experiment Station in August 1944 warned that droughts were inevitable "and the next one appears just around the corner." "With regard to crops," their analyst reiterated in 1946, "it is not a question of 'if,' but only 'when' we will have . . . another year or series of years of low yields."[33]

Few agricultural leaders responded to the changing demands of wartime need as did the county agent of Williams County, North Dakota, who in March 1943 urged farmers to produce as much as they could and to expand production if they had the resources to acquire and handle additional land. Government analysts contemporaneously reporting on the wartime policies of federally sponsored farm mortgage credit agencies described them as "relatively passive," "in keeping with the 'hold the line' program of inflation control." The FSA stipulated that money not be allocated for farm purchase loans if land could not be found at prices justified by "long-term earning capacity," and as early as 1942 reported prices in some areas beyond the range of sound investment. The federal land bank and FCA adopted similar lending criteria. "Long-term," however, generally rested upon crop history derived through the drought decade.[34]

For some six months early in 1943, the Regional Agricultural Credit Corporation was authorized to provide nonrecourse loans to finance production of war crops. Testifying in support of such assistance, John D. Black informed the Senate Committee on the Agricultural Appropriation Bill for the next fiscal year that "large numbers of farmers fully capable of expanding the usual production of vital war crops or of producing such crops instead of other safer but less-needed crops . . . made it perfectly plain that, although they were willing and eager to apply their land, equipment, and labor to the extent necessary for such production, they were not able or were not willing to risk their personal credit standing or their estates by assuming full personal liability beyond the amount necessary to finance what would be their normal operations." Nevertheless, under congressional pressure in behalf of private banking interests, the program of nonrecourse loans was dropped. The total amount of loans for farm production purposes by the Production Credit Associations, Regional Agricultural Credit Corporation, Farm Security Administration, and Emergency Crop and Feed Loan Office, which had risen from $611.3 million in 1942 to $712.5 million in 1943, abruptly declined to $584.1 million in 1944.[35]

Under such constraints and the pressures of forced liquidation, land values in all three states of the region during the mid-forties averaged well below the levels of 1912–1914, while nationally the estimated value of farm real estate rose 26 percent above that index. In Haakon County, western South Dakota, a major district of dry-land development during the earlier period, land prices in 1946 were only 15 percent of their former level. There the amount of farmland transferred during the forties, 78 percent of the total land area, was the largest in an eight-county survey of market trends; but the total land in farms was increased by less than 5 percent between 1940 and 1945. The prices indicated that the sales in large part represented transfers of rangeland. While the amount of harvested cropland was increased by 136 percent, with a very large increase in harvested wheat acreage (168 percent), the total acreage identified as cropland—including idle and failed acreage and plowable pasture—declined 63 percent. It is apparent that farmers were more fully utilizing their cropland—78 percent of the total was harvested in 1945, compared to only 12 percent in 1940; but only 18 percent of the land in farms was listed as cropland in 1945, compared to 51 percent in 1940. With a doubling of the number of cattle and a 25 percent increase in the number of sheep, the pressure for range had been intense. In liquidating their debts farmers surrendered their poorer tracts as grazing land, and land acquisitions were largely at that classification and pricing.[36]

In the state of South Dakota the rate of inflation in land values between 1940 and 1945 approximated the level for the nation generally, 50 percent. That for North Dakota was even lower, 46 percent. By 1945 land values in eight of the twenty-three western counties of the latter state had not yet recovered 1930 values at a rate equal to the recovery rate for the state as a whole. Among those eight, land values in five, all committed largely to range development, either as part of the Federal Land Utilization District or the Cedar Creek Soil Conservation District, were below $10 an acre. In two of the five, Billings and Slope counties, the rate of increase in productivity had not kept pace with their ranking relative to the state average in 1940. All fifteen western counties where average land values had recovered their pre-depression levels were identified as primarily agricultural in development; but there, too, the increase in productivity values lagged in two, Bottineau and Golden Valley. While analysts pointed to the latter trend as indicative of speculation, the return on the dollar investment, whether for rangeland or cropland, was well above the rate for land in the most prosperous Red River Valley counties of the state.[37]

In western North Dakota, as in western South Dakota, there had been a low increase in land in farms between 1940 and 1945 (12–13 percent); but there had been a lower rate of decline in total cropland (9 as compared to 39 percent). Instead, the acreage in wheat in western North Dakota had been increased by 52 percent, together with large increases in the acreage of flax, corn, barley, and oats, while the number of cattle had been increased by 79 percent. Here, too, some of the poorer cropland had been channeled into grazing use, and as in western South Dakota the high rate of increase in harvested acreage as a proportion of the total cropland had necessitated that some accessions to the land in farms be made, primarily as rangeland.

Through this period in both the Dakotas a large number of ownership transfers per thousand farms, compared with the levels for Montana and the United States as a whole, indicated liquidation of debt for taxes or mortgages. This was the case particularly during the early forties, when counties and corporate institutions comprised the largest proportion of the parties in the

TABLE 6.1. Changing Intensity of Land Use, Northern Plains, 1939–1954

Intensity Levels	1939 (acres)	1939 Ratio between Levels[1] (percent)	1944 (acres)	1944 Ratio between Levels[1] (percent)	1944 Ratio between Census Years (percent)	1954 (acres)	1954 Ratio between Levels[1] (percent)	1954 Ratio between Census Years (percent)
Western North Dakota								
1. Land in farms	17,470,940		19,544,297		12	20,171,952		3
2. Total cropland	10,959,997	63	10,013,344	51	– 9	11,366,438	56	14
3. Harvested cropland	5,490,296	50	8,206,115	82	50	7,512,453	66	– 9
Western South Dakota								
1. Land in farms	19,184,983		21,740,130		13	23,361,791		8
2. Total cropland	7,042,639	37	4,300,184	20	–39	4,979,486	21	16
3. Harvested cropland	2,030,229	29	3,753,752	87	85	4,146,540	83	11
Eastern Montana								
1. Land in farms	38,256,041		48,718,618		27	50,863,812		4
2. Total cropland	13,205,588	35	9,710,802	20	–27	11,704,074	23	21
3. Harvested cropland	4,697,459	36	6,161,956	64	31	7,162,098	61	16

Source: Computations from U.S. census of agriculture, vol. 1 (county tables), for the respective years.
[1]Computed as percentages of rows no. 1 by rows no. 2 and of rows no. 2 by rows no. 3.

transactions. Relatively low selling prices characterized these sales, because county governments were eager to return the lands to the tax rolls, and corporate lenders were generally content merely to recoup their investment. By mid-decade, however, as the bulk of such holdings had been liquidated, a larger proportion of the transactions represented sales by private landholders, more speculative in their concern to profit from the favorable conditions. The rate of ownership transfers then declined in the Dakotas but remained at peak level in Montana.[38]

Although the number of ownership transfers had remained somewhat fewer in Montana than in the Dakotas during the early years of the decade, perhaps because so much of the county and abandoned acreage had been transferred earlier to the federal government, inflation in land values was more evident there. The index of estimated values of farm real estate (1912–1914 = 100) rose 62 percent in Montana between 1940 and 1945, and the acreage in farms was increased by over 27 percent. Prices for rangeland rose from an index of 101.3 in 1940 to 201.4 in 1946; but the increase was even greater for nonirrigated cropland, which rose from an index of 91.3 to 225.1 in the same period. Approximately one-third of the area in private or county ownership had been sold in eight sample counties surveyed, and analysts reported competition for land so keen that operators leasing tracts were forced to buy them to maintain control. Here, too, warning was voiced that prices exceeded productivity values.[39]

As in the other states of the region, the proportion of total cropland relative to land in farms fell in Montana from 35 percent in 1940 to 20 percent in 1945, while the amount of cropland declined 27 percent. But here the harvested acreage was increased only 31 percent, and a considerably lower percentage of the acreage from the total cropland was converted to harvested cropping. The number of livestock was increased by 87 percent, a commitment to the livestock industry in absolute numbers about 60 percent greater than in either of the Dakotas. While over one-third of the acreage identified as cropland appears to have been fallow, idle, or pasture, a considerable proportion of the harvested acreage, also, was devoted to feed grains and hay. There had been only a 23 percent increase in wheat acreage, the least expansion in the northern Plains. It was exceeded by a 255 percent increase in the acreage of barley, a 55 percent increase in that of oats, and a 47 percent increase in that of alfalfa.

The war years had not occasioned pressures comparable to those of the period from 1914 through 1918. An adjustment toward increased use of land for livestock had been evident throughout the region, but it continued to be integrated with dry-land cropping. Such operations had potential for expansion, whether for food grains or feed as weather and market conditions might permit. The flexibility that Elmer Starch had identified in the late thirties existed as the chief requirement and primary characteristic of agricultural development in the region.[40]

Favorable precipitation in the northern Plains and food shortage throughout much of the world fostered expansion of dry-land cultivation to new levels during the postwar years. Rainfall was seriously deficient in Montana and North Dakota in 1949 and 1952, in northern and western South Dakota, also, in 1952, and in Montana again in 1956, but it otherwise remained near or above the 15-inch margin of semiaridity throughout the region until the late fifties. It then declined below long-term means from 1958 into the early sixties, but the severe and protracted drought that began in the southern Plains as early as 1950 was long delayed, relatively moderate, and brief in the states to the north.[41]

Wheat yields per harvested acre from 1946 through 1955 averaged 17 bushels in Montana, 12.5 bushels in North Dakota, and 11.4 bushels in South Dakota and were limited more by a serious rust outbreak during the early fifties than by climatic conditions. Production, curtailed by drought in eastern Montana in the census year 1949, exceeded wartime levels there in 1954 and 1959. In western South Dakota it exceeded those levels in 1949 and 1954. Only western North Dakota, stricken by drought in 1949 and serious rust losses, particularly in the north-central durum-producing districts, during the early fifties, showed significant reduction in wheat production.[42]

Dry-land districts retained leadership in such productivity. In 1954 Chouteau and Hill counties of Montana, both of which had been consigned to land retirement in the report of the National Resources Board twenty years earlier, ranked first and third, respectively, in the nation in the amount of acreage devoted to wheat. They ranked third and fifth nationally in the volume of such production, with 10,276,000 and 7,102,000 bushels, respectively. Nineteen other eastern Montana counties, fourteen western North Dakota counties, and Perkins County, in western South Dakota, each produced over 1 million bushels.[43]

Agricultural planners had feared the build-up of surpluses again with the end of the war, but they faced instead a crisis of wheat scarcity by the winter of 1945–1946. Clinton P. Anderson, who had succeeded Wickard as secretary of agriculture the previous July, continued the policy of expanding livestock supplies. Wheat stocks, an estimated 100 million bushels, were again diverted to feed. Pleas on behalf of the United Nations Relief and Rehabilitation Administration were ignored as domestic rationing ended. By December it was recognized that drought had curtailed the Mediterranean wheat crop and that scarcities of seed, fertilizers, and manpower had limited European production. Reports also revealed poor harvests in China, India, Australia, and Argentina. The United States, pledged by Truman to meet the requirements of our allies, was committed to supply 6 million tons of wheat for shipment during the first half of 1946 but by February held only 16.9 million tons for domestic as well as export purposes. To draw deliveries of farm-stored wheat and corn to the Commodity Credit Corporation, bonuses of thirty cents a bushel above market prices were offered. Although this closed the market to millers and processors, the nation met its goal only by supplementing wheat with other grains. Through the summer of 1946 local extension workers emphasized the need for reduction in wheat acreage and increased attention to grass, but until February 1947, the federal government purchased wheat on the open market to meet the export demands.[44]

Prices rose accordingly, for feed grains as well as wheat. By March 1947 wheat was bringing $2.65 a bushel at Williston, North Dakota, a level attained only once before, during World War I. Futures prices for March delivery at Chicago then stood at $3.05 a bushel. The index of prices received for crops (1910–1914 = 100) climbed nationally from 202 in 1945 to 228 in 1946, 263 in 1947, and held at 255 in 1948.[45]

The government resumed large-scale purchases of wheat for foreign shipment in June 1947. That year's crop was the largest on record, 18 percent greater than that in 1946, and 50 percent above the ten-year average. It sold for an average price of $2.31 a bushel, also a new record and $ .48 above the support level. Livestock prices were also high, quoted at $27.80 to $28.80 per hundredweight on the Lewistown, Montana, market that September. Flax, supported at $6.00 a bushel, 50 percent above the parity price for that crop, was twice the market value of wheat. Farm profits for the year were reported as the largest so far recorded.[46]

By February 1948 wheat prices plummeted, falling close to the level of government support for the first time since 1943. Under the Steagall amendment to the Emergency Price Control Act of 1942, however, the wartime guarantee to maintain prices at 90 percent of parity for two years after the cessation of hostilities assured a fifteen- to seventeen-cent a bushel rise in support prices for the 1948 crop. Although production and income declined somewhat in the Dakotas, both rose again to new records in Montana. Per capita income for the state ranked third in the nation.[47]

In 1948, as Congress enacted the first revision of its wartime price-support program, it retained the requirement of 90 percent parity for wheat and corn, among other basic items, through 1949, but in effect lowered the support price somewhat for wheat, while raising it for livestock, by shifting the base period of farm and non-farm price relationships from 1910–1914 to the most recent ten-year period. Nevertheless, legislation the following year continued the formula of support prices at 90 percent of parity in 1950. As the Korean War developed, that level was retained until 1955. Growers had been asked to reduce their wheat acreage for 1950 from 10.8 to 35.5 percent, according to a formula that assigned the largest adjustments to areas of expansion over the past decade. Under this schedule Montana's wartime emphasis upon live-stock resulted in reduction of the wheat allotment by 23.5 percent. North Dakota's was cut only 15 percent. With the outbreak of war, however, allot-ments for all crops were removed, and increases in production generally, apart from livestock, were requested for 1951. Since this action came too late to affect planting of winter wheat, stocks were greatly reduced by late sum-mer. The index of crop prices climbed from 233 in 1950 to 265 in 1951; that for livestock rose from 258 in 1950 to 336 in 1951. Annual production of wheat soared again to new levels in Montana and South Dakota, 95,033,000 bushels and 58,044,000 bushels, respectively; while, at 145,732,000 bushels, the North Dakota crop was the largest since 1947. Net farm income reached levels in constant dollars not again exceeded until the 1970s in Montana and South Dakota and only once so exceeded in North Dakota.[48]

National production goals for corn, wool, and wheat, among other sta-ples, were again raised for 1952. New emphasis was assigned to wheat in South Dakota; little change was anticipated in North Dakota; and some re-duction was requested in Montana, with a large shift to barley. But owing to scarce rainfall, production was greatly reduced in all three states, and above-normal precipitation the following year failed to counteract the decline in the Dakotas, where spreading infection from the rust strain 15B was severe. With wheat supported at $2.20 a bushel for 1952 and $2.21 a bushel in 1953, well above 90 percent of parity, production expanded strongly westward to the semiarid districts of less serious infection. The crop of 1953 in Montana was the largest to that date, 114,232,000 bushels; the crops through 1955, although somewhat smaller, continued to exceed the production of North Dakota, long dominant in the hard red spring wheats. Durum production, too, extended into Montana, a shift stimulated by price differentials elevated as much as $ .90 a bushel above the returns on Northern Spring.[49]

When the Korean War ended, carry-over stocks again soared. Marketing quotas were reinstituted for wheat in 1954, the first since before World War II; acreage allotments were reduced 20 to 30 percent; and the program of flexible price supports, projected initially under the Agricultural Act of 1948, was fi-nally put into operation. Still, however, the support level for 1954 was re-tained at 90 percent of parity, and with the base price for the year at $2.24 a bushel, the season average for Montana farmers, notwithstanding their dis-tance from terminal centers, reached $2.08. The crop of the succeeding year was the state's second largest to date and that in North Dakota was showing marked improvement. Despite further reductions in acreage limits, yields throughout the region were increasing, and so was total production. In 1958 North Dakota's crop of 147,372,000 bushels had not been equalled since the wartime years of 1942–1945; South Dakota's 55,722,000 bushels had been ex-ceeded only once, in 1951; and Montana's 101,882,000 bushels constituted that state's third largest on record and the third bumper crop within the decade. Average net income of Montana's farmers ranked with that of farmers in the top five states nationally every year from 1950 to 1959.[50]

With the continuance of generally favorable climatic conditions and profitable operations, farmers of the region had greatly expanded their holdings. Farm size was increased in eastern Montana from an average of 1,829 acres in 1944 to 2,008 acres in 1949 and to 2,190 acres in 1954; in western North Dakota, from 713 to 784 and to 862 acres in the respective census years, and in western South Dakota, from 1,464 to 1,739 and to 1,924. Over 2,145,000 acres were added to the land in farms in eastern Montana between 1944 and 1954; almost 1,622,000 acres, to that in western South Dakota; and almost 628,000 acres, to that in western North Dakota. Land values rose 40 percent in Montana between 1945 and 1950, 50 percent in North Dakota, and 56 percent in South Dakota, while increasing only 34 percent nationally. From 1950 to 1955 they rose another 47 percent in Montana, 24 percent in North Dakota, and 29 percent in South Dakota. In all three states the level of prices now signified pressure for cropland as well as range.[51]

Sale of state school lands, which had long lagged because of the mandated minimum price of $10.00 an acre, reflected the changed situation. North Dakota, having canceled many contracts during the liquidation period, still held about 3 million acres, the bulk of its grant, in 1940. By 1948 it had sold 1.4 million acres of that amount, over half since 1945. The average selling price was $17.28 an acre in 1946, $19.44 in 1947, $22.63 in 1949. A quarter section in Williams County brought the highest bid in 1950, $37.50, but several others in that county were sold that year for $34.38. The average for the state sales was $25.38 in 1951, $34.15 by 1958, and $55.31, in 1959.[52]

Little regard was given to the prewar commitment to safeguard state-sponsored sales from speculative development. The North Dakota State Land Department still wrote contracts in units of not over 160 acres, but no restriction was placed on the number of contracts held by an individual nor was any applied to the use of the land, except in the leasing of virgin sod. After the decline of corporate sales in the mid-forties marked the diminution of mortgage foreclosure, analysts viewed the trend toward an increasing number of individual land transactions as indicative of better adjustment in farm size. Studies by the U.S. Department of Agriculture and the North Dakota Agricultural Experiment Station concluded that approximately two-thirds of the land buyers were farm owners seeking to enlarge their holdings or tenants advancing to ownership, that three-fourths of the land owners were active or retired farmers, and that there was essentially no change in the proportion of land owners who were nonoperators between 1945 and 1958.[53]

Blatant promotionalism, which had figured so prominently in the settlement booms of the early century and the mid-twenties, was no longer a significant factor in this development. Most advertising was channeled through

the programs of the Greater North Dakota Association (GNDA) and its coun-
terpart organizations in the other states. Their approach was broadly based to
attract industrial, financial, and tourist, as well as agricultural, interest. Statistical data were relatively accurate. Pictured scenes of current conditions need
not greatly exaggerate to convey—as reported of a GNDA film designed for
display via television, libraries, steamship lines, and the Brussels World's Fair
in 1958—"a 'soft sell' of North Dakota as a land of untapped resources and a
pretty good place to live."[54]

Yet speculative activity was evident in the trend of rising land prices. The
rate of ownership transfer remained high throughout the region until the
widespread drought in 1949. During the early forties this trend did, in large
degree, reflect the liquidation process; but its continuance, and in some areas
increase, in the postwar period represented profit-taking on landed capital.
The percentage of resales in total sales climbed in Haakon County, South
Dakota, from 4.4 percent in 1946 to 15.4 percent in 1947, to 21.4 percent in
1948, to 39.3 percent in 1949.[55]

Pressure mounted for counties to cancel long-term leases by which their
tax-reverted holdings had frequently been committed to the federal Land
Utilization projects. Only a veto by President Truman in February 1948 forestalled congressional legislation authorizing sale or transfer to local advisory
boards of 245,000 acres designated as submarginal land, bought under the
land-retirement program for expansion of Indian reservations in Montana
and the Dakotas. In 1953, a Valley County, Montana, planning committee
complained bitterly that farmers of the Fort Peck area had given up "hard-
earned acres of the most fertile land in the nation" at prices that "were far
below the cost of putting the land into production," prices that now appeared
"ridiculous." Strong, but unsuccessful, agitation called for transfer of the
Land Utilization projects as a whole to local control as part of a long-contin-
ued controversy over administration of the grazing domain.[56]

Meanwhile, the annual reports of the Great Plains Agricultural Council
presented a record of large-scale retreat from the commitment of cropland to
regrassing, the so-called restoration program of the thirties. Over 9 million
acres of these tracts were broken in the Plains as a whole between 1944 and
1955, most notably between 1946 and 1949 and in 1951. The heaviest breaking
occurred in North Dakota and Texas, but it was extensive also in Montana,
particularly in Chouteau, Toole, and Hill counties. Some 200,000 acres of sod
land were plowed in the last of these counties, alone, in the single year 1947–
1948.[57]

Despite the activity in the real estate market, the total acreage in farms
was not greatly increased between 1945 and 1955, a mere 3 and 4 percent,
respectively in western North Dakota and eastern Montana and only 8 percent in western South Dakota. Decreases of 3,000 to 4,000 in the number of
farms in each of those areas during the decade significantly contributed to the

acreage available for increased farm size. More notable were the increases in total cropland in all three states and in the harvested cropland in western South Dakota and eastern Montana, the latter category having been reduced in western North Dakota by the severity of the rust epidemic through the early fifties. The effect of the retreat from the restoration program was particularly apparent in the cropland expansion.[58]

Census definitions of "cropland" differed somewhat within the period of this study. Generally it encompassed all pasturelands that "could have been plowed and used for crops without additional clearing, draining, or irrigating," but the definition for 1945 specifically stipulated that the pasture land to be included as cropland must have been plowed within the preceding seven years. Even so limited, the increase in total cropland over the decade amounted to 14 percent in western North Dakota, 16 percent in western South Dakota, and 21 percent in eastern Montana. Harvested cropland declined by 9 percent in western North Dakota, but increased 11 percent in western South Dakota and 16 percent in eastern Montana. The divergence between amounts of total and harvested cropland represented increased acreage in pasture, idle land, or fallow.[59]

There had been a 32 percent increase in the number of cattle in eastern Montana within the decade, compared to declines of 23 percent in western North Dakota and 27 percent in western South Dakota. This expansion in eastern Montana had been counteracted in part by a 41 percent decline in the number of sheep, but the need for greater feed and forage production was evident. It led to increases of 108 percent in the acreage of barley and 32 percent in that of alfalfa. Barley acreage declined somewhat in the western Dakotas, but that in corn was increased slightly and that in alfalfa, significantly, as farmers in those states, too, acted to provide feeds for livestock. While 68 percent of the cash farm income of South Dakota in 1955 was derived from livestock, feed grains being marketed primarily in that form, a majority of the farm income in North Dakota and Montana came directly from crops. The development of the region was agricultural, with production of livestock integrated in the farm economy.[60]

These had been prosperous years in the northern Plains. Notwithstanding lingering conceptions from the drought experience which had strongly channeled wartime agricultural expansion into support for a livestock economy, wheat production had been greatly increased throughout the region. Farmers had garnered record-breaking income from high yields and high prices over most of two decades. They had been able to liquidate their debts and expand their operations.

Yet wariness had prevailed, at both the local and the national levels, to assure that during the war years expansion beyond real values should not generate inflation comparable to that prior to 1920. Conservation themes had been preserved even as acreage restrictions were eliminated. Promotional

TABLE 6.2. Number of Farms and Percentage of Census Change, 1939–1987

	United States	
	Number	Census Change (in percent)
1939	6,102,417	
1949	5,388,437	−12
1959	3,703,894	−31
1969	2,730,250	−26
1978	2,257,775	−17
1987	2,087,759	− 8

	North Dakota (State)		South Dakota (State)		Montana (State)	
	Number	Census Change (in percent)	Number	Census Change (in percent)	Number	Census Change (in percent)
1939	73,962		72,454		41,823	
1949	65,401	−12	66,452	− 8	35,085	−16
1959	54,928	−16	55,727	−16	28,959	−17
1969	46,381	−16	45,726	−18	24,951	−14
1978	40,357	− 3	38,741	−15	23,565	− 6
1987	35,289	−13	36,376	− 6	24,568	+ 4

	Western North Dakota		Western South Dakota		Eastern Montana	
	Number	Census Change (in percent)	Number	Census Change (in percent)	Number	Census Change (in percent)
1939	29,896		16,752		29,797	
1949	25,083	−16	13,460	−20	24,555	−18
1959	20,597	−18	10,711	−20	20,495	−17
1969	18,367	−11	9,386	−12	18,220	−11
1978	16,365	−11	8,488	−10	16,845	− 8
1987	14,871	− 9	8,913	+ 5	16,894	+ 0.3

Source: Data and computations from U.S. census of agriculture, vol. 1 (county tables), for the respective years.

activity was restrained, and during the early forties the extent of liquidation sales had checked rising land values. In the Dakotas the rate of increase was comparable to the national average; in Montana it was somewhat greater, but it was still measured more markedly in rangeland values than in those for cropland.

As prosperity continued into the postwar period and was again stimulated through the Korean War, greater optimism developed. Below-average precipitation in 1949 proved transitory. Yields of wheat were rising, apart

from two years of rust outbreak in the Dakotas, and prices were high. Government grain purchases in 1946 and 1947, to meet needs for foreign assistance,
and congressional action preserving wartime parity levels in pricing through
1955 encouraged farm expansion. It took form in cropland, rather than range,
and inflation of land values exceeded the national average. Increasingly local
sentiment deplored the retreat that had marked agricultural land use in the
thirties, and many contended that the limited impact of diminished rainfall in
1949 and 1952 indicated the effectiveness of better tillage methods and better
understanding of the flexible management systems required for dry-land
operations.

Not all operators in the region shared these views. Some 25,000 farm
enterprises were closed between 1940 and 1960. Absorption of these holdings
accounted for much of the expansion of farm size during the period. Through
the forties the rates of farm abandonment had been somewhat higher in the
dry-land counties than in the states as a whole or nationally, considerably so,
at 20 percent, in western South Dakota. Liquidation of debts from the drought
period had contributed to the differences. During the fifties, however, the
rates of attrition still ranged at 16 to 18 percent, generally within the Dakotas
and Montana as well as within the dry-land areas of the two northern states.
The rate remained at 20 percent in western South Dakota, but nationally it had
climbed to 31 percent (see Table 6.2).[61]

Declining farm parity levels nationally, from 100 in 1952 to 80 by 1960,
pointed to the difficulties. While the reduction in number of enterprises in
part represented profit-taking on appreciated land prices, the competitive
squeeze of commercial operations had forced the action for many who closed
their ventures. Agricultural adjustments throughout the nation in the postwar
years were very largely directed toward meeting that challenge.

That the rates of farm abandonment in dry-land districts approximated
those in less arid areas of the respective states and, indeed, that they were so
far less than that of the nation as a whole, lent substance, however, to the view
that the regional enterprise held advantages in economies of production. For
proponents of such development, the basic consideration appeared to be the
continuing need to safeguard operations from the threat of recurrent drought.
With success in that effort, it was assumed that the marketing specialization
conceptualized under the old Christgau proposal would prevail.

CHAPTER SEVEN

Postwar Drought Adjustments

"Flexibility" became the keynote of regional recommendations for postwar drought adjustment. In the early forties ranchers were still struggling to rebuild herds and grasslands decimated by the droughts of the previous decade. The short drought of 1949 again forced them to cut herds drastically, a reduction that continued to limit output into the early fifties. During that protracted period, wheat acreage provided net returns that exceeded all but those from irrigated sugar beets and alfalfa. In the long run, some analysts contended, livestock and supplementary grass, feed, and forage production would afford greater stability of income; but they recognized that farmers would choose to operate between the two extremes. They would expand production in either channel as circumstances favored.[1]

The kind of diversification that had figured so prominently as a recommended adjustment in earlier periods was now, however, muted. The Farmers Home Administration continued to emphasize it as a protection against sharp fluctuation in income from reliance on cash crops alone, but an analyst of the North Dakota Agricultural Extension Service went so far as to note that the program reduced the possibilities of "cashing in" on exceptionally favorable conditions for grain. The station warned in its annual report for 1953 that overdiversification might split resources so greatly as to be inefficient. Agricultural leaders conceded that certain districts even in range areas might combine production of wheat with raising of cattle: The seasonal labor requirements were noncompetitive; and if wheat failed to yield grain, it usually produced some roughage for livestock. But the aim was to "specialize," developing no more than three operations within the farming program.[2]

With recognition that cropping remained a vital aspect of the regional economy came major changes in the livestock industry. Flexibility evolved in respect to both the age at which animals were marketed and the location of the markets. Since the mid-twenties the marketing focus had shifted from three-year-old steers to younger cattle, so-called baby beef, a trend stimulated by growth of the midwestern feeding industry. In the early fifties supplemental feeding and fattening prior to sale began to develop closer to the breeding area. In South Dakota, where corn figured most prominently in the cropping of the eastern counties, an intrastate linkage formed between this as an area of intensive feeding and the west-river country as a continuing base of range-cattle operations. In North Dakota, too, the number of cattle being fattened increased rapidly, from a record of 80,000 head in 1954 to 110,000 head by

1956. The trend toward livestock feeding was heightened in Montana by a shift to marketing of cattle in West Coast centers, where no adjacent feeding areas comparable to those of the Midwest had been established to serve fast-developing metropolitan communities. To reduce freight rates and animal weight loss on long hauls, Montana stockmen also turned to organization of local markets. By 1949, 70 percent of Montana stock sales were handled through local auctions.[3]

The availability of local feed supplies was crucial to these developments. In the early fifties the experimental substation at Dickinson, in southwestern North Dakota, initiated tests on use of corn silage as a fattening ration. Introduction of the field chopper had facilitated harvesting; and even when drought reduced the quality, a county agent advised, fair ensilage could be produced that was cheaper for roughage than scarce hay. Greater use of alfalfa, also, for silage was stimulated by introduction of sodium metabisulfate as a preservative. Perhaps most significant for the northern Plains was the development of barley pelleting in the late fifties.[4]

In the spring of 1957, a group of Bainville, North Dakota, farmers incorporated as the Little Missouri Feeders Cooperative to initiate what was locally termed a "unique experiment," a feeding lot with space for 2,000 animals, "self-feeding bunks," and use of barley pellets providing "a complete balanced ration." Over the following year, the Farmers Union Grain and Supply Cooperative began demonstrating feedlot operation in the region and marketing barley pellets with additives including hormones, antibiotics, and chemobiotics. While feed grains could not compete with wheat cropping under normal circumstances, they were now achieving a commercial market at a time when acreage restrictions were again limiting wheat allotments. Diversification, as Starch had anticipated a decade earlier, was taking an intra-regional form.[5]

Mixed livestock ranching and small-grain operations were interspersed in many districts. Most of western South Dakota was identified as range country, but almost one-fifth of the farms reported fewer than five head of cattle in the mid-fifties. Corson, Perkins, Lyman, and Tripp counties each reported over 300,000 acres of harvested cropland; Meade and Gregory reported just short of that amount. Southwestern North Dakota and southeastern Montana were even more strongly oriented toward ranching, with only 17 percent of the farms reporting fewer than five cattle; yet here, too, Dunn, Grant, Hettinger, McKenzie, Morton, and Stark counties had over 300,000 acres of harvested cropland, each. North of the Missouri River in northwestern North Dakota and across the northern third of Montana, an area identified primarily for specialized wheat production, nearly two-thirds of the farms reported more than five head of cattle. Irrigated valleys throughout the region were characterized by sugar-beet, alfalfa, and mixed farming, combined with dairying and livestock feeding.[6]

Improved crop varieties contributed to these developments. A remarkable expansion of barley production in Montana had followed the introduction of Compana, developed at the Moccasin, Montana, experimental substation and released to farmers in 1941. Within two years it was the top-ranking barley for dry-land operations. A two-rowed variety, it offered high drought resistance, early maturity, ability to withstand loose smut, low susceptibility to grasshoppers, and excellent yields. While two-rowed were less desirable than six-rowed varieties for malting, Compana provided more digestible nutrients than oats, wheat, or corn as livestock feed. A six-rowed barley, Glacier, released in 1943 for cropping either on dry-land or under irrigation, afforded greater resistance to covered smut and lodging as well as good yield, but lacked the high test weight of Compana. By 1952 Glacier had been dropped from the recommended varieties and replaced by Titan, a six-rowed hybrid developed in Canada and introduced into Montana in 1948. Titan brought the advantages of stiff straw, high test weight, and resistance to both loose and covered smut, but it, too, was marred by a weakness in its susceptibility to shattering. The pre-eminence of Compana as a feed grain for dry-land cultivation was not seriously challenged until the higher-yielding DeKap, imported from the Caucasus, was released by the Montana Experiment Station in 1962.[7]

Meanwhile, in the Dakotas efforts were being concentrated on the development of improved malting barleys. Although the high protein levels of dry-land production contributed to the value of the grain as livestock feed, brewers preferred less of that quality. Kindred, selected by a Jamestown, North Dakota, farmer, had been commercially produced as a malting barley since 1942 and accounted for over 70 percent of that state's crop a decade later. It afforded good yield, rust resistance, and considerable ability to withstand the prevalent root rot, but it was relatively weak strawed. A cross of Kindred with the feed barley, Titan, was released as an improvement by the North Dakota Experiment Station under the name Traill, in 1956; but because of susceptibility to the disease septoria, it, too, proved disappointing. The following year the South Dakota Experiment Station released Liberty, which withstood heat exceptionally well. Again, however, high protein content unless the grain was grown under irrigation made it unacceptable. The search for varieties that combined drought resistance with brewing quality continued into the sixties.[8]

Alfalfa had also benefited greatly from varietal introductions for dry-land cultivation during the period. The yellow-flowered *Medicago falcata*, brought to the United States by Nils Hansen of the South Dakota Experiment Station early in the century, had not attained widespread commercial use. It produced limited seed, with a high percentage of hard kernels that hampered seedling emergence. Efforts to produce a strain for general use, either by selection or hybridization, had failed. Dry-land alfalfa production developed primarily from the distribution of Ladak in the early thirties. Introduced by the U.S. Department of Agriculture from Leh, India, it had been placed in varietal testing at Redfield, South Dakota, in 1920. Offering far greater resistance to wilt than the older Grimm and Cossack varieties and hardiness against both cold and drought, Ladak remained a leading variety throughout the region into the sixties. It was supplemented after 1941 by the release of Ranger, developed at the Nebraska Experiment Station. Both were long-lived, afforded one or two hay cuttings a year, and in some areas yielded a profitable seed crop.[9]

Corn production increased somewhat in the western Dakotas during the war and postwar years but failed to generate interest in eastern Montana. The hybrids coming into use in the region in the early forties were late maturing for grain and offered no great advantage over older varieties for silage or pasture under dry-land conditions. While about 88 percent of corn in Iowa was hybrid by 1940, only 3 percent of that in North Dakota was. Use of hybrids was recommended in Montana only for irrigated lands in the mid- and lower-Yellowstone valleys, and as late as 1952 developmental work on such varieties for South Dakota was directed toward their use in the eastern counties. Several adaptations by the North Dakota Experiment Station were recommended for dry-land use there and in Montana by the mid-fifties, but the only ones that had proved generally satisfactory required too long to reach maturity.[10]

Sorghums—grain varieties such as Altamont, selected from Kaoliang; broom corn; Sudan grass; and the sorgos, Black and Red Amber—were strongly recommended by the South Dakota Experiment Station to provide a supplement to corn for forage in the west-river country. After some interest in their ability to withstand grasshoppers during the late thirties and war years, however, their use sharply declined. Fear of prussic-acid poisoning in cattle grazing on stalks damaged by drought or frost was not greatly alleviated by the introduction of a South Dakota low acid Amber in the early forties or by the release of Norghum, a grain variety, in 1949. Production of the crop was limited even in western South Dakota after the mid-forties and so small in the northern states that it was dropped from the census reports on Montana by the fifties. The Montana Experiment Station concluded in 1959 that they were not adapted to that area.[11]

The best of the tame grasses was Crested Wheatgrass, introduced by Nils E. Hansen in 1898, but not then distributed. Tested at the federal experimental farm at Belle Fourche, South Dakota, from 1908 to 1915 and subsequently distributed by the Northern Great Plains Field Station, at Mandan, North Dakota, it had received little attention until the 1930s, when it proved to be outstanding for regrassing abandoned farm lands. Crested Wheatgrass afforded the basis for the restoration efforts of the Soil Conservation Service in the region. Although it lacked heat tolerance and yielded poorly as hay, it survived cold, competed well against weed growth, and provided desirable forage and seed crops. Sweet clover, highly promoted during the twenties, won scant subsequent favor. It usually required a nurse crop for seeding and, as a biennial, yielded little more than a single year's production. Brome grass, which quickly became sod-bound under good conditions, was even less satisfactory because of low forage value under limited rainfall. A need for better adapted tame grasses remained a handicap to efforts for diversion of land from cash-crop grain.[12]

The principal achievements in varietal development for the region were still associated with production of wheat. Ceres, which had superseded Marquis as the leading hard red spring variety throughout much of the northern Plains during the late twenties, remained strongly favored into the mid-forties. Ceres accounted for over 73 percent of the wheat grown in Adams County, North Dakota, in 1940. But rival strains were introduced because of its poor resistance to smut and black stem rust during severe outbreaks of those diseases in 1935.[13]

Reward, developed from a cross of Marquis and Prelude made in Canada in 1911, was first distributed commercially in 1927 and challenged Ceres in central and western South Dakota by the late thirties. While its early maturity was welcomed, it was criticized in Montana for the same faults as Ceres and for greater shattering. Thatcher, resulting from a double cross of Marquis x Dumillo and Marquis x Kanred, developed as a cooperative undertaking by the U.S. Department of Agriculture and all the spring-wheat producing states, was released by the Minnesota Experiment Station in 1934 and became the principal successor to Ceres. It was early maturing, high yielding, resistant to lodging, and baked satisfactorily, but it proved only moderately resistant to stem rust and susceptible to stinking smut and leaf rust. The fact that Pilot, another product of the cooperative research effort, had the advantage in resistance to those diseases initially won it wide acceptance in North Dakota. In dry-land areas of Montana, however, where rust was a less serious problem, the lower yield and protein quality of Pilot gave the preference to Thatcher or even to the older Ceres and Marquis. By the mid-forties Thatcher, with 32 percent of the spring-wheat crop, had moved well ahead of Pilot even in North Dakota. Rival, introduced by the North Dakota Experiment Station in 1939 as yet another product of the heightened research program, had moved into second place among the state's preferences, with 31 percent of the crop.[14]

Introductions continued to appear in the mid-forties—Mida in 1944, Rescue in 1946, Rushmore in 1948, and Sawtana around 1951. Mida, the product of a double cross of Ceres x (Hope x Florence) and Canadian Rust Lab. no. 625, superseded all varieties in North Dakota, with nearly one-third of the crop by 1949. Yielding over 26 bushels an acre in a test weight of 60 pounds to the bushel, it had overtaken Thatcher. Rescue and Sawtana were developed specifically for resistance to the sawfly and because of the prevalence of that pest in central, north-central, and northeastern Montana had considerable use in that state. Elsewhere their lower yields were limiting.[15]

The introduction of Rescue had demonstrated the potential for rapid response to emergent problems through genetic development. Sawflies had been found only occasionally prior to 1941; by 1943 they had become severe in northeastern Montana; and by 1945 they had spread into central Montana, with losses estimated at some 10,000 bushels, each, in Fergus and Judith Basin counties. A bushel of Rescue seed, imported from the Dominion Cereal Breeding Laboratory, Ottawa, Canada, in October 1944 was sent to Arizona for propagation over the winter. The following May, 35 bushels from Arizona were planted in Montana, from which 877 bushels were harvested. The next winter 100 bushels were sent to Arizona for increase, and 3,870 bushels were returned for seeding together with the Montana carry-over the following spring. Despite late planting, yields of that season were sufficient to permit sample distribution to Fergus County farmers with the privilege of retaining half the yield in 1947.[16]

Similar genetic development was used to counteract the disastrous spread of the rust strain 15B in North Dakota during the early fifties. The disease first appeared in the area in early July 1950 and increased rapidly. Drought may have been the cause of diminished yields in the state by 1952, but despite abundant precipitation in the succeeding years, production of bread wheat was reduced about 25 percent in 1953 and about 45 percent the following year. The damage to durum crops was considerably greater and lasted over a decade.

Progress came first for bread wheats. Rushmore had shown greater resistance to stem rust in 1952 than any other available variety and still ranked second among the North Dakota wheats in 1955. Mida had been second among the state's varieties through 1954 but slipped to fourth by 1955. Neither variety was entirely satisfactory. Lee, selected from a cross of Hope and Timstein, had been released by the North Dakota Experiment Station in 1950. It was also recommended by the Montana station in 1953, with particular reference to its high yield and rust resistance. By 1954 it had become the leading hard red spring wheat grown in North Dakota, covering 31.7 percent of the state's acreage. The following year that amount had been increased to 56 percent. It, too, fell from favor, with but 10.7 percent of the state's crop in 1957, as Selkirk, a variety bred and selected specifically for resistance to the invading rust strain, became popular.[17]

Developed from a cross of McMurachy-Exchange x Redman, Selkirk had been grown successfully for two years in Canada when that government in the autumn of 1953 made 2,500 bushels available to North Dakota and 2,000 bushels to South Dakota for planting in 1954, while an additional 500 bushels were sent to Arizona for propagation over the winter. Increase of the grain was conducted under the auspices of the state experiment stations—in North Dakota by members of the Crop Improvement Association—and focused in the area east of the Missouri River, where the rust problem was most serious. The farmers raised the crop under contract and were permitted to retain only 15 percent of the production. By 1955 about 60,000 bushels were available in North Dakota for distribution upon application to county agents, and seed for an additional 4,000 acres had been brought into the state directly from Canada by seed men and individual growers. So great was the popular demand for the grain that 490 bushels, seized by American customs officers in enforcing regulations restricting wheat importations, were auctioned that spring at an average price of $13 a bushel, to buyers who had traveled snow-blocked roads from as far west as Glendive, Montana, to attend the Pembina sale. With a successful 1955 crop season, the additional purchase of 15,000 bushels from the Dominion Laboratory, and a loan of $65,000 from the Governor's Emergency Commission to finance a large winter increase program in Arizona, the North Dakota Experiment Station was able to announce in January 1956 that Selkirk would be available to seed all the hard red spring-wheat acreage in the main rust area of that state. By 1957, 85 percent of the North Dakota hard red wheat crop was Selkirk.[18]

Efforts to develop new durum varieties progressed more slowly. Under normal conditions durum production was concentrated in half a dozen counties of north-central North Dakota, but it represented about one-third of the state's wheat crop. The danger of intermixture with standard varieties, frequent low quality when grown out of the optimum locale, and the fact that in most years the supply was sufficient to hold the price below bread wheats had limited extension of durums into other areas. Since they require a long growing season and planting had been late in 1950, durums suffered more seriously than standard spring wheats from the invading 15B. By 1952 the durum crop was so short that the price rose to a difference of thirty cents a bushel over standard wheats; and the following year, when the durum crop amounted to only 50 to 65 percent of the ten-year average, the premium ran to a dollar a bushel. By 1954 durum production had fallen to less than 4.5 million bushels, while the domestic semolina industry required 30 million to 35 million bushels annually.[19]

Mindum, distributed as early as 1917, and Stewart, introduced in 1944, were the major durum varieties into the early fifties. In the fall of 1954, Sentry was released by the North Dakota Experiment Station for increase on the basis that it had some tolerance of 15B and was preferable to the older varieties, but it was recognized as merely a "stop-gap," not good enough to withstand the severity of the prevalent disease. About 200,000 bushels were available for seeding in 1956. Meanwhile, four new varieties—Langdon, Ramsey, Towner, and Yuma—specifically selected for resistance to 15B, were under propagation. A mere 242 bushels were available for increase in Arizona over the winter of 1954–1955, from which 8,000 bushels were put out for controlled development in North Dakota the following summer. After further winter increases in Arizona and northern Mexico, some 2 million bushels of seed were available for the spring of 1957. Two additional varieties, Wells and Lakota, both earlier and with still better resistance to the disease, were distributed in the spring of 1960. Only Langdon, a selection from a complicated cross of [(Mindum-Carleton x Khapli)-(Heiti-Steward x Mindum-Carleton) x Stewart] x Carleton, and Wells, a selection from a cross of Sentry x [(Ld. 379) x (Ld. 308 x Nugget)], retained high favor into the early sixties, when durum production finally returned to normal levels.[20]

Stimulated by the pressure of the stem-rust epidemic, varietal development had made remarkable advances. Although the crisis in bread wheats had been alleviated by the latter half of the fifties, the heightened research activity also yielded Conley, released by the North Dakota Experiment Station in 1956, Centana, brought out by the Montana station in 1953, and Sheridan, by the latter institution in 1961 as further improvements. The more limited winter-wheat producing areas of Montana benefited by the introduction of Cheyenne in 1956 and Itana and Rego in 1957. In the late forties, release of Minter as a winter wheat with participation by the South Dakota Experiment Station had brought as much as a 50 percent increase in yields of that grain for the south-central counties of the state and marked the culmination of a 300-mile northward extension of winter wheat acreage on the Great Plains during the preceding half-century. Further expansion of winter wheat cultivation into the northern states was still deterred by the problems of smothering ice cover, heaving soil, breaking of winter dormancy, and spring soil blowing.[21]

Few innovations had marked the adjustment in dry-land tillage methodology during the war and postwar years. The practice of alternating years of grain cropping with fallowing had been among the earliest concepts of dry-farming procedure. Debate on the merits of substituting a cultivated row crop—corn or, in South Dakota, sorghum—had gone on almost as long. Rejection of the fine cultivation of the Campbell system had come with the emergence of soil blowing as drought first challenged the farming effort after World War I. Modification of the operation by alternating strips of crop and fallow had developed in northern Montana in the early twenties. Plowless fallow, subsurface tillage, seeding into roughened stubble, and use of the basin lister were other approaches to the problem introduced in that decade. Contour cultivation and establishment of shelterbelts, the latter being an old conception with renewed support, formed the principal additions to the program through the thirties. Supplemented by a wide range of measures designed to improve pasture and stock-water supplies or to facilitate irrigation, this was the package of dry-land tillage recommendations institutionalized in the operations of the soil conservation districts and the Agricultural Conservation Program of the Production and Marketing Administration (formerly the AAA) over the succeeding decades.

With the dissolution of the Farm Security Administration and the Land Planning Division of the Bureau of Agricultural Economics, the role of the Soil Conservation Service as leader of federal efforts to guide regional readjustment was enhanced. The number of soil conservation districts continued to increase, until by 1955 there were seventy-eight, covering 54,568,000 acres, or 89 percent of the land in farms in Montana; seventy-nine, covering 39,592,000 acres, or 95 percent, of that in North Dakota; and sixty-six, covering 36,011,000 acres, or 80 percent, of that in South Dakota. As the system developed, it tended to become identified more with political than physiographic divisions, and it operated more through contract with individual farmers than with cooperating associations.[22]

The setting up of the Adams County SCD in southwestern North Dakota in 1949 illustrates the approach. Elected supervisors determined that the basic problems to be countered locally were wind erosion, water erosion, pasture management along with water facilities, and shelterbelt establishment. Accordingly, work was undertaken to lay out or construct 2,655 acres of strip cropping, on seven different farms scattered throughout the county; 780 acres of contour stripping, on five other farms in various sections; 4.5 acres of shelterbelt planting, on three farms; and eleven water-stock dams. Such dispersion had been adopted as a method of demonstrating the benefits of the program as widely as possible. In more mature programs the effort was directed toward contractual arrangements by which individual operators received detailed soil and land use analyses with adaptive recommendations covering whole farms.[23]

To promote participation, annual soil conservation achievement contests were instituted in North Dakota through the fifties. Sponsored by the Greater North Dakota Association, the North Dakota Bankers' Association, the North Dakota Press Association, the state extension service, the federal Soil Conservation Service, and the local SCD supervisors, the competition evaluated individual farms on correct land use, scope of participation in the soil conservation program, and quality of work accomplished under it. Districts, in turn, competed with each other for state honors on the basis of the three best farms in their work group. Publicity for such activity was heightened by annual County Conservation Day programs, featuring demonstration plots, educational films, and speeches by conservation specialists, and the whole effort was correlated with celebration of a National Soil Conservation Week. Even church sermons were devoted to the concern. Elementary and rural school teachers were also drawn into it by means of scholarships for their attendance at workshops and camps on conservation.

Technical assistance in developing the conservation plan without charge was one of the major benefits offered under the program, and use of heavy equipment was provided on a "job-cost" basis. The recommended operations, in turn, were largely recompensed through benefit payments under the Agricultural Conservation Program of the AAA. For Williams County, North Dakota, for example, such a schedule of payments in 1944 had provided: $. 75 an acre for summer fallow by "pit cultivation," $1.25 an acre for summer fallow left with sufficient surface stubble to protect against erosion, $ .50 an acre for strip-cropping rows of alternating flax and summer fallow, $ .50 an acre for protective strips of corn on summer fallow, $3.50 an acre for harvesting up to 25 acres of legumes or perennial grasses, $10.00 an acre for weed control, and additional payments for pasture development and construction of stock-water facilities. The Little Muddy Soil Conservation District, covering the western half of Williams County, at that time encompassed 216 agreements, extending to 84,761 acres. The conservation practices reported for 1943 embraced 2,400 acres of crop residues and subsurface tillage, 5,061 acres of wind-strip cropping, 6,766 acres of "approved rotations" under the wartime cropping program, 55 acres of "long-time grass rotation," 6,144 acres of controlled grazing, 387 acres of new permanent grass seedings, and establishment of five new farm shelterbelts. By the end of the decade a total of 235,651 acres, about one-third of the acreage in the district, was under conservation agreements, and the year's accomplishments under the program marked the addition of 6,474 acres of stubble-mulch tillage, 6,316 acres of strip cropping, 2,360 acres of pasture improvement, 422 acres of grass seeding, 35 acres of farm shelterbelt, and, perhaps as a consequence of the rainfall shortage in 1949, 19 stock-water dams and 65 acres of sprinkler irrigation. Progress in adopting the approved practices had been slow and erratic.[24]

Measurement of the usage of summer fallow is complicated by the varying conceptions associated with the designation. Discussing its place in the regional agriculture in 1951, O. R. Mathews, a senior agronomist with the USDA, concluded that census data were fairly accurate where the farming system depended heavily upon moisture-conserving technology, but were unreliable in borderland districts, where much idle or abandoned land was included. So qualified, the extent of summer-fallow use in western North Dakota was recorded as ranging from 2,557,000 acres in 1934 to a peak of 3,188,000 acres in 1939, declining to 1,416,000 acres in 1944, and increasing only to 2,387,000 acres by 1954, representing at the last date a mere 21 percent of the cropland acreage. It had declined in western South Dakota from a peak of 810,000 acres in 1939 to 383,000 acres in 1954, when it amounted to but 8 percent of the cropland acreage. In eastern Montana, however, it had increased from 3,644,000 acres in 1939 to 4,628,000 acres in 1954 and covered at the latter date 40 percent of the cropland acreage. In all three areas it was the most extensively used procedure among the methodological recommendations.[25]

Stubble mulch as an aspect of summer fallow had been supported as an ACP cost-sharing practice in Montana since 1938. The South Dakota program also applied it early to summer-fallow operations, and in North Dakota it had been an approved procedure for both summer fallow and continous cropping since 1942. In 1959 payments approximating $ .50 to $ .60 an acre were made for stubble mulch on some 1,887,000 acres in Montana, 2,808,000 acres in North Dakota, and 331,668 acres in South Dakota.[26]

In 1954 some 2,008,000 acres were strip cropped in eastern Montana, but less than 1 million acres were so handled in western North Dakota, and only 133,000 acres in western South Dakota. The Montana Extension Service conceded in 1945 that use of contour stripping was limited primarily to organized conservation districts and was best suited to "non-glaciated areas," where corn was part of the rotation. Supervisors of the Little Muddy Soil Conservation District in western North Dakota concluded that little land in that area was adapted to the practice.[27]

Moreover, notwithstanding the prolonged emphasis upon such methodology, questioning of the efficacy of individual practices continued among agricultural experts. After a long career of research on dry farming, O. R. Mathews emphasized the value of summer fallow for increased water storage to stabilize returns on succeeding crops; but he recognized that it held limited value on sandy soils, that it might be unsafe where there was great danger of wind erosion, that it maintained less organic matter in the soil than continuous cropping, and that the loss of a crop when wheat prices were high might offset other considerations. There was still much debate on the merits of alternating corn or sorghum with wheat as an alternative to fallow, since it was more profitable to produce some crop than to let the land lie fallow. Research at the branch experiment station for northern Montana, at Havre, where corn had little place in the cropping program, most strongly upheld the merits of alternating fallow, yet with the growth of barley cultivation in the region, a three-year, minirotation of barley-wheat-fallow gained acceptance. Investigations at Froid showed that the latter system yielded as much wheat per acre as was produced by the alternating crop-fallow rotation.[28]

Longer rotations, particularly the inclusion of green manuring as a feature of such cropping, also had little acceptance and, despite its continued encouragement by the SCS, was rejected in research findings of the Montana and North Dakota experiment stations and USDA dry-land specialists. Perhaps even more upsetting to conservationists was the lingering countenance given by several research studies showing higher yields and better protein quality for wheat grown on land where the stubble of the previous crop had been burned prior to seeding. A more compact seedbed, providing easier access to subsurface moisture, appeared under such analysis to offer better returns than restoration of soil nutrients on dry lands, an argument that reverted to the basic conceptualization of the early Campbell Farming System.[29]

This disagreement among the experts was disturbing. Leaders of the Little Muddy SCD protested in 1950 that while cooperation was good among farmers, business interests, and such publicity agencies as the press and radio, that among the USDA agencies in the area "could be improved." Both the ACP and the SCDs institutionally provided for administration by locally elected representatives, who were free to select from a broad range of recommended practices in emphasizing those most applicable to their regional situations. The conflicting advice made it evident, however, that individualized technical guidance rather than formalized regional recommendations was required.[30]

There was a continuing call for more thorough and localized land classification. H. H. Finnell, of the SCS, warned the Great Plains Agricultural Council in 1949 that mapping of "generalized zones" was of little use. Through a borderland, "marginal zone" along the western edge of the Plains, "Second class, fourth class, and sixth class lands lie every-which-way in relation to each other. . . . The land use planner, the conservation planner, the land owner, and the operator," he continued, "have all got to go deeper than that in recognizing various uses and farming methods for different tracts of land in close proximity to each other."[31]

In South Dakota soil surveys were renewed in 1947–1948 to enable the SCS to provide "farm planning legends" for new SCDs and the Angostura irrigation project. The work covered eastern Grant County and was under way in Gregory County of the trans-Missouri region by 1950, but the South Dakota Experiment Station acknowledged that the tendency was to vary the degree of detail according to prevailing use patterns. Rangeland, for example, was not judged to require the intensity accorded districts previously identified for farming or irrigation. Comprehensive surveys dating from the thirties had been prepared for only three counties of western North Dakota—McKenzie, Billings, and Morton. A major survey, based on county reconnaissance—providing descriptions of the soil series and their associations, their structure, permeability, drainage, chemical properties, and rating for productivity—was finally published for North Dakota in 1968. It, too, presented data generalized for relatively large areas and carried a disclaimer: "not suitable for obtaining accurate, detailed information about the soils of individual farms or fields." Publication of the Gieseker mapping of Montana counties had been interrupted in the early forties and was not resumed until the late fifties when it was supplemented by maps according to SCS grading standards.[32]

For cooperators the SCS program promised detailed soil and topographic inventories—field by field survey of soil type, steepness of slopes, extent and degree of erosion, existing land use, and changes necessary for conservation and a well-rounded operation. But the proportion of cooperators within the SCDs and the assistance they were provided increased slowly. Fifteen years after organization of the Little Muddy SCD, only 375,000 acres had been surveyed jointly with the adjacent Lewis and Clark SCD, amounting to less than one-third of the land in farms in Williams County, North Dakota. About half the farmers of the Slope area to the south, another early organized district, were cooperators in the program, but only two-thirds of those had received "conservation plans" for their land by the mid-fifties. Adams County residents, who organized an SCD in 1948, were still on the waiting list for surveys eight years later. Farm planning surveys had been made by 1955 for about 42 percent of the land in farms in North Dakota, 46 percent of that in South Dakota, and only 19 percent of that in Montana. Even for acreage in organized SCDs the surveys amounted to but 52 percent, 62 percent, and 23 percent, respectively.[33]

While there had been little change in the recommendations specifically identified as dry-land methodology, there had been developments in general agricultural technology following the war that were of considerable significance to farm operations in the semiarid region. Fertilizer had not heretofore been used estensively there, partly under the view that the minerals had not been leached from soils deficient in moisture, partly under belief that the available moisture was too limited to sustain heavier vegetative growth. Even decomposed manure had been found not to improve yields, except in very favorable seasons. As early as 1939 the Montana Agricultural Experiment Station had noted that use of phosphate was profitable on two-thirds of the soils tested but limited its recommendation for application to irrigated land. In 1952 they still advised that commercial fertilizers be used only at low rates and on an experimental basis. But in 1949 the South Dakota Agricultural Experiment Station reported that small grains seemed to withstand lack of water better where adequate nitrogen was supplied, and that station moved quickly into the forefront as an advocate of its use. The North Dakota station at the same time pointed to beneficial results from application of phosphates, and by the spring of 1953 was advising heavy applications, alone or in combination with nitrogen. The practice was included as a conservation measure under the Production and Marketing Administration (PMA; formerly ACP) program for dry-farming districts south of the Missouri River in that state. The soils of the northwestern counties, however, had been found better supplied with minerals than other sections of the Dakotas, and studies in 1955 showed that yields from fertilization were not as regularly advantageous in dry-land districts as elsewhere. The practice, nevertheless, was recommended as a measure to reduce risk by hastening crop maturity. In five years after 1949 use of fertilizers doubled in the Dakotas, and after the mid-fifties was greatly increased in dry-land districts of Montana, also. By the late sixties such applications had grown by 363 percent and a decade later, by 966 percent, over the northern Plains as a whole.[34]

The availability of an unspecified product, probably the chemical alticide, for weed-killing was advertised in central Montana as early as 1935, but use of herbicides as a significant feature of agriculture in the region began with extensive field testing of 2,4-D (2,4-dichlorophenoxyacetic acid) in 1945. The development held particular promise for farming practice that rested so heavily upon plowless, stubble-mulch operations, where moisture loss through weed growth was a critical consideration. Like fertilizers, 2,4-D gave less consistent results in areas of low precipitation. It was not effective on grassy weeds and failed to kill the perennial, regional nuisance, Russian thistle; but by the mid-fifties herbicides were estimated to increase wheat yields in many cases by more than one-half. The Great Plains Agricultural Council reported in 1953 that over 1,225,000 acres of cropland in Montana, over 2,016,000 acres in North Dakota, and approximately 810,000 acres in South Dakota were being sprayed for weed and brush control. Testing for the most desirable timing and for the effects of applying herbicides to growing crops continued through the decade. The introduction of amino-triazole in 1957 was reportedly very effective on thistle, and newer products designed for pre-emergent action greatly extended use of weed-killers by the early sixties.[35]

Insecticides also came into wider application following the war. DDT (dichloro-diphenyl-trichloroethane), developed by a Swiss chemist in 1939–1940, had been restricted to military use until 1945. Then it was recommended for a wide variety of farm operations—spraying cattle barns, poultry houses, manure pits, grain bins, and crops. By 1950 alarm was developing about its residual effects. Chlordane, introduced in 1947, had been found superior for grasshopper control and less harmful to honey bees. While DDT was still recommended for mosquito eradication as late as 1958, aldrin, heptochlor, dieldrin, toxaphane, tetra-ethyl-pyrophosphate, and finally malathion were introduced as alternatives during the decade. When grasshoppers recurred in North Dakota in 1957, the most serious threat since the thirties, the progress of insecticide development permitted speedy control of the infestation.[36]

Experiments with using airplanes for crop spraying were under way in western North Dakota in 1939. By 1947 Montana had established an Aeronautics Commission, which served to publicize the value of air service for finding lost cattle, surveying scattered crop fields, and crop-dusting. Annual clinics on those practices were instituted by the North Dakota Agricultural College in 1949. At the same time, the first commercial aerial fertilizer service in the region was established. Three years later 143 planes were licensed for crop spraying in North Dakota, and five years later the number had been increased to 332. Gains were most notable in the western half of the state.[37]

Other transportation developments also enhanced mobility in farming operations. With greater income, farmers increased their purchases of automobiles, trucks, tractors, combines, and other equipment. In Montana in 1954, 87 percent of the farms had tractors and trucks; 84 percent, cars; and 48 percent, grain combines; there were averages of 2 tractors and 1.7 trucks per farm reporting. Similar ratios were reported in the western Dakotas. By 1987 the holdings of combines and tractors remained approximately the same, increasing somewhat more in North Dakota, but in both Montana and North Dakota trucks, including pickups, averaged approximately 3 to a farm. During the thirties the introduction of rubber tires on tractors and combines and of smaller one-man combines that could be hooked to a general-purpose tractor made it easier to move machines from field to field and from farm to farm as operators acquired additional tracts. Sales of the first practical self-propelled combines, released by the Massey-Harris Company of Canada in 1939, were rapidly expanded after the war. As with the earlier introduction of small, light tractors, however, medium-sized implements proved more suitable in the region, and 12- to 16-foot combines were most common. By 1950 the use of diesel fuel reduced the cost for increased power, and by the 1980s more than half the tractors were forty or more horsepower.[38]

As supplementary farm equipment, the furrow drill, introduced in the northern Plains in 1940, was particularly suited to seeding in stubble mulch, and by the eighties, further drill improvements permitted better control of the depth of seed placement and reduction of equipment clogging. A field pick-up baler, appearing with a range of devices for hydraulic lifting in the late forties, reduced by about one-half the cost of harvesting hay. A field ensilage cutter, first used during the 1947 season, eliminated the laborious handling of bundles from the corn binder. As such developments were continued, labor requirements for cropping in the northern Plains were reduced by one-third between the mid-fifties and late sixties, and by almost one-half over the next decade.[39]

The innovations greatly added to the capital investment, however. Long-run USDA studies of the capital requirements for operations on combination wheat-livestock farms in the northern Plains showed a steady increase from an average total of less than $10,000 in 1942 to over $40,000 by 1957, with the cost of machinery rising over that period from an average of less than $1,500 to $7,650. By 1961 the Economic Research Service had raised the estimate of the cost of machinery and equipment on a "typical" Montana wheat farm to $16,000; but the Montana Agricultural Experiment Station three years later placed the amount as high as $46,000, including for production of 820 acres in wheat two 14-foot combines, two tractors, a 24-foot rod weeder, a 20-foot disk tiller, two smaller "toolbar" cultivators, three seed drills, three trucks, and an automobile. By 1987 census count placed the average market value of all machinery and equipment on Montana farms at nearly $61,000, a reduction then from over $66,600 earlier that decade.[40]

While custom service and leasing of machinery developed for some activities, particularly in harvesting, these alternatives entailed greater risk of delay and loss from bad weather. Only large-scale operators, however, could profitably purchase much of the equipment for individual use. Pressures for expanded operations were consequently dictated not merely by speculative urge to profit at a time of favorable climatic and market conditions, but also by necessity for economic use of labor and capital investment. Farm size continued to increase—from an average of 2,190 acres in 1954, to 2,553 in 1959, to 3,023 acres in 1974, to 3,111 acres by 1982 in eastern Montana and from 1,924 to 2,180, to 2,716, and to 2,841 acres for the same years in western South Dakota. The effect of the rust losses in western North Dakota was apparent in the meager increase of average farm size from 862 to 864 acres during the latter half of the fifties, but there, too, the trend toward larger holdings was repeated with expansion to 1,203 acres by 1974 and 1,318 acres by 1982.[41]

Development of stratagems for reducing agricultural risk became even more important, and financial arrangements for stabilizing income became one of the major endeavors of the fifties. Previously, recommendations had stressed the value of carry-over reserves sufficient for a year or two in cash, seed, and feed stocks. Now there was recognition that seasons of productive variability might come in "runs" or "bunches" so protracted that such holdings would be inadequate and even prosperous operators could experience severe difficulty. Large-scale operations permitted not only efficiencies in specialized management but also flexibility in using the resource base. Large holdings presented opportunity for fuller utilization of broken areas by livestock, and since drought seldom embraced all localities equally, benefits might also be derived by cropping dispersed tracts.[42]

Part ownership, combining rented and owned holdings, under the prevalent crop-sharing lease arrangements, shared risk as well as gain. In the northern Plains in 1959 about half of the farm operators were part owners, about a third were full owners, and 16 to 17 percent, tenants or managers. Part owners, however, held as much as 72 percent of the farm acreage in western South Dakota and about two-thirds of that in western North Dakota and eastern Montana. They farmed from 62 to 68 percent of the harvested cropland. Full owners held less than one-quarter of the farm acreage and harvested cropland in western North Dakota and less than 20 percent of those components—in fact, only 13 percent of the farm acreage—in the other areas.[43]

By the late eighties the ownership patterns had not greatly changed in western North Dakota, but part owners had increased their proportion of the harvested cropland to 71 percent. Elsewhere in the region the prosperity of the seventies had brought some expansion of land—including cropland—ownership, but part owners continued to hold more than half of the former and nearly two-thirds of the latter acreage. Tenantry had declined slightly but had not been a significant element in either period. Renting was primarily a management decision for flexible operational expansion while minimizing risk.

Like most regional operators, Thomas D. Campbell, whose mammoth enterprise had contributed greatly to advertise dry-farm development in the twenties, had known lean years through the early thirties but survived to

TABLE 7.1. Farms and Farm Acreage, as Related to Tenure of Operator, 1959 and 1987 (number and percentage of component)

	Western North Dakota		Western South Dakota		Eastern Montana	
	Number	Percent	Number	Percent	Number	Percent
1959						
Full owners						
Number of farms	7,052	34	3,276	31	6,996	34
Total acres	4,597,407	23	2,983,158	13	7,032,430	13
Harvested cropland acres	1,703,597	24	571,973	18	1,330,855	19
Part owners						
Number of farms	10,283	50	5,673	53	10,066	49
Total acres	12,546,698	63	16,940,418	72	34,910,813	67
Harvested cropland acres	4,383,432	62	2,128,744	68	4,482,114	65
Managers and tenants						
Number of farms	3,250	16	1,722	16	3,433	17
Total acres	2,704,675	14	3,425,207	15	10,376,641	20
Harvested cropland acres	1,003,510	14	441,469	14	1,085,206	16
1987						
Full owners						
Number of farms	5,225	35	4,132	46	7,267	43
Total acres	4,164,503	21	8,928,193	37	11,961,250	24
Harvested cropland acres	1,186,010	17	832,893	26	1,803,093	23
Part owners						
Number of farms	7,278	49	3,461	39	7,104	42
Total acres	13,171,256	66	12,724,156	52	33,051,373	66
Harvested cropland acres	4,782,380	71	2,029,355	63	4,979,037	63
Tenants						
Number of farms	2,333	16	1,320	15	2,521	15
Total acres	2,638,242	13	2,651,124	11	5,207,022	10
Harvested cropland acres	794,928	12	343,623	11	1,065,199	14

Sources: Computations from U.S. census of agriculture: 1959, vol. 1, pt. 18, pp. 121–23; pt. 19, pp. 130–33; pt. 38, pp. 133–35—county table 3; 1987, vol. 1, pt. 26, pp. 224–31; pt. 34; pt. 41, pp. 235–43—county table 10.

become highly successful through the forties and fifties. Identified in 1959 as America's "largest wheat grower," he relied primarily on leased holdings throughout his career. Of about 45,000 acres then plowed by the Campbell corporation, it owned only about 15,000 acres prior to Campbell's death.[44]

Skillful use of public credit was another risk-reducing measure contributing to the success of these operations. Interviewed in 1959, Campbell explained that he had obtained loans from the Commodity Credit Corporation on his grain for twenty-six years, but that all but five of those loans had been repaid with interest. The credit had enabled him to hold the grain off the market until rising prices permitted more profitable sales. Except for those five years, the firm had sold its wheat at from seventeen to seventy cents a bushel above the loan rate.

During the fifties agricultural economists initiated proposals on a variety of ways to supplement privately developed reserves with a combination of regionally adapted public programs—amortized credit, tax adjustments, crop insurance, as well as price support and grain storage provisions. Credit and tax arrangements centered on flexibility in payment according to seasonal conditions. Farmers in the postwar years remained heavily dependent upon private lenders, who supplied the bulk of the mortgage credit. Life insurance and investment companies, the federal land banks, and the federal Farm Mortgage Corporation extended real estate loans at such limited appraisal values that their role declined sharply during the expansion of sales activity in this period. Federal lending institutions withheld all credit from North Dakota farmers for nearly fourteen years, until 1951 when the state legislature modified its statutes on foreclosures and deficiency judgments that conflicted with federal regulations. By the early fifties, the Farmers' Home Administration (FmHA), which was designed to extend credit where private funding was not available, had made only 403 farm ownership or housing loans in western North Dakota.[45]

Intermediate credit, necessary for machinery and livestock purchases, was also largely dependent on commercial dealers and private banks, frequently on the basis of real estate collateral. The FmHA primarily extended this type of credit. Its resources were limited, however, and as a policy determination its lending centered upon livestock promotion. Moreover, if at any time FmHA borrowers were able to refinance their loans at rates and terms prevailing in their communities, they had to resort to private commercial channels. The Production Credit Associations, with authorization to make loans up to seven years, were intended to meet this demand, but their activity was largely confined to annual arrangements, with limited renewal.[46]

Many insurance companies and private lenders applied to mortgage loans amortization arrangements and longer terms, such as had been introduced under the federal lending program during the Depression, but they rarely permitted the flexibility in annual payment rates needed for adjustment to variable income levels. In 1945 individuals and commercial banks were still writing mortgage loans predominantly for five-year terms; insurance companies, for fifteen years; the federal land bank, for thirty-three years; and the FmHA for forty years. In the northern Plains 35 to 40 percent of the mortgages ran for five years or less; 20 to 25 percent, for six to ten years; and 30 to 40 percent, for longer. Interest rates on real estate mortgages ranged from 4 to as high as 7 percent; those on commercial loans for livestock or machinery were commonly 6 to 7 percent but in many cases higher; short-term commercial loans for current operation were predominantly at 7 percent or more. The need for cheaper and more available credit was a problem identified in most planning efforts of the period.[47]

Since the thirties, federal farm credit policy, where applicable, had been coordinated with USDA programs. That conservatism and the heavy reliance accorded to private lending institutions, much criticized by regional analysts during the fifties, reflected the department's concern over mounting crop surpluses. The view markedly changed as export markets developed and grain stocks dwindled during the seventies. Under Secretary of Agriculture Earl L. Butz, agricultural expansion was encouraged and credit was extended more abundantly. Real estate debt in the Great Plains increased by 189 percent and non–real estate debt, by 259 percent between 1973 and 1983.[48]

Assumed at a time of rampant inflation, these obligations carried nominal interest rates as high as 12 to 14 percent, a service charge that posed serious difficulties as crop prices declined from 1976 through 1978. In the latter year Congress authorized the secretary of agriculture to forego foreclosure on loans of the FmHA and to insure or guarantee privately issued loans in order to refinance outstanding debt, reorganize farming operations, purchase essential livestock and agricultural supplies, even to provide family subsistence. Through the early eighties the program was repeatedly liberalized, until by 1985 the federal farm credit system held some $76.8 billion in outstanding loans, amounting to over one-third of the nation's total farm debt. Following a presidential veto of legislation further funding these activities, the Farm Credit Administration was then reorganized. The secretary of the treasury was authorized to take over delinquent loans, to renegotiate them where feasible, and where that was impossible to institute foreclosure. Debt referral and debt adjustment under federal auspices had represented disaster relief, but not an institutional adjustment for long-run application. The focus of credit extension was re-directed to private lending agencies.[49]

Property-tax reform also remained largely related to disaster relief. North Dakota voters considered but repeatedly rejected proposals pressed by the Farmer's Union that these taxes be graduated according to levels of farm income. The South Dakota Agricultural Experiment Station suggested, as an alternative, that they be scaled to the parity price ratio, building public revenue when income was high to balance a decline when it dropped. But moratoria—provision for installment payment of back taxes without penalty, even in North Dakota, by legislation of 1943 that canceled accumulated interest on such debts beyond a single 5 percent charge—merely helped to alleviate specific emergencies. There was no assurance of flexibility in meeting possible recurrence of such problems.[50]

Federal income tax regulations did, however, provide a modicum of stability for farm accounts through authorization of income averaging over a five-year base. Moreover, by allowing generous deductions for depreciation of equipment and exclusion of income-in-kind and realized capital gains from farm revenues, the federal program taxed farm income considerably less than non-agricultural or business earnings. During the seventies, it has been estimated, approximately 86 percent of farm income could be sheltered. Unfortunately, these benefits were at the same time conducive to speculative investment during a period of severe inflation.[51]

All-risk crop insurance, covering crop loss resulting from weather, insects, plant disease, and other stipulated unavoidable causes, was available on wheat as a federal program after 1938, except for a one-year interruption during 1944. Throughout much of the period, North Dakota and Montana also continued to offer state protection against hail loss. The federal program was designed to compensate only for production expense and was limited to a percentage of the average yield as calculated for individual counties over the past twenty years. State programs afforded reimbursement at a stipulated scale, ranging from $8.00 to $12.00 an acre for regular policies or, in North Dakota, $13.00 to $22.00 an acre for special policies requiring a higher premium.[52]

Because of the extreme variability of yields and climatic conditions actuarial data for the programs was limited. Until the late fifties the North Dakota system levied charges at the end of the crop season according to rate schedules based upon the amount of total loss incurred. Those with special policies paid in advance at a fixed rate averaging 15 percent of their insured risk, but if a surplus remained at the end of the crop year, it was refunded to the participants. Since no capital fund was accumulated, little compensation was available in periods of stress. In 1957 an effort was made to establish a more equitable basis by differentiating three major risk areas and assigning heavier premiums accordingly. Williams and the southwestern counties— Stark, Slope, Hettinger, Adams, and Bowman—constituted the highest risk area, with premiums ranging at 14 percent of the insurance carried. Most of the remaining western counties were grouped with the central district, paying premiums of 8 to 12 percent. In the wake of continuing dissatisfaction with the program, it was discontinued in the sixties.[53]

Despite a noteworthy actuarial record in the region, the federal all-risk insurance program was also poorly adapted to the farmers' needs. Premium payments in North Dakota from 1939 through 1951 exceeded indemnities every year but one, 1949. During the rust outbreak of the next few years deficits were incurred, but by 1957 a reserve had been re-established. The loss ratio for North Dakota counties over the whole period 1939 through 1957 was a mere 0.56; that for Montana counties was the same; and that for South Dakota counties, 0.70—when the ratio for Great Plains counties, generally, amounted to 0.99. Reserves accumulated on wheat insurance continued to be high in the region, amounting to over $15.7 million for Montana and over $11.1 million for North Dakota. Yet the premiums under the federal program ran to over 10 percent of the coverage for many counties of western North Dakota. Credits were allotted to those who had consecutive years without loss; but that arrangement, designed to invite long-run participation by those with good records, penalized the unfortunate in their time of hardship. Moreover, for want of adequate funding, the program was closed to new contracts during periods of crisis—in 1955, because of heavy rust losses, and in 1960, because of drought losses the previous year. Under such limitations the insurance won little support, notably so in districts of dry-land wheat cropping.[54]

Skeptical of the benefits of formal insurance programs, farmers of the northern Plains relied on the tacit insurance available through federal price supports, conservation payments, and disaster relief loans. Their rate of approval of these programs was among the highest in the nation. While governmental payments represented a small part of farm cash income in the region through the forties and fifties, amounting to less for that area than for the nation generally during the immediate postwar period, they provided assurance of relief in emergencies. Following a severe winter and spring drought in 1949, thirty-seven North Dakota counties, including sixteen of the twenty-three in the western region, were designated disaster loan areas, entitled to 3 percent production loans under the Farmers Home Administration for the following crop season. All North Dakota counties and several counties of north-central South Dakota were again given such assistance, primarily as a consequence of rust losses, from 1953 through 1955. Largely as an aspect of a new land-retirement program enacted in 1955 but also as a response to drought that spread northward over the Plains at the end of that decade, government benefit payments were rapidly and greatly expanded through the sixties.[55]

Disaster relief was formally established as free insurance under the Agriculture and Consumer Protection Act of 1973. With a deficiency in yield below 60 percent, producers of wheat, corn, barley, and grain sorghum were thereby to be compensated at the rate of one-half a target price designed to cover costs of production. Under amendment in 1977 the program also assured payment at the rate of one-third of the target price for three-fourths of the normal yield if farmers were unable to plant their eligible acreage because of "natural disaster" beyond their control. The secretary of agriculture was at this time directed to conduct a study of alternative programs for general all-risk, all-crop insurance.[56]

A major revision of the federal crop insurance program was enacted in 1980. The capital stock was raised from $200 million to $500 million. The administering board was altered to include a person experienced in crop insurance and three active farmers, none of whom was to be otherwise connected with the government. Committees of producers were to be used to the maximum extent possible, and contracts were to be negotiated with private insurance companies where actuarial data made such action feasible. Insurance was to be offered against losses covering up to 75 percent of average yield, with lesser levels of coverage as adjusted. Premium allowances were to be provided for those carrying alternative protection against hail, fire, or disaster loss. In order to encourage "the broadest possible participation" in the program 30 percent of each producer's premium, reduced as applicable for the alternative insurance arrangements, was to be paid by the administering body for coverage up to a maximum of 65 percent of the average yield. Sixty days before the beginning of the planting season of 1981, farmers were to be given the option of filing for disaster payments or covering the acreage with insurance under the new legislation. Significantly, however, the Food and Security Act of 1985, in denying eligibility for emergency loans to farmers if crop losses were incurred when insurance under the federal program had been available, indicated a continuing widespread reliance upon the relief alternative.[57]

Farmers of the northern Plains showed brief interest in rain-making experiments as an approach to risk reduction when diminished rainfall in the late forties appeared to presage the long-predicted recurrence of drought. Research begun in 1943 by Dr. Vincent J. Schaefer, as a meteorologist at the General Electric Company had shown promise when a method was developed for changing super-cooled clouds to ice crystals. The U.S. Office of Naval Research and the Army Signal Corps joined the experimentation in 1947. By 1951 Dr. Schaefer enthusiastically reported to the Great Plains Agricultural Council that sufficient knowledge existed about "precipitation mechanisms . . . eventually [to] do most anything we want with weather." Although meteorological scientists were still unable to determine where their cloud-seeding chemicals would go or how best to distribute them under varying conditions, in 1950 a Colorado group had formed a commercial enterprise, the Water Resources Development Corporation, to continue the tests in large-scale farm and ranch application.[58]

A young farmer of northern Williams County, North Dakota, had undertaken cloud seeding with dry ice as early as 1948. His results were of questionable significance, since precipitation over the region generally was only half an inch below the long-term average, but over the next several years farmers in western North Dakota, the Grand River district of South Dakota, and along the Montana High-Line west to Cut Bank and Shelby contracted with several firms for cloud seeding. Despite claims of improved pasture conditions in the area, the fact that rainfall for the region as a whole was well above normal during 1950 and 1951 left doubt of a causal relationship. The Grand River Valley Weather Improvement Association reported 23 percent greater rainfall in the district covered during the very dry season of 1952, but during a three-week period of cloud seeding over the adjacent North Dakota–South Dakota border counties that July results varied widely, with between 60 and 150 percent of normal precipitation. With well-above-average rainfall in 1953, the program was postponed so that farmers could complete seeding. Although a contract was renewed for continuation of the project over north-central North Dakota in 1954, interest in the effort was waning.[59]

At most, proponents claimed merely to enhance the activity of existing rain clouds. Results had varied greatly in both scope and duration. Scientists of the U.S. Weather Bureau were highly skeptical of the reported successes: "Much more experimentation and analysis" were needed "to separate the magnitude of the precipitation attributed to seeding from that of the background noise caused by natural variability of precipitation." Addressing the Missouri Basin Inter-Agency Commission in 1957, a bureau spokesman reported that results had been "relatively limited and specialized," "least promising . . . in drouthy flatlands where rain is needed most." The South Dakota Legislative Assembly at that time joined a number of other western states in funding further research under the National Science Foundation. A decade later that body reported that redistribution of precipitation and organization of cloud systems into more productive units remained ends to "be pursued." In 1976 Congress requested that a Weather Modification Advisory Board review the possibilities. That board subsequently outlined research proposals to span another twenty years, but they were not funded. Speakers at a symposium sponsored by the Research Committee of the Great Plains Agricultural Council in 1982 noted continuing frustration "because of the lack of clearcut evidence of rainfall increases." While experiments in seeding cumulus clouds at Miles City, Montana, had developed ice particles into "precipitation embryos," they had dissipated too quickly to produce rainfall. Understanding was still lacking on the mechanisms to link such seeding to ground-level precipitation.[60]

The prospect of expanded irrigation as a stabilizing factor in agricultural development of the region was also pursued. Despite wartime pressures, construction of small dams, reservoirs, and dugouts had continued under programs of the AAA, SCS, and the FSA, in conjunction with the state extension services, primarily to provide better stock-watering facilities but also to permit small-scale irrigation for home gardens, feed-grain production, and improved pasture. Between 1936 and 1955 the Agricultural Conservation

Program in Montana had assisted in the construction of some 1,500 dams specifically for irrigation, plus approximately 45,000 for erosion control and livestock water. Through farmer and district cooperation, the SCS by the latter date had brought irrigating water to approximately 40,000 acres in that state. Although upkeep on many of these structures was permitted to deteriorate during the years of ample precipitation, it was believed that public funding could bring them back into service quickly during drought. Congressional legislation in 1954 expanded the water-facility loan program to provide credit for small-water development, up to $25,000 to individuals for as long as twenty years and up to $250,000 to incorporated associations for terms running up to forty years.[61]

In planning for postwar employment relief, state and local communities had listed numerous projects to use the waters of the Missouri basin. State water conservation agencies kept agitation for this work active, and promptly brought them forward as peace returned. Ultimately, many of them were incorporated in the headwaters phase of what came to be identified comprehensively as the Missouri Basin Development Program (see Map 7.1). That program evolved out of more than a decade of bitter rivalry and strong controversy on a variety of government fronts—opposition by state authorities based on fear of surrendering local control when presidents Roosevelt and Truman sought to establish a Missouri Valley Authority; division between the states of the upper valley, who sought to preserve the waters for irrigation, and those of the lower valley, who wanted to utilize them for navigation; and within the federal establishment itself, in struggles between the Army Corps of Engineers and the Bureau of Reclamation over the design and construction of the facilities and between the Interior and Agriculture departments over guidance of settlement under them. As an aspect of irrigation development, however, the conflict was resolved in the compromise Pick-Sloan plan, a consolidation of design by which the Army Corps of Engineers, under the proposals of Colonel Lewis A. Pick envisaged construction of a number of small reservoirs upstream, five large dams on the Missouri itself, and a system of levees, primarily to afford flood control and navigation, while the Bureau of Reclamation, under the planning of Glenn Sloan of its Billings, Montana, office, promoted a series of some ninety small reservoirs to expand irrigation along the headwaters.[62]

Authorized under the Flood Control Act of 1944, the work embraced as major facilities the already completed (1938) Fort Peck Dam and Reservoir in Montana; the Garrison Dam and Reservoir (Lake Sakakawea) in North Dakota, begun in 1947 and operational by 1953; the Oahe Dam and Reservoir, extending from near Pierre, South Dakota, north almost to Bismarck, North Dakota, begun in 1948 and structurally completed by 1956; and the Fort Randall and Gavins Point Dams and Reservoirs (Lakes Francis Case and Lewis and Clark) in southern and southeastern South Dakota, begun in 1946 and

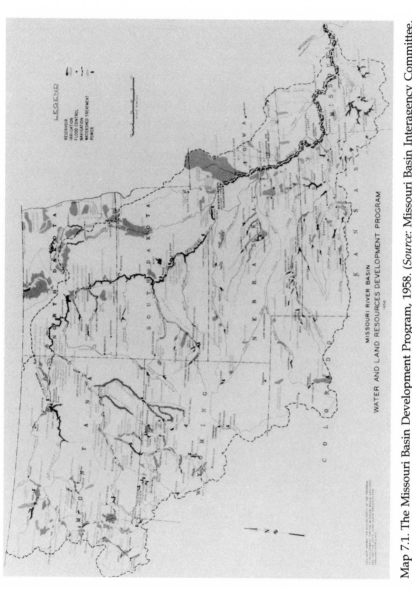

Map 7.1. The Missouri Basin Development Program, 1958. (*Source*: Missouri Basin Interagency Committee, *The Missouri: A Great River Basin of the United States, Its Resources and How We Are Using Them*, U.S. Department of Health, Education, and Welfare, Public Health Publication no. 604 [Washington, D.C.: Government Printing Office, 1958], endpaper) This map encompasses the modification under the Garrison Diversion Plan.

1952, respectively, and also completed in 1956. Subordinate structures included in Montana the Canyon Ferry Reservoir on the upper Missouri, near Helena; the Yellowtail Dam on the Big Horn River; the Tiber Reservoir on the lower Marias River; in North Dakota the Dickinson and Heart Buttle Dams on the Heart River and the Broncho Dam on the Knife River; and in South Dakota the Angostura Dam and Reservoir on the Cheyenne River, Pactola Reservoir in Rapid Valley, and Shadehill Reservoir on the Grand River. In conception the Missouri Basin project was to have encompassed 4,760,000 acres of new irrigation and supplemental water for an additional 520,000 acres. The irrigable areas provided under new works ranged from as little as a few hundred acres at Dickinson to some 15,400 acres along the Knife River in North Dakota and from 1,200 acres at Stanford and 3,000 acres at Ross Fork in central Montana to 45,000 acres on the Big Horn, 59,000 acres between the Milk River and the Fort Peck Reservoir, and 100,000 acres by diversion below Fort Peck into the Medicine Lake area of Montana. At the same time work was completed and in some cases extended for earlier-authorized projects on the Sun, Milk, Lower Yellowstone, Tongue, and Powder rivers in Montana, at Buford-Trenton in North Dakota, and at Angostura in South Dakota.

Most of the dams were completed by the early fifties. The multipurpose system provided large-scale generation of electric power, vast improvement in downstream flood control, numerous recreational lakes, and improved municipal water supplies, but relatively little increase in the acreage of irrigation for areas of dry-land cropping. The core of the proposed development under the last category had centered in an effort to bring water to 1.3 million acres extending across northeastern Montana and northwestern North Dakota.

The idea of constructing a canal from the Missouri River in Montana to the Red River of the North, the eastern border of North Dakota, had been presented to Congress by North Dakota leaders as early as 1889. During the drought of the thirties appeals for such a federal project had multiplied. In 1940, Glenn Sloan had developed a proposal for a diversion canal from behind the Fort Peck Dam across northwestern North Dakota into the Souris River Valley. The Army Corps of Engineers had reported unfavorably on the project, but it was incorporated in the Pick-Sloan compromise package authorized under the legislation of 1944 (see Map 7.2). In 1949, the Bureau of Reclamation discovered that the soils through much of the proposed route were so compacted by an underlay of fine-grained, clay-like sediment, the result of glacial action, that adequate drainage could not be provided for irrigation.[63]

North Dakota government and business leaders were not prepared to accept abandonment of a proposal that might be extended into the long-contemplated linkage of the Missouri River with the Red. Commercial interests of the Devils Lake area were particularly active in pursuing the development in conjunction with Congressman Usher Burdick, of Williston, and Harry Polk,

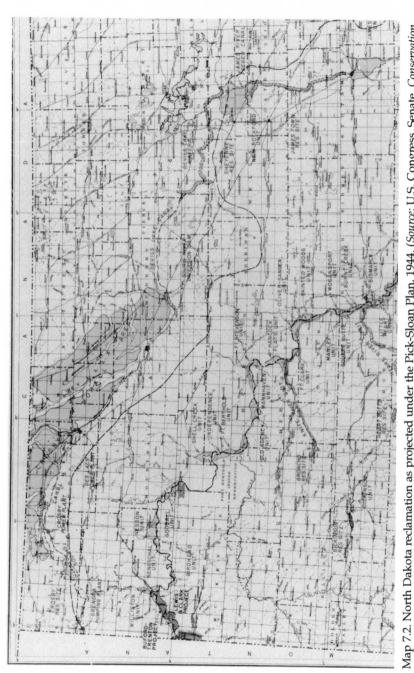

Map 7.2. North Dakota reclamation as projected under the Pick-Sloan Plan, 1944. (*Source:* U.S. Congress, Senate, *Conservation, Control, and Use of Water Resources of the Missouri River Basin*, 78th Cong., 2d sess., April 1944, Sen. Doc. no. 191, serial 10856 [Washington, D.C.: Government Printing Office, 1944], Appendix 3, plate 58-D-492)

editor of the *Williston Daily Herald* and four-term president of the National Reclamation Association. State reclamation associations, organized throughout the region in the early forties to press postwar developmental efforts, rallied to the cause. The North Dakota State Water Commission, the Greater North Dakota Association, the State League of Cities, local chambers of commerce, the North Dakota Press Association, agricultural agents of the transcontinental railroads, the North Dakota Stockmen's Association, the Farm Bureau Federation, leaders of the state experiment station, even the American Legion and its Auxiliary joined in public meetings of support for it. Surveying farmers on the Missouri-Souris segment in 1944, the Bureau of Reclamation claimed that 80 percent "were outspoken" in favor of the proposal.[64]

With the collapse of its plans for irrigation by canal from the Fort Peck Reservoir, the Bureau of Reclamation developed a new proposal, submitted under a feasibility study in 1957, calling for diversion of water from the Garrison Reservoir for development of 1,007,000 acres in central and eastern North Dakota. The first phase of this project, again designed to provide irrigation for 250,000 acres in the Souris River Valley, was authorized in 1965. Within three years it was in large part suspended because of opposition from landowners along the route, wildlife and conservation groups, and the Canadian government, which in the last case submitted a diplomatic note expressing environmental concern over undesirable drainage into the Hudson Bay Basin. Since this segment of the project was the only part that fell within boundary limits definable as dry-land cultivation, congressional support for the massive investment eroded while negotiations were conducted to relieve the dissatisfaction (see Map 7.3).[65]

Opposition to the proposal had increased locally as well as nationally over the years. Indians of the Ft. Berthold and Standing Rock reservations and farmers in many districts along the route were angered when their lands were engrossed for reservoir or canal construction. Three small towns and a branch of the Great Northern Railroad had had to be relocated as under the original design, the Medicine Lake Reservoir was constructed in Sheridan County, Montana. Settlers on part of the recently completed Lewis and Clark and Buford-Trenton projects were also forced to surrender their holdings when the Corps of Engineers proposed to raise the pool level of the Garrison Reservoir to accommodate the expanded design. This alienated some of the most ardent of the early proponents of the development. Even the Williston editor protested the threatened high pool level for "what now seems conclusively to be a very presumptive basis that farmers in the less arid sections of the state want and are ready to pay for irrigation." When a spring ice jam washed out the Buford-Trenton Canal and bridges in 1952, forcing irrigators to retreat to dry-land benches, fears were engendered concerning irrigation elsewhere. Only two North Dakota state senators opposed organization of a Garrison Diversion Conservancy District in 1955, but they spoke for four western coun-

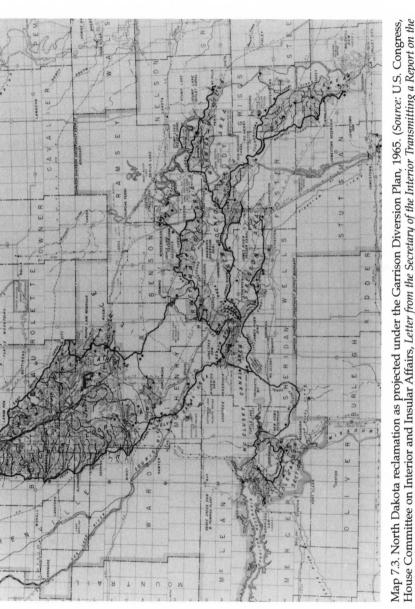

Map 7.3. North Dakota reclamation as projected under the Garrison Diversion Plan, 1965. (*Source:* U.S. Congress, House Committee on Interior and Insular Affairs, *Letter from the Secretary of the Interior Transmitting a Report on the Garrison Diversion Unit,* 86th Cong., 2d sess., February 4, 1960, H. Doc. no. 325, serial 12275 [Washington, D.C.: Government Printing Office, 1960], facing p. 88) Note that the projected development has been shifted eastward out of Burke, Renville, and Ward counties into central and eastern Bottineau, with major areas of extension in the vicinity of Devils Lake and along the James River.

ties as one of them protested that with "huge amounts" of silt filling up the dams, living behind dikes would be "like putting prisoners in death row."[66]

Realization that acreage limits under the federal reclamation program would require reduction of holdings increased the protest. While average earnings per irrigated acre were about double those under dry-land cultivation, the 160- to 320-acre allotment for a man and wife on the projects set income limits not constraining dry farmers. As the years of adequate rainfall continued, question was increasingly raised on the desirability of relinquishing large, highly mechanized operations that were profitable under dry farming to assume the smaller holdings, higher production costs, and unfamiliar techniques of irrigated enterprise. By 1955 two farmers from the Souris district, submitting petitions bearing 675 signatures, representing holdings of 400,000 acres, testified to a congressional subcommittee that the vast majority of farmers living in the area proposed for irrigation wanted "no part of it."[67]

Local government leaders also began to question the desirability of the development as federal land purchase reduced their tax base. Environmental groups who had initially favored reservoir construction as an enhancement of fish and wildlife habitat very early found that the canals were disrupting waterfowl sanctuaries and wetlands, some of them developed at the cost of resettlement only a few years earlier. Finally, old quarrels revived on the potential diminution of upstream water supply to maintain downstream navigation.

By 1982 Senate votes no longer supported the necessary authorization for federal appropriations to continue the work. A reformulated project was adopted in 1986, which restricted the potential irrigation to a mere 130,940 acres, of which but 38,500 would be located within the dry-land region west of the 100th meridian and 17,580 acres of that amount would be allotted to Indian reservations. Construction to the eastward into the Devil's Lake area and the James and Sheyenne river valleys would be continued but under a design intended primarily to provide water for municipal and industrial use. Some $275,240,000 had already been expended for the Garrison Diversion Project; an estimated $902,240,000 more would be required to complete the reformulated program.[68]

Despite the heavy promotion of irrigation and completion of numerous small-water projects under the Water Facilities and Water Conservation and Utilization programs during this period, the amount of irrigated harvested cropland was increased by less than 100,000 acres in eastern Montana, by less than 25,000 acres in western North Dakota, and by less then 34,000 acres in western South Dakota between 1939 and 1959. Only 11 percent of the harvested cropland was irrigated in eastern Montana, less than 1 percent in western North Dakota, and 2.6 percent in western South Dakota at the latter date. Irrigated pasture had been increased by less than 100,000 acres in the region as a whole. There were only about 8,000 acres irrigated from ground-water sources in eastern Montana, where a bulletin of the state experiment station early in the decade had featured sprinkler development. Such irrigation was still in the "talking stage" in North Dakota, an extension specialist reported. Only 363 acres in the western counties of the latter state were irrigated from ground-water sources and only 2,453 acres in western South Dakota.[69]

Nearly thirty years later, when pumping of ground water for irrigation had become a major development elsewhere in the Plains, it still had little role in the northern states. Irrigation of harvested cropland had increased by about 25,000 acres in eastern Montana and about 10,000 acres in western South Dakota by 1987, but these amounts did not offset the very large expansion that had occurred in dry-land cropping. The percentage of harvested cropland that had been irrigated remained nearly the same as in 1959 in both the Dakotas and had declined from 11 to 10 percent in eastern Montana. The amount of irrigated pastureland had declined in all three states.

In anticipation of the proposed development, agricultural experiment stations during the fifties and sixties had sought to compare the advantages of irrigation and dry-land cropping, particularly on the new, integrated projects. In general it was found that farmers who combined the operations through this period made faster progress and experienced less financial difficulty than those who relied on irrigated enterprise alone. Although irrigated farms showed few years of negative return and less variability in the return, their long-term average of losses, because of heavier costs, was greater than for operations under dry farming. Besides the larger investment entailed in land, construction, and developmental expense for irrigation, irrigated cropping required more inputs of machinery, labor, fertilizer, and other production factors. While cash farm income per acre was also higher under irrigation, analysis suggested that if the additional capital were put into expanded dry-land operations, less labor time would be required, higher farm income would be received, and a very much better rate of return would result from the investment.[70]

Given the cold climate and distance from markets, cash crop production under irrigation faced difficulties in this region not found in most other areas of such development. Potatoes were already in surplus, and sugar beets, under quota limitation. Studies indicated that dry-land wheat gave average net income returns per acre higher than those from other dry-land crops and higher, also, than those from irrigated feed grains. Net returns from irrigated wheat, corn silage, and alfalfa exceeded those from dry-land wheat; those from fed livestock were also slightly better; and irrigated pasture provided nearly triple the cash return of native, unirrigated pasture. Livestock, generally, however, showed much greater rates of variability in net income than either dry-land or irrigated wheat. Range cattle brought only about one-third as much as dry-land wheat, and native pasture cut for hay gave the lowest net income of any crop. The research emphasized the importance of wheat, whether dry-land or irrigated, in the agricultural economy of the region, showed the disadvantages of dependence upon the old, range-cattle operations, but pointed to some potentiality for profit by feeding livestock on irrigated corn silage, alfalfa, and pasture.[71]

Settlers on the irrigated projects had not turned extensively to the last alternative. Although most combined operations with livestock raising, their irrigated land was devoted primarily to seasonally maturing wheat and other small grains, less than 20 percent to the slower-developing alfalfa, and little to pasturage. Of the farmers surveyed on the Lower Yellowstone project in 1950, 69 percent preferred pasturing cattle on dry land and finishing them with feed grains raised under irrigation. Dry-land farmers adjacent to projects had not been found eager to acquire irrigated tracts, and analysts questioned whether those who held them had benefited greatly by the integrated enterprise. Thirty-eight settlers under the Lewis and Clark facilities and seventy-two at Buford-Trenton were judged reasonably successful with one-half to two-thirds of their acreage in dry-land cropping and range. Settlers on the Kinsey and Buffalo Rapids projects, on the other hand, then farming for an average of six years, had acquired a total net income averaging only about $7,000. A Montana study reviewing a variety of developmental plans for farms on the newly irrigated Lower Marias project in the mid-fifties foresaw financial problems under any of the proposals if there were declining farm prices in the future. Unless a more profitable management program were introduced, the principal advantage of irrigation appeared to rest in reduction of risk.[72]

Optimism among regional analysts during those years centered upon utilization of the dry-land areas. Studies by the South Dakota Agricultural Experiment Station indicated that of some fifty farm families interviewed, from 75 to 80 percent who had remained in farming over a lifetime owned their land. Research sponsored by the Great Plains Agricultural Council found markedly higher median incomes and a larger percentage of families with incomes greater than $5,000 in the western Dakotas than in eastern districts. Elmer Starch controverted those who argued that the Plains could no better withstand strain than in the thirties, that the pattern of "boom or bust" was destined to be repeated. He conceded that the region was subject to great droughts that affected the whole area occasionally and also to localized droughts that occurred somewhere almost every year. He contended, however, that residents had learned to figure in terms of long-time performance, that over time they would have better than the national average income. "While those of us who work in the Plains area are not so sanguine as some writers about having achieved control of wind erosion," he wrote in 1951, "we do know that we have reduced the susceptibility of many areas to a marked degree, and we have evolved practices of soil management which are effective."[73]

H. H. Finnell, of the Soil Conservation Service, shared these views. Reporting in 1947 on tests in the southern Plains, he concluded that soil damage had not been nearly so serious or extensive as had been thought in the thirties. Much of that area was producing more wheat than ever before. A large percentage of good land had not been seriously injured; only about 2 percent that was heavy textured had suffered enough erosion to require removal from production, and that could have been prevented with conservation measures.[74]

This optimism was challenged, however, when in September 1949 Secretary of Agriculture Charles F. Brannan belatedly announced the USDA's comprehensive recommendations for development in conjunction with the Missouri Basin program. He called for retiring some 20 million additional acres of cropland to grass and legumes; reseeding 17.5 million acres of depleted range; constructing 245,000 miles of shelterbelt and 750,000 acres of windbreaks, 407,000 stock ponds, and 14,000 to 16,000 reservoirs on headwater streams; increasing irrigated acreage from the existing 5 million to 12.5 million acres; assisting to rehabilitate existing irrigation projects and develop new ones; and continuing emphasis upon terracing, contour farming, and laying out grassed waterways.[75] It was a program that in large measure reverted to prewar adjustment planning.

Discussion of the proposals in the Great Plains Agricultural Council brought serious questioning. The expense of regrassing and introduction of livestock had raised concern as crop diversion was reinstituted earlier that year. A regional agronomist for the SCS warned that in areas of light rainfall three or four years were often required before grass could provide a substantial return. Other speakers argued that far more acreage than the customary half-section prevalent in eastern districts would be needed to provide a living from grass and livestock production. Starch noted that income from hay, even on irrigated land, had been unstable: when climatic conditions were favorable, the farmer found scant market for it; when the entire area was short of feed, prices were high, but the producer then had little surplus available for sale. Even a representative of the U.S. Bureau of Agricultural Economics questioned whether seeding the poorest land to grass would be profitable and whether supplies of such seed were adequate for planting more than a small part of the acreage proposed to be shifted out of wheat.[76]

The importance of maintaining feed reserves had been repeatedly stressed. How much was practicable now raised query. Under the prolonged weather cycles experienced in the region, reserves for a year or two, as generally recommended, would have aided little. Indeed, one analyst concluded that they would have worsened the stockman's plight by deferring his sale of cattle until the period of lowest prices. Longer storage entailed costs that ought to be weighed against the practicality of more flexible herd adjustments.[77]

Moreover, even operators in hazardous areas would resist substitution of grass for wheat. To be usable, grass must be accessible to livestock operations, be supplemented by other feeds, and be located where stock water was available. Areas remained, particularly in Montana and Wyoming, where water was inaccessible for livestock. For many farmers such a conversion of their enterprise would require large expenditures and better credit for purchase of special equipment, fencing, and foundation stock, as well as for water development. Land values and tax structures had been set on the basis of wheat income. Often families had moved to town and conducted farming from a distance, while enjoying the school and community services of town residence. Livestock production, one speaker commented, "is not very feasible without someone living on a farm."[78]

Starch protested that short-term productivity records should not be used to determine "submarginality"; and a conference on credit needs, sponsored by the Great Plains Council in 1954, echoed the view that "too much weight . . . [had] been given to price and weather conditions of the thirties." By the fifties the postwar population upsurge was also bringing recognition that the earlier population estimates had been "overly conservative, to say the least." In the thirties Warren S. Thompson and Pascal K. Whelpton had forecast a population of about 155 million in the United States by 1980. These estimates had formed the basis for reports of the National Resources Committee and for the census projection in 1941. Revising the forecast in 1946, Whelpton still placed the total at approximately 167 million in 1980, and the census estimate in 1946 for the later date was somewhat lower at about 165 million. Joseph S. Davis of the Stanford Food Research Institute, writing in 1949, questioned these projections and advised that the rate of growth had achieved proportions "only beginning to be recognized."[79]

Conceding that improvements in agricultural technology were expanding productive capacity, regional planners of the fifties emphasized program development that embraced expansion to meet potential need while it also permitted flexibility to respond to alternative circumstances. M. P. Hansmeier, of the North Dakota Agricultural Experiment Station, cited "flexibility," as a summation of the discussions in a planning committee of the Great Plains Agricultural Council in August 1949.[80] Later that year Starch, whose emphasis upon integrated development had long embraced this conception, provided a multifaceted interpretation of it in a paper titled *The Future of the Great Plains*, which reappraised the 1935 report of the Great Plains Committee.

Great Plains farmers, Starch wrote, must employ management principles that would enable them to "shift quickly and roll with the punch." All cultivation must be adapted to the requirement of moisture conservation. Use of stubble mulch was important not only as a technique for conservation, by increasing the soil's ability to absorb water and affording protection from wind action, but also as a means of adjusting the timeliness and scale of operations. If the spring season were dry, farmers might need to delay soil preparation and to feed cattle on stockpiled hay through a longer period. They might later, perhaps, supplement feed reserves with quick-maturing grains and silage. If drought persisted through July, however, Starch advised that they cull their herds rather than attempt to carry larger reserves or ship the stock to distant feeding areas. Integrated livestock and crop production and restoration of grass on less productive lands facilitated adaptive responses but also required increased size of units. In this connection, partial ownership of land provided a means of sharing risk. Financial arrangements—credit, insurance, and tax support programs—should be adjustable. While Starch regarded the statutory rigidity of zoning as inhibiting to flexibility, he noted that grazing districts in "areas quite obviously and permanently given over to range use" could promote it by providing for rapid changes in the level of stocking. Similarly, he added, conservation districts might through experience evolve controls "for a systematic shift in land use that will no doubt some day include the measures which will enable manpower to come into an area and quickly convert it to wheat; but with certain protective restraints."[81]

The basic expectations in this regional analysis were that cropping would remain central to the land utilization and that wheat was to provide the primary focus. "In the long run," Starch had commented in responding to remarks by a representative of the USDA on the prospective program changes, "wheat is about as good as anything, and adjustments ought to keep that in mind."[82]

Wheat specialization in this view combined drought resistance with marketing preference. The adaptive concerns of farmers in the northern Plains over the succeeding decades were, however, to be shaped more as responses to national problems of surplus and scarcity in wheat supply than as adjustments to drought. Under that consideration, regional production faced an uncertain future.

CHAPTER EIGHT

The "Yo-Yo" Marketing Problem

Problems of unstable market demand were not new to operators in the northern Plains. Concerns other than drought and soil conservation had been evident in the initiation of land-retirement proposals as early as the late twenties. Eastern and, in turn, midwestern grain producers had long deplored the opening of new lands to cropping. Their voices had joined those of western stockmen in calling for a halt on the basis of contracted wheat exports and falling grain prices after World War I. The Christgau proposal, in turn, had represented an effort to support a competitive base of regional claim over the remaining segment of the trade. Recurrent drought had supplemented the counteractive proposals, but it had not generated them.

Similarly, the aborted crop-diversion program recommended by Secretary of Agriculture Charles Brannan and partially incorporated in the Agricultural Act of 1948 had been dictated by the redevelopment of agricultural surpluses, not by pressure of drought. Food and feed requirements of the war and immediate postwar years had brought a lifting of restraints during the forties, and the reinstituted proposals were held in abeyance through the Korean War period. By 1954, however, accumulating grain surpluses again fostered demands for lower price supports, production quotas, and land retirement. At the same time, severe drought in the central and southern Plains gave focus to proposals for agricultural contraction, and by association they were expanded regionally to the northern states as well. But drought—defined as generalized, prolonged, and sharply diminished precipitation—was not a widespread problem in the northern Plains again until the early sixties, *after* the renewal of the program for "conservation reserve" in 1954 and the approval of the Great Plains Conservation Program in 1957.

Montana precipitation records, weighted according to the area of harvested cropland acreage within the various reporting districts, averaged only 14 inches from 1950 through 1986 and fell below that level 50 percent of the time. The diminished precipitation occurred consecutive years, however, only four times, including twice, from 1958 through 1961 and from 1983 through 1985, for longer periods. Such records for North Dakota averaged nearly 18 inches for the thirty-six-year period and fell below 15 inches only five times, none of them consecutive. For harvested cropland in South Dakota the average precipitation was 20.36 inches and fell below 15 inches only three times, none consecutive.[1]

State averages masked numerous divergences, but detailed mapping of

July 1 "Crop Prospects," as reported by the USDA Agricultural Marketing Service, did not present a repetitive record of localized hardship. Crops in 1955 were "Good to excellent" over all Montana and North Dakota, although only "Fair" in northern South Dakota. The following year crops were "Near failure" in central South Dakota and the tri-corner border district of southeastern Montana and the Dakotas; elsewhere through the region, they were "Poor" to "Fair." They were again "Good to excellent" over the whole region in 1957 and over most of South Dakota, southwestern North Dakota, and central Montana in 1958. Only the extreme northern border areas of northeastern Montana and northwestern North Dakota were "Poor" the latter year, the first of four recording critically low precipitation. By July 1959, "Poor" crops characterized southern North Dakota, the vicinity of Bismarck, and most of South Dakota, but they were listed as "Fair" for eastern Montana and "Good to excellent" for the central district of that state. They were "Good to excellent" for all North Dakota and northern and central Montana in July 1960, and "Fair" over the remainder of the region. Only the report for 1961, the last year of the most extended drought period recorded during the above-noted, thirty-six-year precipitation history, described the crops as "Short" throughout the region. Thereafter through the sixties, mapping showed the prospects generally "Good to excellent."[2]

"Pasture Feed Conditions," on the other hand, ranged from "Fair" to "Extreme drought" over all northern and western North Dakota and east-central and northeastern Montana in 1958; "Very poor" over most of South Dakota, central and much of western North Dakota, and northeastern Montana in 1959; only "Poor to fair" over western South Dakota and the southeastern quadrant of Montana in 1960, and virtually lacking throughout the region in 1961. After culling herds since 1956, ranchers were beginning to rebuild them in 1962, but pastures were still under "Severe drought" in north-central Montana again in 1963, and only "Poor to fair" in northeastern Montana in 1964. Low precipitation running from October to April throughout the period, excepting 1954, 1969, and the early seventies, retarded development of spring ranges, while peak rainfall from April into August concentrated moisture during the cropping season. The drought experience had most severely afflicted livestock producers, not grain growers.

Droughts, as C. W. Thornthwaite and J. R. Mather commented in the *Yearbook of Agriculture* for 1955 yet required study. They were "hard to measure." Little progress had been made from 1940 to 1960 in defining the limits to long-term, successful cropping. Notwithstanding the generally favorable conditions of the period, the mapping of average annual precipitation as published in the *Yearbook* for 1957 used the same chart presented in 1941 as a forty-year average covering the period 1899–1938. The slow progress in soil classification has been previously noted. New recognition had, however, developed concerning the complexity of "consumptive use" of water in plant growth. In the late thirties O. R. Mathews and John S. Cole had recognized that "up to a certain point" wheat yields increased proportionally to the quantity of water available above a minimum requirement. Subsequently, Cole had concluded that a minimum of about 8 inches of water, either as available soil moisture or precipitation, was necessary in the Great Plains for any development of spring wheat and that yields were increased at a constant rate of 2.2 bushels an acre for each additional inch of moisture. A Saskatchewan study placed the minimum as low as 5 to 6 inches and the yield at 14 bushels an acre with 10.5 inches of moisture. But research was clarifying that the minimums, as well as the accretions, varied greatly according to a broad range of factors—temperature, humidity, wind velocity, amount of sunshine, terrain, and soil fertility and structure, as well as crop and varietal selection and tillage methods in the agricultural operation.[3]

The increased attention given to irrigation in the postwar years stimulated efforts to measure the water requirements of comparative crops in varying locales. In western South Dakota, study revealed that alfalfa required 86 percent more water to produce a pound of dry matter than did wheat. Research in Colorado showed that the "consumptive use" of water for irrigating an acre of small grain during a growing season was 14.87 inches; that for corn, 19.66 inches; that for grass hay, 23.34 inches; and that for natural vegetation, 37.34 inches. Grasses and legumes not only had the higher water requirements but, accordingly, left the soil in much dryer condition for subsequent cropping. Those who on the basis of observation and experience had questioned the value of rotational diversification and green manuring in dry-land cultivation were acquiring statistical verification. It was also evident that, while such cropping might improve the condition of the soil in time, it would prove a slow and difficult operation under dry-land conditions. Discussion in the Great Plains Agricultural Council in the early sixties, at the conclusion of the most severe and prolonged shortage of rainfall in the past quarter century, continued to point to wheat and other grains as both the most economically competitive and the best adapted for agriculture in the region.

Speaking before the council in 1961, William R. Bailey and Dana C. Myrick of the Economic Research Service, USDA, questioned whether there was not a tendency to "sell the Plains short, looking for and emphasizing the chances for failure." They noted that Plains farmers in 1957, at a time of rapidly rising land and machinery costs, held about 90 percent equity in their assets, "exceedingly high equities" when compared with most other kinds of business. Wheat was their chief money crop, and at the current price relationship of wheat and feed grains they would produce wheat to the full extent of their allotments. For the northern Plains, in particular, where sorghum was not successful, there was no incentive to shift to forage crops. The primary forage for livestock in the area was native range, "almost entirely a residual of land unsuited to crop farming." Most farmers had some rough, poor land that they utilized as range, but to expand livestock as necessary for income equivalent to their wheat would entail much larger holdings, combination of existing units, and great reduction in the number of farms. Additional fencing, water facilities, and buildings would be required, while the investment in existing holdings of land and machinery would be devalued. If regrassing were involved, there would be several years of delay and increased risk pending its re-establishment. "We can only conclude that under most foreseeable circumstances, the grain farmer of the Plains will remain essentially a grain farmer."[4]

The experience of livestock producers in the northern Plains during 1961, when shortage of forage and grass, as well as feed grain, forced drastic herd reduction, re-enforced such views. Bailey and Myrick recognized that wheat occasionally failed completely, but noted that "at such times production of other crops is small too." Again addressing the council in 1966, Bailey conceded that wheat and barley under dry-land cropping gave highly variable returns, but he attributed that variability largely to the occurrence of years of exceptionally high yields. Combinations of grain crops and forage or grain crops and livestock presented less variability, but the "moderate reduction" was achieved only by sacrifice in income. A better strategy, he proposed, might be to accept variability as inevitable and seek ways to blunt the effect of poor years. "Exploitation of the latter surely warrants inquiry," he argued, "as a farm occasionally gets an average 3-year income in 1 year. One cannot imagine the Great Plains without wheat," he concluded.[5]

By the mid-fifties, however, U.S. farmers operated in an economy marked by surpluses of both wheat and feed grains. Between 1953 and 1961 allotments of wheat acreage were reduced about one-third, land diverted in Montana and North Dakota primarily to barley, but in 1961 the latter crop was also brought under the crop-reduction program. Wheat production nationally in 1958 was the largest crop to that date, and in the northern Plains it was among the largest. Although there was a sharp decline in Montana and North Dakota in 1961 and in South Dakota in 1962–1963, it was followed by surging increases during most of the remainder of that decade and still higher levels through the seventies and early eighties. Yields were reduced in Montana from 1969 through 1970 and from 1972 through 1973, in North Dakota in 1966 and 1970, and in South Dakota in 1970 and 1976, but they were severely contracted only in 1985 and 1988 in Montana and in 1988 in the Dakotas. As Bailey had argued, from 1974 through 1984 production in Montana and North Dakota was annually two and sometimes more than three times that of the subsequent poor years. Farmers had placed their less productive land in reserve under a variety of set-aside programs, and at the same time enlarged their productivity on fewer acres by fertilization, use of weedkillers in place of summer fallow, adoption of improved, higher-yielding crop varieties, and greater mechanization. Montana wheat yields, which had been considered profitable in the twenties at 12 bushels to the acre, averaged 27.7 bushels in 1968, 31.3 bushels in 1975, and 33.6 bushels in 1982.[6]

From 1940 until 1956 direct government payments designed to cope with crop surpluses were only a small proportion of gross farm income in the region, less than 1 percent in North Dakota and Montana and approximately that level in South Dakota. By 1970 they had climbed to 11 percent in Montana, over 17.5 percent in North Dakota, and about 8 percent in South Dakota as a variety of programs were instituted to alleviate the problem. In the mid-seventies the percentages of those payments fell as low as they had been two decades earlier, but by 1984 they had again risen—to 14 percent in Montana, 13.5 percent in North Dakota, and 6.6 percent in South Dakota. Government assistance took form in part as price support through purchases by the Commodity Credit Corporation as the farm parity ratio slipped from 100 in 1952 to 58 by 1984; but a large proportion of the direct payments also represented compensation to farmers for diverting acreage from cash crops to forage production and regrassing of pasture and range lands. Such expenditures, and vast sums spent in food stamp distributions and subsidized exports of surplus stocks, were dictated less by regional production handicaps than regional production capabilities, less by microeconomic than macroeconomic considerations, compensation for individual restraints entailed under national policies.[7]

After the Korean War, Secretary of Agriculture Ezra T. Benson reinstituted quotas intended to limit production of wheat beginning with the crop of 1954. The national base, an acreage rather than a production standard, was again set under a formula establishing a minimum at the prewar allotment of 55 million acres and remained at that level through 1963. Quotas for individual states, as before, were apportioned by the USDA according to their production history over the past three years, and within those allotments, state production and marketing (formerly ACP) committees apportioned county shares under the same standard. Wheat acreage allotments between 1939 and 1961 were accordingly reduced over 11 percent in North Dakota and 7 percent in South Dakota but increased 17.5 percent in Montana—and more than 40 percent in New York, Vermont, Michigan, Alabama, and Mississippi among the eastern states and in Colorado, New Mexico, and Utah to the west.[8]

Under a marketing exemption for small producers, some 4 million acres, largely in the Corn Belt, were also cultivated in wheat during the period, independent of historical production rights. Regional wheat growers tended to attribute the mounting surpluses to such production, but analysts argued that it added little to the commercial supply, since it was used largely as stock feed on the farms where it was grown. Commercial stocks were derived primarily from production on the Great Plains—61 percent of the national wheat crops as an average from 1946 through 1955; 64 percent from 1958 through 1962, after the quotas were reinstituted; and as much as 67 percent in the bumper-crop year of 1958. In structure, the quota arrangement had not imposed geographic criteria as direct restraints upon production. Indeed, the program evidenced some shifting of allotments toward the states of the western Plains and within North Dakota, at least, a disproportionate decline of such cropping in eastern counties.[9]

Meanwhile, under a shifting system of price supports, all wheat growers who accepted their acreage allotments were assured of seasonal minimum prices according to their yields. If, under referendum, quotas were approved by two-thirds of wheat growers, supports for 1954 were guaranteed at 90 percent of parity. Flexible lowering to a range between 82.5 and 90 percent of parity was authorized for the crop of 1955 and between 75 and 90 percent subsequently. For 1956 the support level was set at 76 percent of parity; for 1957, 83.7 percent, when in consequence of election-year political pressures only a presidential veto had forestalled enactment of legislation restoring the 90 percent minimum. Thereafter the support base held at 75 percent of parity until 1964.[10]

The Agricultural Act of 1961, however, had required approval of a 10 percent reduction in allotments, as carry-overs of wheat and feed grains in the United States remained at record levels. Because of low yields in the region that year and rapidly expanding exports beginning at the turn of the decade, market prices continued to be strong. The proportion of the crop placed under the support program declined from 42 percent in 1958 to only 22 percent in 1961 and about 27 percent in 1962. Feeling less dependence upon the arrangement, wheat growers rejected quotas for the following year. The problem of disposing of surplus stocks through export while maintaining domestic support prices well above world levels led Congress to institute a multi-price support system. In 1964 participating growers were guaranteed supports at 65 to 90 percent of parity, in operation about 79 percent, on their proportion of the grain used domestically, about 45 percent; support of 0 to 90 percent of parity, in operation about 61 percent, on their estimated proportion of the exports, also about 45 percent; and 0 to 90 percent of parity, in operation about 52 percent, for the remaining 10 percent, the estimated surplus. Only those growers who remained within their allotments and agreed to participate in a land-diversion program would receive payments.[11]

Under the still-applicable Agricultural Adjustment Act of 1938, farmers continued to receive payments for practicing cover cropping; cultivating in strips and according to contours; establishing terraces; planting trees; seeding range and pasture lands; constructing dams, small irrigating works, and grassed waterways; providing stock-water ponds and wells; and a variety of other approved measures deemed desirable for improved land utilization. While focus on such activities had centered in the program of the soil conservation districts during the forties, PMA committees remained a vital force in local administration of agricultural policy. In 1953 the two programs were brought together within one agency for agricultural stabilization under the departmental reorganization of Secretary of Agriculture Ezra T. Benson. At the same time the regional offices of the Soil Conservation Service were eliminated. The move enhanced the role of local committees and their programmatic activities through the succeeding decade, just as efforts to restrict surplus cropping were generating increased focus on conservation as an alternative to production. Between 1954 and 1958 annual payments for ASCS programs increased from about $1.5 million to $3.7 million, for a five-year total of $14.1 million in eastern Montana; from $2.9 million to nearly $4 million, a five-year total of $20.6 million in western North Dakota; and from $2.6 million to $4.8 million, a five-year total of over $22.4 million in western South Dakota.[12]

As acreage restraints on cash-crop production were reinstituted, planning for use of diverted cropland directed renewed emphasis to cover cropping and pasture development. Such a program derived additional impetus from the fact that after 1955 national surpluses of feed grains were even larger than those of wheat. Under the Soil Bank program established by the Agricultural Act of 1956, farmers were compensated to divert land below the level of their established allotments or, as applied after 1957, their base acreage—technically defined as nonconserving crop acreage; more broadly stated as all crops excepting hay harvested on the farm during the two preceding years— and to devote the diverted cropland to "conserving uses." Until 1958 farmers could choose to place the diverted acreage in either an "acreage reserve," which permitted planting the land in alternative crops as long as they did not exceed the allotments, or a "conservation reserve," which required that the overall crop acreage be reduced, with the excess taken out of production completely, including use for pasture or grazing. Under the latter arrangement farmers contracted to provide a cover crop adequate to give "good protection from wind," primarily legumes or perennial grasses but, in fewer instances, also trees. The program of acreage reserve contracts, drawn for periods ranging from one to ten years, was dropped in 1958 because of the expense for so limited a benefit. Conservation reserve contracts, running from a minimum of three years to a maximum of ten years, were continued. Payments encompassed two sums: rent under a government lease arrangement at sums recommended by the local committees under stipulated guidelines and a 50 percent share of the cost of any remedial work in cover cropping. Participating farmers who brought noncropland into production during the life of a Soil Bank contract were subject to loss of payment according to the number of acres of new land.[13]

Proponents of the program assessed it enthusiastically in 1960, the final year under its original authorization. They anticipated that vast amounts of the acreage would remain in grass after termination of the contracts. Nationally, they claimed, 28.6 million acres had been taken out of production, about one-third of it in the Great Plains as a whole. Of the latter, only 2.8 million acres were in the northern states. A total of 1,820 farms had been brought under the program in eastern Montana, 6,013 in western North Dakota, and 3,964 in western South Dakota. The average rental cost for the Montana land, at an average of $9.04 an acre, amounted to $5,257,334 annually; that for North Dakota holdings, averaging $9.15 an acre, came to $13,580,159; and that for South Dakota tracts, at $8.69 an acre, $6,681,185. The bulk of the land also required "conservation cover," financed under government ASCS payments, averaging from $4.11 an acre in Montana to as much as $11.04 an acre in South Dakota, on a cost-share basis. Even under limitation of maximum payments to $5,000 a farm, it was an expensive operation. Moreover, the General Accounting Office at that time reported that 23 percent of the land listed in the "conservation reserve" had never been in production or had no cropping history. Analysts of the Economic Research Service of USDA estimated that the reduction in acreage of all harvested crops amounted to only about one-half of the acreage placed in the reserve.[14]

The program had not resulted in the intended reduction of crop surpluses. As the United States entered the harvest of 1959, a surplus of over 1.25 billion bushels of wheat had already accumulated, enough to fill the domestic requirements for two years. The government had some $3 billion invested in stocks held off the market under price supports, and storage costs amounted to about $150 million annually. Seeking expansion of the program in 1960, the administration urged that at least 45 million acres be retired. Analysts of the Agriculture Committee of the National Planning Association estimated that 60 million acres would need to be withdrawn to eliminate the surpluses.[15]

Farmers had pragmatically diverted their poorest cropland to the reserve. Allotments assigned on the basis of crop acreage, coupled with support prices based upon yields, had led to intensified inputs for maximized returns from the better quality lands. Even summer-fallow acreage, for which diversion payments were only one-half those for cropped land, was being brought into production through use of herbicides to reduce moisture loss in weeds. One strong advocate of soil conservation noted, moreover, that much acreage committed to programs designed for "cropland" withdrawals in actuality fostered plowing up areas that had earlier been regrassed, so that after a short period of cultivation, the tracts might again qualify for government rental payments.[16]

The conservation reserve had also engendered strong regional dissatisfaction. To encourage large-scale land retirement, annual payment rates might run as much as 10 percent higher for contracts covering whole farms than for those on part farms. Whole-farm contracts permitted operators to live on the land and use the permanent pasture, woodlands, and orchards, but barred all cropping. Since the arrangement supported traditional range operations, a large segment of the contracts in the Plains area as a whole, but especially in the northern states where forage and feed grain production af-

forded little alternative as diversion cropping, were whole-farm contracts—
75.5 percent in eastern Montana, 71.7 percent in western North Dakota, and
60.5 percent in western South Dakota. But the $5,000 limit to individual diver-
sion payments and the restrictions against running stock on the diverted acreage were unpopular. Studies by the Montana Agricultural Experiment Station
showed that whole-farm contracts were generally a less profitable adjustment
for the dry-land grain-livestock operators than limiting their acreage in conservation reserve to cropland normally in feed grains and summer fallow,
while combining livestock with a full wheat allotment.[17]

The program was most strongly opposed on the ground that it promoted
depopulation, thus curtailing community and business activity and reducing
the tax base for local government. The managers of the Williston, North Dakota, electricity and telephone cooperatives predicted a need to raise rates.
The operator of the local grain elevator argued that as farmers relinquished
their holdings, they would crowd the town labor market and drive down
wages. The program would "hit the implement dealer, gasoline and tire dealers—then the grocer and clothier as unemployment grows." Farmers, too,
pointed out that idle land was a nuisance to the community, a source of weeds
and insects and a fire hazard. The combination of heavy cost and constituent
protest led Congress in 1960 to reduce funding for the program, so that few
new contracts could thereafter be initiated. Those already operating were
continued and in some cases ran as late as 1972, but the acreage in the reserve
steadily declined through the sixties. By 1969 it amounted to only 3.4 million
acres nationally.[18]

While the exemptions authorized under the whole-farm contracts and
the imposition of acreage, rather than yield, limits on crop production were
features of particular relevance to agricultural operations on the Plains, the
ACP/PMA/ASCS and Soil Bank programs were national in application and
were intended to extend environmental considerations generally. The drought
in the central and southern Plains during the mid-fifties however, led to development of more specifically oriented measures under what was known as
the Great Plains Program. This undertaking, launched in March 1957, had
been proposed nearly two years earlier at a meeting in Denver called by the
secretary of agriculture in conjunction with the Great Plains Agricultural
Council. Enacted under Public Law 1021, and occasionally so identified, the
program authorized expenditure of $150 million over fifteen years to make
long-term land use changes possible in that region.[19]

Administrative leadership and responsibility in the effort were assigned
to the SCS, and in large measure the programmatic activities of that agency
were emphasized in it. Where "land adjustments" were deemed desirable,
the program was very similar to "Conservation Reserve." It, too, called for
negotiation of contracts ranging from three to ten years, for application of
conservation procedures, and for federal cost sharing, ranging from 50 to a

maximum of 80 percent, in application of measures for soil and water improvement. Again the hope was that such funding would encourage farmers and ranchers to convert cropland to perennial grasses and permanent cover.

Important distinctions, however, emphasized that the program was applicable only in counties specifically designated by the secretary of agriculture as requiring conservation measures, improved cropping systems, and land-use changes; that plans must be developed to encompass entire farms or ranches; and that a system of tiered planning effort be formulated, extending upward from county and state advisory committees to the federal administering agency in the manner of the local planning activity of the late thirties. As projected by a planning committee of the Great Plains Agricultural Council, acting in an advisory role, county committees were to be established at the call of the county agricultural extension agents and to be comprised of local representatives of the SCS, ASCS, Farmers Home Administration, and other action agencies, together with "key agricultural leaders" representative of laymen and organizations in the community. Representation was to be "adequate in terms of problems and interested leadership" and to encompass spokesmen for farmers, ranchers, members of SCD boards of supervisors, county ASCS committees, county FmHA committees, county extension boards, boards of county commissioners, business organizations, credit agencies, farmers' organizations, commodity groups, and agricultural service bodies. The committees were to formulate their own rules for "operation and continuation." State committees, in turn, were to be composed of representatives of the federal and state agencies, the county advisory groups, and "such other . . . state and local organizations, agencies and groups as may be needed to provide effective representation within the state." Officers of the state bodies were to be elected from the lay membership.[20]

County advisory committees were to be charged with promoting community participation in development of local programs, with identifying the local problems and deciding on the priority in approach to them, and with establishing a land capability classification system which local operators would "understand, accept and use." State committees were to provide "leadership and guidance" to the county committees, to review problems of state and regional scope and offer proposals for their resolution, and to afford liaison between the various levels of program development.

There was some mention within the Great Plains Council that this organizational structure replicated a planning program of considerable local involvement at an earlier day, but local newspapers present little indication of any comparable community activity under the Great Plains Program. As described by the assistant administrator of the SCS, the administrating agency, county committees were, in fact, involved mainly in details of processing applications, assembling cost-share rates for approved practices, establishing standards for plans and contracts, coordinating local projects in the work plan of the SCS, and arranging for cooperation among the involved local agencies. That the committees operated in the planning process to shape the focus of development under the program is questionable. That they provided widespread popular participation in the process is still more doubtful.[21]

Work under the program centered on planned, whole-farm management with the technical guidance of the SCS as that agency had hoped to develop its program with cooperators in the SCDs. Farmers were given the option whether or not to participate; but when they accepted a contract, they agreed to conduct for its duration the full range of recommended conservation practices. They were not free, as under the ACP/PMA/ASCS programs to choose only part of the proposed measures or to commit only a limited segment of their operation. Unlike the restrictions under the "Conservation Reserve," however, these contracts included tracts other than cropland and, where conditions were satisfactory, they could be used as pasture or range. Indeed, under the regional focus of the SCS program, development of grasslands for such usage received primary attention; annually recurring expenditures relating to reformed tillage practice were to be left to the program of the ASCS. The large share of the expense that the federal government was prepared to assume and the length of the contract period were designed to assist farmers in bearing the cost of the slow process of re-establishing grass on depleted range and broken sod lands. The SCS estimated that 11 million to 14 million acres of cropland in the Plains region were unsuited to cultivation. While acreage allotments and marketing quotas as established under the AAA were to be reserved for resumption at the expiration of the contracts, it was anticipated that the Great Plains Program would constitute a new approach to cropland retirement.[22]

But the program served more to provide for rangeland improvement than cropland adjustment. In the congressional hearings on proposed enactment of the program, Assistant Secretary of Agriculture Ervin L. Peterson had projected that 75 percent of the funds, $112 million, would be spent to convert 10 million acres of cropland to grass; but by 1964 only a little over 1.1 million acres out of over 33.8 million acres under contract in the program represented cropland, and USDA analysts estimated in 1970 that only about 15 percent of the program funding had been devoted to cropland conversion. Although 774 contracts, covering some 3.4 million acres, had been signed in Montana by June 30, 1964, they encompassed conversion of only 116,000 cropland acres. In South Dakota 688 contracts covered 2.4 million acres, but entailed conversion of only 93,000 cropland acres. More numerous contracts, 1,842, in North Dakota covered 2.4 million acres, including conversion of 229,000 cropland acres.[23]

The Great Plains Program has been continued under various extensions, and by 1979 the chief of the SCS reported that 9 million acres of cropland had been converted to grass. Two years earlier, however, the General Accounting

Office reported that 26 percent of the farmers under the program had already plowed their new grassland or planned to do so when their contracts expired. A recent survey by the SCS showed that 30 percent of the participating farms and ranches had been primarily livestock operations; 53 percent, combination livestock and crop enterprises; and only 10 percent crop and cash-grain farms. Even doubling the number of counties originally projected, thus drawing under the program districts averaging as much as 20 inches of annual precipitation, more prairie than plains in environmental character, had not shifted the nature of the participation.

Expanded soil mapping was a crucial aspect of the planning effort. The Denver meeting had called for detailed soil survey and land classification for each locality in the Great Plains but conceded that reconnaissance work would have to be used until such analysis could be accomplished. Despite repeated emphasis upon this need over the years, relatively little had been done. In 1955 only 42 million acres of the 122 million acres in SCDs in the critical wind-erosion areas of the Great Plains had been surveyed. During the last six months of that year the place had been quickened to cover 1.75 million acres, a 78 percent increase over the comparable period in the previous year. Under the Great Plains Program the annual rate was increased from a total of 12,750,000 acres in 1955–1956 to 24.3 million acres in 1960–1961. At the latter date it was estimated that mapping of the Plains region as a whole would be completed in sixteen to eighteen more years, at the current rate of progress.[24]

The situation also led to a comprehensive effort to inventory the nation's soil and water resources. The SCS had initiated less inclusive surveys in 1943, 1948, and 1952, to assess the capabilities for meeting the food requirements in those crisis years. As the Great Plains Program was being developed in 1955, the agency proposed that a review of current land and water conservation problems, including the extent of acreage needing treatment, be compiled on a county basis for the nation as a whole. It was to cover all nonfederal land in use or available for production of food and fiber, with separate data on irrigated and nonirrigated cropping, and also to provide information on a watershed basis covering the nature and scope of water management problems suitable for treatment under the legislation for small-water development. Procedures were to be instituted to keep the record current.[25]

The Conservation Needs Inventory (CNI) was under way in 1957 and continued over the next five years. It rested on the system of land-use classification introduced by the SCS in the late thirties, but in the absence of detailed soil mapping, utilized sampling, at the rate of 160 acres to the county. Largely accomplished by workers described by the SCS as "technicians well acquainted with local conditions," these surveys were presumably the work of the county committees operating under the various federal action programs. Critics have noted that the attitudes of collecting and analytical personnel influenced the report and have commented that the hazards may have been "slightly overrated." At best, the mapping only broadly defined the land within the eight basic categories of the SCS classification, four identifying quality of cropland and four, the worth of grazing land.[26]

The extent of cropping in Class IV, the marginal grade for agriculture, and upon land under the grazing categories signaled the focus for readjustment efforts. The acreage under individual grazing classes seems not to have been delineated; it was certainly not publicized. Of slightly less than 43 million acres of Class IV cropland in the Great Plains as a whole, about 16,116,000 acres were being cultivated—scarcely a surprising amount for a category classified as basically agricultural, notwithstanding that it was identified as marginal. Only a small fraction of this acreage was in the northern Plains: about 440,000 acres, primarily in the southern counties, of western North Dakota; about 1,650,000 acres in eastern Montana; and approximately 1,520,000 acres in western South Dakota. More disturbing was the fact that an additional 10,625,000 acres, 9.6 percent of the cropland in the Great Plains as a whole, were being cultivated in areas classified for grazing. In eastern Montana about 990,000 acres, 8 percent of the cropland; in western North Dakota about 910,000 acres, 6 percent of the cropland; and in western South Dakota about 630,000 acres, 9 percent of the cropland fell under grazing classifications. Those classes, however, included areas cultivated in alfalfa and perennial grasses as pasture or range.[27]

Many in the Great Plains Agricultural Council during the early fifties had argued that dry-land cropping techniques could adequately meet the environmental problems of wind erosion. Reputable agricultural scientists had emphasized, too, the particular adaptability of wheat to the conditions of the region and the difficulty and expense of regrassing rangelands. Such arguments were less frequently voiced as the federal program was promulgated, but they were not absent. In 1956 the Tenure Committee of the Great Plains Council questioned whether the wheat allotment program was not shifting such production from areas where the crop was well adapted and where alternatives were limited to other, less suitable areas of the country. A speaker at the council meeting in 1959 devoted his whole paper to the propositions that the Plains could better than any other region supply the hard wheats needed for bread flour; that wheat cropping contributed to soil maintenance and the availability of stock feed; that the area, unlike many others, had few alternative enterprises; and that this land and its people were an important segment in the national economy whose interests ought to be considered.[28]

Environmental concerns calling for land retirement had less support among regional agricultural leaders than was operative in Washington as an approach to reduction of crop surpluses. The CNI made evident, too, that the acreage requiring such conversion was only a small fraction of the amount necessary to be taken out of cropping if surpluses were to be notably reduced. Other programs, developing changed crop disposal as well as production patterns, now began to command greater attention as measures to alleviate the problem.

Use of surplus commodities for purposes of domestic relief had been started in 1934, when President Franklin Roosevelt authorized distributions from these stocks by the Federal Emergency Relief Administration. A year later Congress assigned to the USDA, as an amendment to the Agricultural Adjustment Act, 30 percent of all customs receipts, the so-called Section 32 funding, for use in supporting farm prices, either by subsidizing exports or by diverting agricultural commodities from domestic channels of trade through payment of benefits or indemnities or "other means." By 1939 some 13 million people were receiving assistance in school lunch and other relief activities through use of the federal holdings. A food-stamp program begun at this time introduced normal marketing channels into the process but still, directly as well as indirectly, rested upon the availability of the government stores. Eligible relief recipients were entitled to buy orange stamps to a value approximating their normal food budget and to receive as a bonus blue stamps equal to about half that amount. The orange stamps could be applied to purchase of any food items; the blue ones were to be used only for commodities identified by the USDA as surplus.[29]

These programs had been discontinued by 1943, as the prosperity of the war period reduced both the accumulation of food surpluses and the need for public welfare assistance. With the end of the war, however, a National School Lunch Act was adopted, financed on the basis of three dollars in state funds to every dollar from the federal government. Amended repeatedly through the sixties, the legislation was designed to reach needy school children and expanded to encompass school breakfasts and food service at such nonprofit institutions as preschools and day-care centers, settlement houses, and recreational facilities. The federal aid took form in monetary outlays as well as donated food stocks. By 1969 such disbursements amounted annually to nearly $230 million from the federal treasury but also some $224 million worth of food from the stores of the USDA.[30]

In 1949 direct distribution of commodities from government holdings acquired under price-support programs was authorized for charitable purposes on a permanent basis. The USDA liberalized administration of the program in 1953 by agreeing to pay shipping costs to state welfare agencies, and legislative amendments later in the decade expanded the categories of eligible recipients, permitted the Commodity Credit Corporation to process the grains into flour and meal, and increased the range of available items. By 1959 flour, corn meal, dried milk, rice, butter, and cheese to a total value of $75 million monthly were being distributed in some 1,200 counties of the nation.

With record carry-over wheat stocks amounting to more than 1.3 billion bushels, including CCC holdings of nearly 1.2 billion bushels, on July 1, 1960, the Kennedy and Johnson administrations of the sixties gave new emphasis to food relief activities.[31] Besides liberalizing the school lunch program, Kennedy, in his first executive order, doubled the range of commodities available to welfare recipients. Under persistent pressure by a group of congressional liberals and spokesmen for ethnic minority interests, Johnson ultimately expanded this list to twenty-two commodities. Meanwhile, however, the focus of the public welfare effort shifted to a renewed food-stamp program.

Again by executive order during his first month in office, Kennedy revived use of section 32 funds to finance a three-year trial of the program. Expenditures increased from $3.7 million during the partial fiscal year ending 1961 to $45 million during fiscal 1964. Under President Johnson's urging Congress finally specifically funded the program in 1964, largely as an alternative to direct distribution of surplus commodities and under a bargaining arrangement with those who sought preservation of high commodity price supports. Since local governments had to bear only minor administrative costs under the arrangement, rather than the cost-share obligation of direct commodity distributions, they very generally embraced the new plan as a replacement for commodity distributions. For what they would normally have spent on food, families whose need was certified by local administrators according to state welfare guidelines were authorized to purchase stamps of greater value. The cost averaged $6.00 to $7.00 for every $10.00 worth of stamps, and even the poorest recipients were required to pay at least $2.00 monthly for the stamps. The difference between the purchase price and the stamp value represented the federal contribution.

By 1967 nearly 1.4 million individuals were receiving the stamps monthly in forty-one states and the District of Columbia. At that time the program was renewed for another three years, and the minimum purchase requirement was reduced from $2.00 a person to $ .50, with a $3.00 payment stipulated for households of more than six members. Under subsequent administrations the program was still further liberalized. The number of recipients was increased from 2.5 million in 1968 to 14.3 million by 1974 and to 17 million by 1977, when the public expenditure for the program amounted to $5.1 billion annually.[32]

Although initiated as an important phase of national effort to reduce poverty, politically and administratively the emphasis rested upon the availability of crop surpluses and the coalescent desire to reduce them. Responding to senators who sought greater variety in the distribution of commodities for domestic relief, Assistant Secretary of Agriculture Clarence L. Miller explained in 1959 that the program applied only to surplus commodities: "If it is not coming into the inventory of Commodity Credit Corporation it is not available. We are happy when that situation arises. Our prime responsibility is the operation of the price support." Secretary of Agriculture Orville Freeman, in particular, has been severely critcized on the ground that, in his efforts to placate conservative southern congressional leaders, he lagged in liberalizing the food programs. Advocating adoption of the food stamp program before a Senate committee in 1964, Freeman submitted studies showing that food stamps generated greater consumption of food and feed grains than did commodity donations. The farm value of food consumed under direct donations, it had been found, was only $1.75 per person per week, while that for foods acquired under food stamps came to $2.01. Explanation for the difference raises questions, but the statement was indicative of the motivating concern. President Carter's later efforts to link such distributions with a Better Jobs and Income Program were rejected.[33]

Farm surpluses were also reduced through export development. In the immediate postwar years, U.S. exports of wheat to Europe had been large—some 44 million bushels to France, 39.5 million to Italy, and over 15 million, each, to Belgium, Yugoslavia, and Greece during the crop year 1945–1946. Following the severe winter and spring of 1946–1947, aid to France and Italy was also again large, amounting to nearly 34 million and over 35.5 million bushels, respectively. In the latter year nearly 26 million bushels were also sent to West Germany, with additional shipments of 92.6 million bushels in 1947–1948, 115.6 million in 1948–1949, 77.6 million in 1949–1950, and 61.7 million in 1950–1951. Beginning in 1947, large shipments, amounting in 1949–1950 to nearly 57.9 million bushels were also sent to Japan.[34]

Most of these shipments represented relief which was sent in fulfillment of international commitments, to an extent that depleted the available grain reserves. By the early fifties, however, trade with most western European states stabilized at less than 15 million bushels a year, individually, except for that with West Germany and the United Kingdom which ranged moderately higher. As surpluses again began to accumulate, an International Wheat Agreement (IWA) was put in operation in 1949–1950, which with periodic revisions continued into 1968. Under this arrangement, the signatory exporting countries agreed to supply specified quantities of wheat at a stipulated maximum charge, and importing nations agreed to buy set quantities at a price no lower than an established minimum. Between those limits, a spread of about thirty cents a bushel, negotiation would presumably operate. Lack of participation by the Soviet Union and, until the sixties, by Argentina posed serious limitations to the usefulness of the arrangement as a market regulator. The actions of individual governments in subsidizing domestic shipments at prices above the established international trading levels still more seriously and persistently counteracted the stabilizing effort.[35]

Most nations subsidized a guaranteed price to domestic wheat producers. This was notably the case among importing countries, which sought to encourage home production. It was less common among exporting nations, but of those who did, the United States offered much the highest subsidies among the major exporting nations. Designed to meet competition at the port of destination, they varied according to the price by class and grade of wheat at the port of shipment and averaged $ .55 a bushel in 1949–1950, ranging to $ .67 in 1950–1951, $ .75 in 1954–1955, and $ .81 in 1956–1957. With U.S. exports under the IWA amounting to nearly 266 million bushels in 1950–1951, subsidy payments by the U.S. government that year came to $178,179,000.[36]

Such payments inhibited not only adjustments in domestic production and use of the grain but also in bargaining within the range of treaty commitments under the IWA. During the first four years the maximum prices had been set too low; market levels were so much higher that subsidization was inevitable if the export commitments were to be honored. Revised in 1953, the minimum price was set so high that the United Kingdom withdrew from the agreement; underdeveloped importing nations would not, and frequently could not, meet their purchase contracts; and the United States, in an effort to move mounting surplus stocks, introduced a variety of alternative purchase and transfer arrangements. This country's shipments under the IWA, which in 1950–1951 had amounted to about 70 percent of total U.S. wheat and flour exports, fell to only 20 percent in 1960–1961, and transactions under the agreement declined from 40 to 25 percent. It was generally concluded that the IWA had been a failure.[37]

While wheat production in the United States during 1952 reached the second highest level to that time, total exports of wheat and flour by that year had declined 36 percent below the surge of the postwar relief effort. This situation had given rise to such measures of production control as re-imposition of wheat acreage quotas in 1955 and establishment of the Soil Bank in 1956. It also stimulated aggressive efforts to increase domestic exports. Since Canada, which was the principal wheat-exporting country next to the United States, had extended subsidies to export sales outside her commitments under the IWA, the United States in 1953 adopted a similar policy. Enactment of the Mutual Defence Assistance Act, otherwise known as the Mutual Security Act, Public Law 665, and the Agricultural Trade Development and Assistance Act, Public Law 480, in 1954 introduced or expanded other trading mechanisms by which the U.S. share of total world exports of wheat and flour increased from about one-third in the early fifties to 38 percent at the end of the decade.[38]

In addition to provision for military and development assistance for non-Communist nations throughout the world, the Mutual Defence Assistance Act included authorization, under Title IV, for furnishing commodities to the Allied joint-control areas and, under section 502 of Title V, for utilizing foreign currency in development of new markets and stockpiles of defense materials. Under these provisions sales of wheat and flour were relatively heavy through 1960, but with authorization for acceptance of foreign currency as provided in Title I of the Agricultural Trade Development and Assistance Act, they quickly rose to a peak of nearly 446 million bushels by 1964. Title I under the latter measure likewise permitted credit sales of agricultural exports, first extended through the Export-Import Bank in 1934, but now to be offered by the Commodity Credit Corporation. Other sections provided for donations to assist in disaster relief or programs of economic development undertaken on a government to government basis or through voluntary relief agencies and for barter exchanges in development of stockpiles of strategic materials.[39]

Arrangements under this last provision for several years represented a major channel of trade, accounting for the transfer of over 86.7 million bushels of wheat and flour at the peak in 1956. Such transactions, however, proved highly controversial in relations between the United States and Canada, the principal competitor for American grain exports. The U.S. government did not directly negotiate such transfers, but operated through private traders, who at first were given great latitude to arrange transactions offering the most profitable commissions to the intermediaries. Third-party countries were accordingly drawn into arrangements by which the suppliers of the bartered materials were frequently not the recipients of the agricultural exchanges. Contracts became increasingly distributive. Canadian authorities protested that such trafficking, often developing as "tie-in" arrangements, presented competition that damaged their wheat trade and their sales of strategic materials such as lead and zinc. Some Americans, too, claimed that barter was being used for transactions that might have been handled through regular commercial channels for money. In April 1957, the CCC tightened restrictions on barter contracts, and under congressional mandate the following year the USDA was directed to "cooperate with other exporting countries in preserving normal patterns of commercial trade with respect to commodities covered by formal multilateral international marketing agreements." Thereafter, the volume of such exports declined sharply and by 1967 had ended.[40]

Wheat and flour exports of the United States under the special programs of Public Law 480 and the Mutual Security Act amounted to approximately half a billion bushels during the fiscal years beginning 1961 through 1965, and at their peak in 1964–1965 accounted for 78 percent of the country's total wheat and flour trade. During the mid-fifties such shipments to countries linked in the various defensive alliances were already large—to the United Kingdom, Japan, Taiwan, Turkey, and Brazil; but under the Marshall Plan they had also been extended in considerable amount to borderline Iron Curtain states, such as Yugoslavia and Poland. Public Law 480 assistance became particularly notable with negotiation of an arrangement for supply of 3.5 million metric tons of surplus wheat, rice, cotton, tobacco, and dairy products to India in 1956. Although not nearly so large, exports to Pakistan were also extensive beginning in 1957, as were those to Egypt from 1959 through 1962. The trade to India remained very large and from 1962 through 1967 was massive, ranging to 11.7 million metric tons during the fiscal year 1966–1967. Exports to Korea and Japan were also increased markedly under the impact of the Food for Peace Act of 1966, amending and supplementing Public Law 480, as an aspect of the mounting international tensions in the Far East.[41]

Marketing of American wheat from the West Coast, long retarded because of the unfavorable structuring of transcontinental freight rates, now opened new trading opportunities for trans-Missouri grain growers. As when livestock marketing had shifted westward, the relative absence of established terminal markets in the region contributed to innovation. Aggressive marketing was not limited to federal programs focused on disposal of CCC holdings. Most of the wheat-producing states established wheat commissions in the late fifties to assist in the search for new markets. Cooperating through organization of regional associations serving growers of the Great Plains and the Northwest, such agencies maintained offices in Lima, Karachi, New Delhi, and Tokyo, as well as Washington, by the summer of 1960.[42]

For producers in the inland Plains, transportation costs were a persistent concern, and the problem was magnified in the effort to compete internationally. Hard Spring Wheat growers of Montana and North Dakota complained that because the Canadian government subsidized its railroads the costs of shipping the American crop to the West Coast were three to four times higher than were those of their Canadian rivals. Hauling by truck, which for many Plains producers opened direct access to terminal river ports and downstream water transportation, was of limited importance for the farmers of the northern Plains. Meanwhile, the accumulation of massive surpluses at local elevators, because the development of mechanized harvesting and farm trucking brought the crop to sale within a few weeks, led to emphasis upon provision of farm storage facilities, financed by government construction loans and maintained by government storage payments. Railway companies, after long denouncing the competitive inroads of motorized hauling, found that the combination of local trucking with large-scale farm storage opened possibilities for massive bulk shipments. By the mid-sixties trainload shipments directly from loading point to domestic destination or port of export, without intervening transit stops, were by-passing older terminal centers even in marketing eastward. Over the next two decades, some forty so-called rail ports were established in Montana. "Unitrains" as long as fifty-two cars, hauling 176,000 bushels of wheat on one run, made the trip directly from local markets to shipboard, at a saving of fifteen cents or more a bushel for shippers.[43]

Total U.S. wheat and flour exports, which had averaged some 321.6 million bushels annually from 1950 to 1954, increased to an average of 449.3 million bushels over the succeeding five years, to 716.1 million through the period 1960–1964, and to 696 million as an average from 1965–1969. By July 1, 1967, the national wheat stocks had been reduced from a carry-over of 1,411 million bushels in 1961 to 424.4 million bushels. The amount was so low as to threaten disaster if domestic production should be curtailed by damage from insects, disease, or drought. Although discouraging fears of possible shortages, officials of the USDA increased the authorized planting allotments by 15 percent for the crop of 1967. Secretary Freeman, who estimated that a stock of some 600 million bushels was desirable, sought authority to build such a stockpile; but Congress, responding to fears of farm groups that the reserves would hold down prices, rejected the proposal. At the same time, however, congressional leaders denied USDA requests to continue the "cropland adjustment," land-diversion program, reinstituted in the Food and Agricultural Development Act of 1965.[44]

TABLE 8.1. U.S. Exports of Domestic Wheat and Flour Equivalent

Fiscal Year Beginning[1]	Thousand Bushels[2]	Fiscal Year Beginning[1]	Thousand Bushels[2]
1945	261,362	1968	539,946
1946	255,483	1969	605,552
1947	407,116	1970	728,235
1948	435,082	1971	621,228
1949	287,433	1972	1,167,494
1950	345,558	1973	1,141,545
1951	455,649	1974	1,030,231
1952	316,854	1975	1,154,933
1953	216,381	Transition	323,276
1954	273,634	1976	919,995
1955	345,565	1977	1,215,837
1956	548,197	1978	1,192,602
1957	401,762	1979	1,369,159
1958	442,106	1980	1,599,533
1959	509,006	1981	1,683,219
1960	660,857	1982	1,428,224
1961	717,779	1983	1,586,305
1962	637,698	1984	1,087,447
1963	849,497	1985	993,793
1964	714,717	1986	1,103,323
1965	858,657	1987	1,554,361
1966	734,081	1988	1,446,078[3]
1967	742,156	1989	1,076,576[3]

Source note: Based on totals by country of destination as reported in USDA, *Agricultural Statistics.* Where data were subsequently revised, the latest figures were used. Where identifiable, total flour, including semolina, was restricted to milling from U.S. grain. Shipments for relief or charity were included. No consistent table for the extended periods has been found. These amounts are somewhat lower than those covering 1907–1960 in USDA, *Agricultural Statistics, 1962,* p. 2 table 1.

[1]The year ran from July 1 until 1976 when it shifted to October 1. The transitional quarter from July 1–September 30, 1976, is listed below the total for 1975.

[2]Equivalency rates for flour as bushels of wheat varied somewhat: 100 pounds of flour equaled 4.57 bushels of wheat from 1944 through 1946; 4.31 bushels of wheat during the summer of 1946; 4.57 bushels of wheat from November 1946 through June 1957; 2.3 bushels after June 1957. Where data were given in metric tons, they have been converted to bushels at the rate of 36.7437 bushels per metric ton.

[3]Preliminary, based on USDA, *Agricultural Statistics 1991,* p. 13 table 15.

As advertisements were being run in the *Wall Street Journal* and the *New York Times* warning of the need to control the population explosion if famine were to be averted in the developing countries, agricultural leaders recognized a newly emerging outlook in national land policy. Addressing the National Plant Institute at Chicago in 1967, Professor Don Paarlberg, then of Purdue University, soon to become chief economist in the USDA, noted growing awareness of a need for reversal of the programs which in the early sixties had projected retirement of some 51 million acres by 1980. He cited the development of an expansionist rather than a restrictive attitude toward agriculture and the emergence of a market-oriented rather than a government-dominated policy. The United States would be committed to the alleviation of hunger in those nations willing to join in an effort to solve their problems of food scarcity, but the capacity of American agriculture for economical production in large volume would be channeled to foreign assistance through exports of technical expertise as well as food. Alluding to Paarlberg's remarks, the Great Plains Agricultural Council centered its sessions of 1967 on the "Changes in the Economic Climate Affecting Great Plains Agriculture."[45]

The new outlook had been evident in passage of the Food for Peace Act in November 1966. Adopted as an amendment of Public Law 480 of 1954, the program now stressed the importance of relating American assistance to efforts of countries "to help themselves toward a greater degree of self-reliance" in food production and population growth. Besides continuing the provisions for sale of agricultural commodities on credit and for foreign currencies, as measures designed to develop new markets for American products, the legislation called for direction of relief donations "insofar as practicable . . . toward community and other self-help activities . . . to alleviate the causes of the need for such assistance." A program of "farmer-to-farmer" guidance was to be established through the USDA to help farmers in beneficiary countries "in the practical aspects of increasing food production and distribution and improving the effectiveness of their farming operations." The Peace Corps and research culminating in the Green Revolution thus emerged as outgrowths of efforts to meet the crisis marked in the dwindling American food reserves.[46]

The International Grains Arrangement, which revised the International Wheat Agreement in 1968, reflected the duality of this approach. It called for cooperative effort by which the signatory nations agreed to provide some 4.5 million metric tons, 165.3 million bushels, of wheat and other food grains annually for assistance to developing countries, but it also raised the price limits for contractual obligations as assumed under the earlier program. Linked during the negotiating process with the so-called Kennedy round of talks on reduction of world trade barriers, approval of the revised arrangement became a condition for American reduction of industrial tariffs. The development of grain shortages in the Soviet Union, marked in 1963 by the transition of that nation from the position of exporter to importer in the world grain market, removed one of the major competitive influences that had weakened American support for the IWA. Under the new arrangement exports of the United States as a participant rose to approximately 53 percent of the total transactions and amounted to 74 percent of this country's total wheat and flour exports in 1972–1973.[47]

As an effort for international price stabilization, however, the program was quickly in difficulty. Grain dealers contended that because of the weather-induced grain shortages in India and the Soviet Union during the early sixties, the lower base, fixed at $1.955 a bushel for Manitoba no. 1 spring wheat, was abnormally high. The United States, Australia, and other exporters began subsidizing sales. While Canada at first resisted it, too, eventually dropped its price to $1.70.[48]

American efforts from 1977 to 1979 to bring about international agreement for new price stabilization through use of coordinated grain reserves as buffer stocks were likewise unsuccessful. During grain shortages in the early seventies, the United States complained that import levies set by the European Economic Community (EEC) established a domestic subsidy so high as to close those markets even from exporting for relief of world need. When surpluses once more began to accumulate during the early eighties, Canadians, in turn, protested that subsidies by the United States were only moderately less than those of the EEC as inhibitions to international trade. Through recent bilateral agreements with Canada and Mexico, the United States has again attempted to reduce or eliminate the trade barriers; but as the discussions were renewed in the current Uruguay round of international tariff and trade negotiations, the American proposal, backed by the other major grain exporters, for a reduction of from 75 to 90 percent in the subsidies has been rejected by the EEC. Against this background American trading activity has been pursued vigorously.[49]

The opening of U.S. trade with the Soviet Union in the winter of 1963–1964 had signaled an increased focus upon commercial arrangements, rather than foreign assistance programs, in the disposition of American grain surpluses. Despite strong criticism from congressional and labor leaders, who questioned support for a weakened national adversary, President Kennedy and Secretary Freeman took the measures necessary to enter the booming market already opened the previous September with negotiation of Canadian, Australian, French, and West German grain sales to the Soviet Union following failure of that nation's crops. For subsidies later acknowledged to have amounted to $1.7 million, higher American transfer costs were balanced to develop a market for large sales of wheat, corn, rice, and soybeans during the mid-sixties.[50]

Responding to the export demand, American farmers increased wheat production markedly in 1967 and 1968, but as the volume of exports fell sharply in 1968, carry-over stocks again rose. Grain prices declined, and the farm parity ratio fell to the lowest point since 1933. Price supports on the 1970 crop amounted to 56 percent of the farmer's gross return on a bushel of wheat. Seeking to stimulate agricultural exports and reduce the cost of farm programs, the Nixon administration relaxed some of the crop acreage controls while lowering the farm subsidies. These changes, incorporated in the Agricultural Act of 1970, freed farmers to choose their cropping programs as long as they remained within general acreage limits. They responded by reducing the amount of fallow and producing even more wheat. Although authorization was continued for cropland conversion to long-term land retirement, together with an appropriation for the purpose, the program was not implemented. In July 1972, as surplus stocks reached the highest level in nearly a decade, the USDA announced a record high farm subsidy, as much as $100 million more than the current billion-dollar payments, for reduction of wheat acreage in the coming year.[51]

Already, however, a reversal was again emerging in the marketing situation. Drought reduced yields from the Soviet Union eastward into Australia during 1972 and 1973. As Russian buyers purchased massively on the world market, including from the United States an amount equal to more than half the normal American grain sales to all countries in a year, prices rose phenomenally, from around $1.90 to $4.82 a bushel for North Dakota wheat between 1972 and 1973. While harvests in eastern Europe and Asia improved somewhat in 1974, scarce petroleum supplies limited the availability of fuel, fertilizers, and pesticides for agriculture. That year's crop still failed to meet requirements, and that of the following year yielded too little to replenish stocks. American exports of wheat and flour equivalent averaged over 1 billion bushels annually from 1972 through 1976, from 17 to 26 percent higher than the highest previous level. Carry-over stocks in this country dwindled from 863 million bushels in 1972 to 438 million in 1973 and to a mere 247 million in 1974.[52]

Contributions to the World Food Program under U.S. Food for Peace legislation dropped from close to 36 million bushels in 1972 to less than 11 million in 1973 and to 2.75 million by 1975. Total wheat and flour exports under Public Law 480 declined from 227.6 million bushels in 1971 to 56.8 million in 1973. Commercial sales made at inflated prices had drastically reduced the availability of food stocks for less developed countries, now grown dependent upon imports. Delegates from 130 nations meeting in Rome at the urging of the United Nations Food and Agriculture Organization in November 1974 confronted an immediate need for 8 to 10 million metric tons, over one-third of a billion bushels of grain, to meet such requirements. Projected trends pointed to a potential cereals deficit of nearly 85 million tons under normal production in world agriculture by 1985. If conditions were unfavorable, the deficiency, as calculated, could reach as high as 120 million tons.

American delegates, pointing to the problem of worldwide population growth, called for cooperation by all the world's governments to stimulate food production and recommended establishment of an international stockpile of food reserves. For too many years, they argued, the world had depended on North American surplus stocks to make up for continuing shortages and seasonal crises. Secretary of Agriculture Earl Butz noted that the United States had contributed 46 percent of all food aid provided to the developing countries since 1962. Secretary of State Henry A. Kissinger specifically urged the oil-rich countries to assist with investments and contributions, as well as to keep the cost of petroleum within reasonable bounds.[53]

These were not popular views. Leaders of the Arab states, Cuba, and China criticized the agricultural and financial policies of the United States. The president of Algeria pointed to the "world inflation, caused and maintained by industrialized countries" and proposed that the "concept of production for market purposes" be replaced by "production for humanitarian purposes." Libya's minister of agriculture condemned the "excessively luxurious life led by the USA and its allies, as well as its deliberation to raise the prices of agricultural production inputs such as machinery, fertilizers, and chemical pesticides." Iran blamed the West because of its "past food policy" that "created the present food shortage." Programs that had called for crop diversion to feed grains, land returned to grass, and enhanced production of livestock were especially denounced as promoting inefficient nutritional use of resources.[54]

In the end the conferees agreed to establish a grain reserve of some 10 million metric tons and an international fund to help the less-developed nations work toward self-sufficiency. They also decided to set up an early warning system on incipient food disasters, to organize a program for worldwide fertilizer production and distribution, and to intensify agricultural and weather research. Cooperation under the project was, however, left to the determination of individual nations. American farm spokesmen, fearful of the impact on seasonal markets, had yielded reluctantly to the call for a reserve.[55]

Under the Food and Agriculture Act of 1977 the U.S. secretary of agriculture was directed to administer a farmer-owned wheat reserve of 300 to 700 million bushels, with the upper limit adjustable according to any commitments made for an international reserve. President Carter was authorized to negotiate for establishment of such a reserve, and as early as August 1977, Secretary Butz committed 30 to 35 million metric tons (1.1–1.3 billion bushels) of American grain for the purpose. By legislation of 1980 Congress formally approved establishment of a five-year wheat reserve. Meanwhile, American aid under Public Law 480 programs was again stabilized at about 8 to 11 million metric tons, 294 to 404 million bushels, of agricultural produce annually, at an annual cost of nearly $1.5 billion.[56]

The domestic agricultural program of the United States had been markedly changed by legislation of 1973. Price-support payments had been supplanted by a system of target pricing, under which farmers were to be paid the difference between market prices and target levels. Targets were intended merely to cover production costs, and at $2.05 a bushel in 1973 were well below the market averages of $4.24 and upward for hard red spring wheat in the northern Plains. Export subsidies had been dropped the previous autumn, notably, however, after grain traders had already collected them on the Soviet contracts. Limitation of producer loans to levels below market prices also placed greater emphasis upon market forces. While provision remained for the secretary of agriculture to condition loans and crop payments upon farmer acceptance of allotments and acreage set-asides for conservation uses, no diversions were called for. American wheat production, beginning in 1973, climbed to greater amounts than ever before; and since then, excepting only 1985 and 1988, farmers of the northern Plains have produced the most valuable wheat crops ever grown in the region.[57]

Nationally the index of crop prices (1910–1914 = 100) reached 504 in 1974. It was a period of high inflation generally, but the parity ratio of farm prices to farm costs in the mid-seventies reached 91 for the first time since 1953. Between 1969 and 1982 the price received for a bushel of wheat increased from $1.23 to $3.65 in Montana and from around $1.35 to $3.69 in the Dakotas, while average yields increased from 25.3 to 30.0 bushels per acre in eastern Montana and from 22.5 to 28.6 bushels an acre in western South Dakota, although declining slightly, from 27.8 to 26.9 bushels an acre, in western North Dakota. Notwithstanding higher production expenses, these were prosperous times.[58]

Farmers accordingly expanded their cropping. The SCS estimated that during 1974, 150,000 acres in Montana, 250,000 acres in North Dakota, and 290,000 acres in South Dakota were converted from grass to crops. Of these amounts a mere 20,000 acres in Montana but 105,900 acres in South Dakota and 125,000 acres in North Dakota were classified as marginal for long-term cropping. From 1974 to 1982 the area of total cropland was increased by nearly 1 million acres in eastern Montana. In the Dakotas, however, the severity of drought in 1976 checked the expansion, limiting it to less than 7,000 acres in western South Dakota and even contracting the area by nearly 500,000 acres in western North Dakota. Still the rate of decline in number of farms stabilized in western North Dakota and decreased in western South Dakota and eastern Montana during the period. In all three areas the size of the average farm was expanded—from 2,868 to 3,111 acres in eastern Montana, from 1,161 to 1,318 acres in western North Dakota, and from 2,616 to 2,841 acres in western South Dakota.[59]

The value of farm lands rose in Montana from $56 an acre in 1969 to $271 an acre by 1982, in North Dakota from $96 to $455 an acre during the same period, and in South Dakota from $33 to $349. Economists have maintained that even through the expansion of the period, land was not over-valued, that returns on the investment were "much more favorable than on major alternative investments," and, indeed, that investors minimized estimates of net cash rent to levels that would have given a negative rate of increase over the sixties if interest had continued at former rates. But as farmers borrowed to enlarge their operations, real estate loans negotiated at 5 to 8 percent in 1970 cost 11.5 to 13 percent when renewed in 1981, and production loans obtainable at 7.5 percent in 1970 carried 18 percent interest in 1981. At the same time, machinery and operating costs were rising as well as costs for land. Machinery prices increased 300 to 400 percent during the seventies; fertilizer, 250 percent; gasoline, 440 percent; oil, 335 percent; and diesel fuel, used in most farm work, 730 percent. Debt was growing faster on the Plains than for the nation as a whole through this period. Many farmers were so deeply in debt that survival of their operations was jeopardized if their level of income declined.[60]

Wheat prices fell sharply in 1975 and remained low until 1979. As early as 1975 wheat stocks had again begun to rise. By 1976 they amounted to over a billion bushels; by 1982, over 1.5 billion; by 1985, nearly 2 billion. The Food and Agriculture Act of 1977 restored the system of acreage allotments, abandoned in 1973, and raised the target prices—from $2.47 to $2.90 a bushel, still below the seasonal average market price. The parity ratio of prices received to prices paid by farmers sagged in 1977 to 66. That winter farmers of northwestern North Dakota, responding to a call by the American Agricultural Movement, withheld grain and cattle from market and joined in a farmers' march on Washington, demanding establishment of parity at 100. Congress responded with emergency legislation authorizing the secretary of agriculture to increase the target prices, allotting $4 billion for an emergency farm-loan program and declaring a moratorium on foreclosures by the Farmers' Home Administration.[61]

Wheat prices climbed rapidly to nearly peak levels in 1979 and 1980, and production over the next two years set new records, both nationally and regionally. Meanwhile, however, inflation permeated the national economy, with high interest rates, high production costs, high price supports, and a high dollar in international trade. As foreign buyers found it increasingly expensive to purchase American wheat, governmental policy further disrupted the export trade by applying embargoes as an instrument of diplomatic action.

Public Law 480 credits had long been used to support or inhibit the actions of nations and their leaders in accordance with objectives of American foreign policy. Termination of such funding to Iran in 1975 as a protest against increased oil prices caused the loss of wheat and flour trade that had amounted to nearly 64 million bushels the previous year. Embargoes on wheat exports to the Soviet Union—from July to October 1975 and again from January

1980 to April 1981, the first as a bargaining maneuver to effect a petroleum purchase at discounted price and the latter as a protest against Soviet invasion of Afghanistan—suspended an arrangement under which the Soviets entered a long-term agreement to buy 6 to 8 million tons of wheat and corn (the equivalent of 220 to 294 million bushels of wheat) annually. Exports of hard red spring wheat, which had averaged only about 40 percent of the production of that variety from 1964 through 1971, had risen to 72 and 70 percent with the opening of the Soviet market in 1972 and 1973. They fell to a mere 30 and 39 percent, respectively, in 1976 and 1977, rose to about 60 percent from 1978 through 1980, but dropped again to 44 percent in 1981.[62]

For wheat producers of the northern Plains much of the effect of export development, generally, was residual. Durum wheat, a specialty crop used particularly for semolina in pasta and cous-cous, was not extensively exported but did find a growing market during the late seventies and eighties in Italy, Iran, North Africa, and Pakistan. Grown primarily in north-central North Dakota, it represented, at its peak in 1981, a quarter of the regional wheat crop. Hard red winter wheat, which normally constituted between 40 and 69 percent of American exports, comprised in the mid-1970s about 27 percent of the crop in the northern Plains, most of it grown in the foothill areas of central Montana. Hard red spring wheat, accounting for a half to more than three-fourths of the regional production, provided less than 10 percent of the national exports through the mid-sixties and only 15 to 16 percent in later years.[63]

Because of their higher protein percentages, the hard northern spring wheats had long held strong demand for domestic milling. Generally their prices were considerably higher than those for hard winter wheat, as much as sixty-seven cents a bushel higher in Montana in 1975, when Soviet demand, notwithstanding the embargo, had pushed the spring wheat exports up to 49 percent of their production. Because of the higher price, the spring wheats were less marketable for export under Public Law 480 to newly developing nations. They had not, for instance, shared in the acceleration of exports during the mid-sixties, when a drought in India had resulted in large increases in exports of hard red winter wheat. Where the latter varieties could be grown successfully, they yielded more than the spring wheats, and geographically their producing area in the United States was about 55 percent greater. With the development of improved strains and increased use of nitrogen fertilizer, protein levels of the winter wheats were greatly improved by the late seventies. Also, development of mechanized baking technology reduced the importance accorded to elasticity in bread dough, which was dependent upon protein strength. Millers who blended wheats for satisfactory flour quality became less demanding in protein requirements. To farmers of the northern Plains, foreign buyers prepared to pay for superior quality bread wheats had become a competitive concern, a market not to be rejected lightly.[64]

American farm legislation in 1981 reflected recognition of farmers' dissatisfaction with embargo diplomacy. Under that measure, if an embargo were initiated against a country receiving over 3 percent of American exports of a particular commodity and the action did not extend to all U.S. exports, the secretary of agriculture was required to provide compensation to the producers, either by raising the rate for commodity loans to 100 percent of parity or by paying the difference between that parity level and the market price of the crop over the sixty days following the embargo. While Congress continued the program of acreage reductions, it elevated target prices still higher and required that the secretary of agriculture provide support subsidies as needed to meet similar action by foreign competitors. A revolving export credit fund was provided for use by the Commodity Credit Corporation in developing markets, and the ceiling was raised on donations under Public Law 480.[65]

These measures and the lifting of the Soviet embargo may have contributed to the record-breaking export level of 1981; but the effect was short-lived. The problems in the export trade, generally, had opened a door for competitors to enter markets long dominated by American dealers. Australian grain sales were expanded in the Far East. A number of importing nations, responding to programs that generated scientific and technical advances, became self-sufficient, even, as in the case of India, surplus-producing exporters. The nations of the European Economic Community, a major market for northern spring varieties, heightened trade and tariff barriers. European farmers were producing three times as much wheat in 1985 as in 1950. Diplomatic crises, at first with Iran and most recently with Iraq, curtailed the large and profitable American grain traffic to the Near East. Between 1981 and 1985 the U.S. share of world wheat exports declined from 49 to 38 percent. The volume of domestic carry-over stocks again soared, most notably for the hard red spring wheats, which in 1985 and 1986 reached more than 108 percent of the year's production of such wheats.[66]

The period became a time of severe adjustment for American farmers. Wheat prices declined from a national average of $3.91 a bushel in 1980 to $2.42 a bushel by 1986. The parity ratio of farm prices to farm costs fell from 61 in 1981 to 56 by 1983. For operators in the northern Plains the decrease was even sharper. The season average price received for all wheat in 1980 was $4.14 a bushel in Montana, $4.24 in South Dakota, and $4.40 in North Dakota, with prices for durum wheats because of unusually low yields averaging $5.90 a bushel for the 7.6 million bushels produced in Montana and $5.35 for the 73.1 million bushels produced in North Dakota. By 1986 the season average price received for all wheat in Montana and North Dakota was about $2.53 and in South Dakota, $2.42, and the average price of durum wheat had dropped to $2.66 in North Dakota.[67]

Declining land and farm equipment values reflected the difficulties in the Great Plains, with a drop of about 20 percent between 1983 and 1984 and about 33 percent from 1981 to 1985. Nationally the debt-assets ratio of the farming sector rose from about 16.5 to 20.8 percent. In North Dakota the increase represented a shift from 16.4 to 23 percent; in Montana, from 17.7 to 22.5 percent, and in South Dakota from 22.2 to as high as 29.5 percent. Moreover, liquid assets as a percentage of the base had fallen sharply from 20.4 percent in 1974 to 12.9 percent by 1984. Demands for cash increasingly necessitated liquidation of productive capital. This situation prevailed widely as the relief activity of the Federal Farm Credit System was constricted during the mid-eighties. Farm liquidation rates in the Ninth Federal Reserve District, which included the Dakotas and Montana, together with Minnesota, northwestern Wisconsin, and the upper peninsula of Michigan, increased from about 1 percent in the early seventies to 3.2 percent by the winter of 1984–1985. A year later the American Bankers' Association reported the rate at 4.9 percent, and it rose to 5.6 percent in 1986.[68]

Drought or poorly timed precipitation had added to the problems. Production was low in South Dakota in 1976, in North Dakota in 1980 and 1983, in Montana in 1979, 1980, 1984, and 1985, and throughout the region in 1988. Rainfall was scant in North Dakota in 1976, but production there remained high. When it declined in that state in the early eighties, moisture was more than adequate through the growing season. Despite the fact that drought was widespread in 1988, with 94 percent of the commercial farm sales of wheat and 70 percent of those of cattle derived from drought areas, net cash income for the year was better in Montana and South Dakota than it had been through the early years of the decade. The USDA reported that drought losses in that segment of the country were largely offset by high prices on the production available. By 1989 net cash income was close to record levels in the region, exceeded in Montana and South Dakota only in 1973. The primary problems of the period were still centered in the instability of the market situation.[69]

Federal farm programs of the eighties were again designed to reduce production. To be eligible for price supports, farmers were required in 1982 to cut their acreage by 15 percent. The following year they were given the option of diverting 10 to 30 percent more in return for payments in kind. They could also bid to divert their whole base acreage from production with payment amounting to 95 percent of normal yield. For the next two years acreage reductions were set at 20 percent, with compensation for an additional diversion of 10 percent. By 1988 the acreage to be set aside was initially raised as high as 27.5 percent. Meanwhile, the Great Plains Program was renewed in 1981 for another decade, and in a special subtitle of the Food Security Act of 1985 provision was made for reinstitution of a massive cropland retirement program to place from 40 to 45 million acres in reserve for minimums of ten to fifteen years.[70]

Cropland reductions were, however, linked with aggressive efforts to expand exports. Payments in kind not only reduced the direct financial cost of the benefit programs, they shifted the domestic subsidies from fixed target levels to market pricing. Abandoning the payment-in-kind arrangement because CCC holdings had proved insufficient for the requirement, the 1985 farm bill restored cash subsidies to farmers but at a reduction of 25 percent, with smaller cuts scheduled for subsequent years. Meanwhile, fiscal policy served to lower interest rates and reduce the value of the dollar, thus offering increased inducements to foreign buyers of American products.

As American wheat prices declined, exports rebounded. Large shipments were continued to Japan and, beginning about 1979, were extended to mainland China and Korea. The reopened trade to the Soviet Union ranged from as little as 5.6 million bushels in 1985 to 321.5 million bushels in 1987 and will no doubt increase under current circumstances. Wheat exports to South American states, notably during the late seventies and early eighties to Brazil and Chile, were expanded during the mid-eighties to Central America and the Caribbean area. Drought in northern and eastern Africa through much of the last decade brought Morocco and Algeria into the market and led to heavy relief shipments elsewhere in that region. Exports of American wheat and flour equivalent exceeded 1.1 billion bushels in 1986. The following year they amounted to over 1.5 billion bushels and in 1988 approximately 1.4 billion.[71]

Concern over the prospect of a world food shortage has again developed. As drought became widespread in the spring of 1988 Secretary of Agriculture Richard E. Lyng lowered the set-aside requirement for American wheat producers from 27.5 to 10 percent, but the following December the United Nations Food and Agricultural Organization reported that cereal stocks had fallen below the minimum requirement for food security. In 1990, Congress, considering agricultural legislation for the next five years, cited findings of the National Research Council and the National Academy of Sciences that food aid throughout the nineties would need to be twice the 10 million metric tons a year previously projected to meet global food requirements.[72]

Under the new legislation, the president was called upon to reenforce the program of sales of agricultural commodities on credit or for local currencies or barter to developing countries; to provide at a maximum annual expenditure of 1 billion dollars not less than 9.88 million metric tons over the next five years for food assistance to foreign countries through governmental and private agencies, including the World Food Program; and to establish under bilateral governmental arrangements a grant program for food-deficient countries qualifying as "least developed" under poverty criteria established by the International Bank for Reconstruction and Development. The secretary of agriculture was directed to develop a "long-term agricultural trade strategy... designed to promote the export of United States agricultural commodities" and, among other goals to ensure "the provision of food assistance and the improvement in the commercial potential of markets for United States

agricultural commodities in developing countries." This linkage of philanthropic with commercial interests for the development of the export market
was at the same time accompanied by provisions for continuing domestic
programs of crop limitation. While the secretary of agriculture was allowed
considerable flexibility for adjustments, the legislation provided for continued conservation land retirement and wheat acreage reductions ranging up
to 20 percent of the base acreage for each producing farm.[73]

The policy duality represented in this combination of export "enhancement" with production curtailment reflected continuing fear of the surpluses
so long a dominant consideration in shaping the national agricultural agenda.
But in making available deficiency payments based on a minimum established price of $4.00 a bushel for the next five years as compensation to pro-
ducers for relinquishing government loan or purchase agreements, as well as
in the leeway accorded the secretary of agriculture in setting acreage reductions, the legislation also indicated renewed uncertainty of market requirements. The "yo-yo or come again, gone again change" in world demand for
U.S. farm commodities, noted in the mid-seventies by agricultural economists
Earl C. Heady and John F. Timmons, is again on the upswing.[74]

Earlier demand periods dictated by foreign events—during the two
world wars, the Korean War, the mid-sixties, and the mid-seventies—were
followed by large surpluses and depressed prices. Dependence upon the in-
stabilities of foreign markets inevitably adds greatly to the agricultural risks.
Over the past thirty years, however, the industrialization of the less developed states, partly as a consequence of commercial penetration, has brought
about disruption of traditional food chains, rising popular agricultural requirements, and a growing inelasticity of world demand.[75] The responsibilities as well as the risks for the United States as a major agricultural exporting
nation can no longer be assessed merely as a regional or even a national concern. The policy dilemmas of farm production require global perspectives.

CHAPTER NINE

Policy Dilemmas

"The national climate of opinion, the national mood and temper, has become strangely sullen, silently embittered, inordinately selfish in both a personal and a national sense of the term."[1] In this judgment Professor Ross Talbot of Iowa State University highlighted the undercurrent of public querulousness relative to national agricultural policy during the early eighties. For farmers the prosperity of a booming market had collapsed as the parity ratio dropped from 91 in 1973 to 65 in 1980 and then declined to 56 by 1983. In 1978 and 1979 the tractorcades of the American Agriculture Movement (AAM), centered in the southern Plains and the Southeast, once more threatened a farm "strike" in support of a demand for 100 percent parity. In Senate hearings around the country, farmers complained of increasing land, machinery, and cropping costs; of a widening income gap between large and small producers; of public commitment to consumer interests in structuring price supports more closely to market influence; of grain embargoes and sharply fluctuating prices associated with export trade. To counteract operations of the Arab oil cartel, seen as the cause of rising fuel prices, the AAM called for export development directed to exchange of wheat for oil: "A barrel for a bushel"; "cheap crude or no food."[2]

Political support for such protests had dwindled as Americans became more urban oriented, more removed from even the sentimentality of rural ties. Price supports, export subsidies, cropping quotas, and land retirement translated into higher grocery bills for city dwellers, who were also suffering from the inflationary economy. In reference to Plains development, more attention had been directed to the relief assistance of the crisis years than to the fact that much of the "relief" had taken form as loans that were quickly repaid when prosperity returned. Few residents outside the region recognized that annual journalistic reports of spring dust storms or congressional laments of suffering farm constituents related to localized areas. Pictures of grain mounded on the ground for want of storage facilities told a story of accumulating surpluses, but few accounts cited the importance of American agricultural abundance in relief of major world need through the forties and early fifties, during the mid-sixties, the early seventies, and much of the eighties.

Until the mid-fifties government payments regularly supplementing farm income had been associated with the national price support (AAA/ACP) program, applicable generally. In spite of the fact that dry-land crops were produced far from the terminal markets to which support prices were pegged, the value of such grains continued to bring farm prices well above the national average. Moreover, cropping quotas and cropland diversion were particularly restrictive in relation to the regional productive advantages of dry-land operators. Study by USDA economists indicated in 1961 that, in view of limited production alternatives, wheat in Montana was the most profitable crop, even at a feed price. With no production controls and no price supports, net incomes throughout the Plains would decrease, they concluded, but the decrease would be least (at 35 percent) in north-central Montana and most (at 66 percent) in Oklahoma.[3]

Reinstituting land diversion and "conservation" programs may have seemed necessary as a response to drought in the lower Plains in 1955, but regional agricultural analysts had found them of questionable desirability over the broad area to which they were extended. By 1972 such payments amounted annually to over $100 million in Montana and South Dakota, respectively, and over $200 million in North Dakota. For the ten Great Plains states that year they amounted to over $1.5 billion, 43 percent of the national total and 31 percent of the realized net farm income in the region.[4]

Amid waning public support for such expenditures, proponents of land retirement found new impetus in the expanding environmental movement of the mid-seventies. When long gasoline lines reminded city dwellers that natural resources were limited, they learned, too, that ground-water tables were falling on the Plains and assumed that agricultural production there was delimited by irrigation. As suburbs, asphalt, and shopping centers spread over the green spaces around metropolitan centers, urbanites thought, ever more appreciatively, of grasslands and wilderness areas, of roaming deer and buffalo. The outcry against USDA budgets was silenced by protest that the nation was mining its soils and underground water to meet an ever-growing world demand, that "the current generation of farmers ha[d] no right to engage in the agronomic equivalent of deficit financing, mortgaging the future of generations to come." From this perspective, the AAMs proposed Arabian barter arrangement took form as an exchange of oil for soil.[5]

While Washington policymakers were not prepared to forego trade development, they embraced the environmental movement and in the process initiated the most expensive agricultural program yet authorized. Until this period environmental considerations as a focus in advocacy of soil conservation had served very largely as a formula for reduction of crop surpluses, and that was to remain the predominant consideration in the new coalition. Hugh Henry Bennett, as the first head of the Soil Erosion Service and its successor, the Soil Conservation Service, had indeed initiated the program from an environmental perspective. That approach had been evident, too, in the views expressed by H. H. Finnell in 1949 and in the lingering efforts of the SCS to attract cooperating farmers who would commit whole farms to a management program based upon identifying differentiated agricultural conditions and adjusting their operations accordingly. With the absorption of soil-conservation measures as a basis for general agricultural adjustment payments under the revised AAA program of 1936, however, environmental operations had been codified to a degree that failed to give recognition to individual circumstances and needs. Reports of the SCS became primarily a talley of construction procedures performed under ACP referrals. The struggle to obtain funding for and to accomplish the requisite soil mapping on which individual SCS management programs could be designed took so long that the "acreage planned" was a small fraction of the "acreage enrolled" and the "acreage treated" was far less. The ACP, in turn, presented a smorgasbord of generally desirable practices from which state and local committees selected those deemed most widely applicable in the area and from which individual farmers chose those most profitable, not necessarily the most needed from an environmental standpoint for their particular operations.

As the cost of the program mounted, public criticism had developed over the apparently ineffective policies. Critics wanted to know why it was necessary, year after year, to renew payments for establishing contours, grassed waterways, and improved tillage and cropping practices. This questioning became the theme of the volume *Soil Conservation in Perspective* by E. Burnell Held and Marion Clawson, published under the auspices of the Resources for the Future in 1965. The conservation movement, these authors argued, had been generated in the belief that land capability afforded "the chief, or only, means of increasing agricultural output to meet growing needs." The SCS had called for comprehensive land classification as the basis for its work, but progress in whole-farm analysis had been so slow that the PMA (AAA) and state extension services had promoted less sweeping measures as a speedier approach to increased productivity. Land management "for greater output and greater profitability," that is, marketing considerations, had replaced erosion control as the dominant goal. For the past thirty years, these analysts noted, public expenditures had supported "a curious mixture of output increasing and output restricting activities, with no real reconciliation of their purposes and terms." The budgetary process, they concluded, should be directed so that public funds were "made available only for investment in soil maintenance."[6]

A wide range of surveys and legislative responses centering on land-use planning developed over the next two decades. Passage of the Federal Water Pollution Control Act Amendments and the Rural Development Act in 1972 initiated the action. The legislation induced states, under threat of federal intervention, to strengthen their laws relating to zoning and the authority of soil conservation districts. In Montana county governments, by acts of 1971 and 1973, were given authority to plan and zone on a county-wide basis. In North Dakota a State Planning Division was organized with eight regional planning offices throughout the state, under mandate to assist counties in development of comprehensive plans, with particular intent to fit federal programs to local needs while developing plans to satisfy the requirements of the federal agencies. South Dakota, also, had a state planning agency responsible for receiving and coordinating local plans with those of the national government. A South Dakota act of 1974 specifically required county planning commissions to prepare comprehensive plans, including zoning ordinances. Much of this interest centered, however, on problems related to mining development and the spread of urban subdivisions. A proposal in South Dakota to require permits from local soil conservation districts before engaging in "land disturbing activities," a precursor of the later federal "sod-busting" legislation, was defeated. The machinery for local land-use control was in place, but public participation was questionable.[7]

Regional speakers at sessions of the Great Plains Agricultural Council in 1973 and 1974 were concerned that if directed from Washington, the planning effort would "steal away the police powers of the state." If left to local initiative, however, they believed that it would prove fruitless. "In rural development we've messed around with the grass roots process for many years and I suggest that we have little to show for the effort and the expenditure," one commentator complained. His criticism of the past governmental guidance of local planning was similarly sharp: "We cannot just ask the local citizens to develop a plan and then put it on paper, and then send it to Washington or some state planning office and get it reviewed. No one knows what that review is going to embody—and those who are put on the panels to do the reviewing don't even know what they are supposed to review!"[8]

During the mid-seventies Congress repeatedly debated proposals calling for grants to the states for development of land-use plans and programs; requiring inventory, classification, and planning for utilization of public holdings; and finally in 1977 extending the latter requirements to the nation's soil and water resources generally. Extended hearings on these proposals evidenced, however, great and widespread opposition to such efforts. Appearing before the House Subcommittee on Energy and the Environment in the spring of 1975, a spokesman for the National Cattlemen's Public Lands Committee opposed much of the proposed legislation as a threat of federal control. The American Farm Bureau Federation and the National Grange expressed similar objections. Of the major farm groups only the National Farmers' Union voiced general approval, but was critical of abuse in the exercise of powers of eminent domain against private land owners. Even the National Association of Conservation Districts, while endorsing federal land-use legislation, urged greater consideration for local concerns and capabilities, for the importance of economic as well as environmental values, and for minimum intervention in state jurisdiction.[9]

As food stocks dwindled under pressure of increased export demand, proponents of conservation rediscovered an inherent concern for protection of the environment based on regard for the national interest. While the Conservation Needs Inventory conducted by the SCS in 1967 had been interpreted as showing an adequate base for domestic needs, the analysts had concluded that 97 percent of the inventoried soils evidenced "a dominant soil limitation or conservation problem," that treatment was needed on 63 percent of the land in the nation. During the early 1930s some 360 million acres had been cropped in the United States. Over the next forty years the land in cultivation had been reduced by over 60 million acres, but from 1970 to 1980 that trend was reversed, with approximately 357 million acres under crop. Lands environmentally marginal for cultivation, it was assumed, were being brought back into production. What role, then, should the United States undertake in meeting the needs of other countries? Could or should the American resource base support the rising demand for food and fiber in developing nations around the world? Proponents of environmental concerns demanded that policy makers "at the highest level" clarify the relationships between policies for food production, land use, and rural development.[10]

A series of surveys over the next decade emphasized the deficiencies in the existing program. The Renewable Resources Planning Act of 1974 led to a review and programmatic recommendations on the handling of the public lands, primarily the national forest holdings. That same year a study by the SCS on the effectiveness of the Great Plains Conservation Program concluded that better targeting of funds among the states and increased attention to the combinations of practices funded on individual farms could appreciably improve the results without increasing the cost. Three years later the comptroller general reported to Congress the results of surveys by the General Accounting Office that were highly critical of most governmental conservation programs. All were found deficient in targeting for areas and individuals with high priority needs in conservation. The ACP cost-sharing operations, as Held and Clawson had reported, tended to enhance production with little concern for soil conservation. The Great Plains Conservation Program still applied only 27 percent of its funds to establishing vegetative cover on cropland, with most of the remaining expenditures devoted to construction of irrigating systems, fencing, and other facilities for livestock and water-fowl. Much of the land supposed to have been restored to "permanent" grass cover had been returned to crops. The SCS commitment to full-farm planning was criticized for being both slow and expensive. A single operative could prepare only about thirty-six plans in full-time service during a year. The SCS in 1975 had spent $50 million formulating 83,180 plans, at an average cost of approximately $600 a plan.[11]

The capstone of the legislative response to the agitation was legislation in 1977 that called for a Resources Conservation Appraisal (RCA). First approved by Congress the preceding session, the measure had been rejected by President Ford through pocket veto, because it called for intervention by a federal bureaucracy in concerns he thought better left to state and local governments and individual landowners. To circumvent the veto the Senate agriculture committee had called upon the USDA to proceed with a broadly based evaluation of all existing soil and water conservation programs. This survey, the National Agricultural Lands Study (NALS), carried a mandate to assess the potential utilization of the entire national agricultural land base, including pasture, range, and forest, as well as cropland. Its focus was directed to the "conversion patterns affecting high quality agricultural land—in particular, how the diversion of rural and urban development onto land less suited for agriculture could reduce the impacts of agricultural land conversions on U.S. agriculture." While its emphasis was thus limited to the loss of prime farmland to urban spread, construction of water impoundments, transportation facilities, and mining development, it stressed an anticipated need for increased agricultural productive capacity at 55 to 80 percent above the 1980 level to meet domestic and foreign demand over the next twenty years. "After four decades of agricultural surpluses, U.S. agriculture has moved away from underused production capacity," the study noted. It projected a large shift of forage land into crops, more confinement feeding of livestock, and a decline in domestic meat consumption attributable to higher cost. Moreover, it anticipated higher food prices generally as incentives to attract the needed agricultural expansion. Although the report rejected the idea that land conversion out of agriculture amounted to "a present-day 'crisis,'" it concluded with a plea for intervention at all governmental levels to control land development and protect agricultural resources as a national policy.[12]

While the NALS was under way, Congress reconsidered the RCA proposal under the Carter administration. The measure then enacted called for the secretary of agriculture to appraise the soil and water resources of nonfederal lands as the Renewable Resources Planning Act had earlier dealt with the public domain, to develop a program for conservation on the private holdings, and to report the findings to Congress by January 1980. Under the original bill the SCS was to have conducted the survey, and that agency had initiated the work with a National Resources Inventory (NRI), begun in 1977, which had already established the basis for the survey and for its computerized model. The new legislation less explicitly assigned the leadership role to the SCS. When other government agencies—the Office of Management and Budget, the General Accounting Office, the Agricultural Stabilization and Conservation Service, and the Senate Agriculture Committee—criticized the SCS dominance, Secretary of Agriculture Bob Bergland set up an interagency coordinating committee, with the Office of Environmental Quality operating as executive secretary. By November 1980 the committee had met some 109 times, but because of limited staff assistance and bitter internal controversy, it had not provided the report by the requested date. With the election of a new presidential administration the operations of the coordinating committee were suspended, and in June 1981 they were terminated. The SCS continued as the primary agency in the proceedings.[13]

Under John Crowell, assistant secretary of conservation, research, and education in the USDA, a draft report was finally announced in August 1981, and a second review draft was released in October. Major changes had been introduced with the new administration. In keeping with President Reagan's emphasis upon federalism, states were to be given a new block-grant program for cost sharing in conservation improvements by farmers. Priorities in the apportionment would no longer be shaped by the federal government alone but by joint state-federal boards, with half the members selected locally. Theoretically the arrangement would meet the objections voiced against either Washington dominance or local inactivity.

From the viewpoint of concerned environmentalists, this arrangement was a fatal flaw. Political pressures had been cited repeatedly as one of the predominant factors in the cost and inefficiency of the ACP and SCS approaches. Every president after Harry Truman had sought to reduce such funding, effort which under President Nixon had extended to impoundment of $222.5 million of the appropriations congressionally committed to the programs. In converting financial support of conservation measures into a form of "pork-barrel" distribution of public largesse, politicians, whether at the national or the local level, responded to demands of their constituents that too frequently mirrored personal rather than public interest. The General Accounting Office in 1977 had complained of the inconsistent priorities established by ACP committees. Critical analysts on the implementation of the RCA commented that local meetings designed to develop recommendations gave high priority to what was already being done, but sought higher payment for the practices.[14]

Through the summer and fall of 1978, the SCS held some 9,000 meetings, reportedly attended by "more than 164,000 people." The following year the agency commissioned Louis Harris and Associates to survey public concerns on conservation policy through a statistical sampling based on 7,000 interviews. A draft report circulated in early 1980 received 65,000 responses, "signed by 118,213 people." Much of the *1982 Final Program Report* consisted of testimonial acknowledgments of the consultation. The public meetings had served well to develop a constituency broadly committed to "protecting the environment," but they had provided little program analysis and little discussion of the difficult decisions requisite to the process. They had not served as a forum to question the assumptions of projected resource needs or the data-gathering process.[15]

The theme of the RCA report was U.S. capability to meet national and world requirements over the next fifty years. Departing from the pessimism of the NALS, the USDA in 1982—and still more decisively in the *Second RCA Appraisal*, updating the analysis in 1989—offered strong reassurance while pressing the call for conservation. The updated *Appraisal* projected large decreases in cropping on both dry-land and irrigated acreage by the year 2030 in the Plains and mountain states. Graphically, the model called for reductions in cropland acreage amounting to approximately 60 million acres in the northern Plains (defined as covering the two Dakotas, Nebraska, and Kansas), another 20 million acres in the Corn Belt, 30 million acres in the mountain states (including, in the latter case, Montana, Idaho, Wyoming, Colorado, New Mexico, Arizona, Utah, and Nevada), and 25 million acres in the southern Plains. No conversion from acreage potentially cultivable into cropland was anticipated, and 40 million acres proposed as "conservation reserve" were not expected to be returned to cropping even under the highest stress of demand. This program stressed expansion of livestock and, again contrary to the NALS, a return to range-fed production. For agricultural interests in the related areas it was a distressing scenario.[16]

Critics of the RCA early pointed to the weaknesses of the inventory on which the projections were based. During an ACP survey in 1980 it was found that the data of the universal soil-loss equation measuring sheet and rill erosion were unreliable for a considerable number of the sampled counties, and the SCS was still unable to provide dependable wind erosion estimates even as late as the 1982 report. The inventory was based on sampling much as the Conservation Needs Inventory had been compiled in the fifties, in most cases throughout the Midwest, West, and South 160 acres to the county. Within those units the SCS examined sample areas selected randomly at three points. State SCS staffs and the Iowa State University Statistical Laboratory checked that data, and the SCS re-examined a little over one-third of them in the field. Within those Primary Sample Units the record was detailed—size of area, identification of water holes, extent of perennial streams, soil name and land-capability class, acreage that had been "modified" by man, areas lacking vegetal cover, kinds of crops and amounts of residue maintained over the past four years, kinds of conservation applied, and type of ownership. These details were then generalized for regional summaries, but the individual farmer outside the Primary Sample Unit could not rely on the applicability of the data to his particular holdings. An *Interim Report* for the NALS, prepared by the National Association of Conservation Districts, had emphasized the limitations of such an approach: "The relation between erosion and productivity can only be quantified with one soil, in one place, and with given climatic conditions. The same relationship may or may not occur elsewhere when one or more of these factors change [*sic*]."[17]

Whatever the shortcomings of the data collection, it did not on environmental grounds point to the programmatic emphasis of the analytical recommendations. The erodibility index (calculated by multiplying together erosion factors representing soil, topography, and climatic conditions and dividing the product by the soil-loss tolerance assigned to the soil) showed only 2.1 million acres in the whole state of Montana, 313,500 acres in North Dakota, and 343,700 acres in South Dakota as "very high" and consequently requiring permanent conversion to sod or trees or long rotations with several years of sod. Less than 4 million acres in the two Dakotas, jointly, and 9.5 million acres in Montana were rated as "highly erodible" and consequently requiring a combination of two or more measures such as maintenance of 50 percent crop residue, contouring, terraces, strip cropping, and sod-based rotations. Sheet and rill erosion amounted annually to only 1.6 tons per acre in Montana, 1.9 tons per acre in North Dakota, and 2.6 tons per acre in South Dakota, while the soil-loss tolerance level (T-value) was estimated at approximately 5 tons per acre annually before productivity would decline significantly over a hundred years. The wind erodibility rate found for Montana was marginally high at 8.3 tons per acre annually, but those rates were also low for North and South Dakota, at 3.1 and 2.7 tons per acre, respectively.[18]

Calculated on a broader, regional basis, about half the cropland in the mountain states was tabulated as "highly erodible" and less than one-quarter of that in the northern Plains, but only an estimated 28 percent and 12 percent, respectively, were found to be deteriorating at double the rate necessary for maintenance of soil quality under 1982 farming practices. Computer simulations of erosion's effect showed an estimated loss of productivity on cropland in the states of either region under such management practices over the next hundred years at approximately 5 percent. Current erosion rates greater than 2T as percentages of total croplands were reported higher in all other regions than in the northern Plains and higher (with 36 percent eroding at double the rate necessary for maintenance of soil quality under 1982 farming practices) in the southern Plains than in the mountain states. The estimated losses of future productivity ranged to 71 percent of total cropland in the Northeast, 65 percent in Appalachia, 39 percent in the Pacific states, 27 percent in the Corn Belt and the Delta region, and 11 percent in the Southeast. Based on current erosion rates and future productivity, the targeting of reform was pointed to very different foci of concern than conservation (see Table 9.1).[19]

Program development as proposed in the *1982 Final Program Report* was vague. Three alternatives had been presented for public "review": (1) continuation of current programs with current funding; (2) redirection of federal programs to target critical programs; (3) expanded emphasis on state and local action through a block-grant arrangement plus redirection of federal support as under the second alternative. Nationally, 43 percent of the respondents reportedly supported the third, or USDA, preference; 48 percent opposed it. Majorities in the western and midwestern states were said to favor the plan, but whether that can be said of the states in the northern Plains is questionable.[20]

The governors of all three states in the area, while agreeing on the need for a stronger conservation effort, were also in agreement that targeting

TABLE 9.1. Erosion Problems on Cropland (million acres)

	Total Cropland	Highly Erodible[1]	Erosion Rate >2T[2]	Productivity loss >5 percent[3]
Northeast	17	7	3	12
Appalachia	23	10	6	15
Southeast	18	3	3	2
Lake States	44	6	6	1
Corn Belt	92	23	23	25
Delta States	22	3	3	6
Northern Plains	93	21	11	5
Southern Plains	45	19	16	<1
Mountain	43	22	12	2
Pacific	23	4	4	9
Total	421	118	87	77

Source: USDA, Second RCA Appraisal, p. 29 fig. 19.
[1]"Highly erodible" cropland has an erodibility index of 8 or greater, calculated by multiplying together the erosion equation factors representing soil, topography, and climate and dividing the product by the soil-loss tolerance assigned to the soil.
[2]"Erosion >2T" means that the soils had either sheet and rill erosion or wind erosion rates greater than twice the estimated soil-loss tolerance under existing management.
[3]"Productivity loss" is estimated by the Erosion/Productivity Impact Calculator (EPIC) Model.

should not be directed in such a way as to reduce funding for existing programs. They were all opposed to introduction of the block-grant proposal if it were to lessen the federal contribution; they all opposed the provision for newly established local administering boards, on the ground that this would reduce the authority of the existing ACP and SCD committees. Similar views were expressed by the National Cattlemen's Association (NCA), the American Farm Bureau Federation, and the National Farmers' Union (NFU). The NCA, which deplored the failure to extend the proposed funding of conservation measures to rangelands, termed the presentation "slanted" and "misleading" in its advocacy of the third alternative. The NFU elaborated its criticism with "a casual comment" concerning the USDA's "response leaflet:" "It's loaded! It does not take a discriminating eye to ascertain that the leaflet assures favorable comments about the Secretary's 'preferred' program, and blatantly downplays the other two alternatives (as weak as they may be). We encourage more respectful solicitation of public opinion in the future."[21]

Because of the delay in completion of the report, its findings were not available for Congress to use in preparation of the agricultural legislation of 1981, which in acreage reduction continued much the same format as the program of 1977, including voluntary diversion of grain crops to conservation uses as a qualification for income supports while prices were upheld through the acreage reserves and adjustments as necessary in the rates on commodity loans. An approach toward geographic targeting was, however, introduced in a program for cost-sharing of technical and financial assistance for conservation measures by farmers and ranchers in special areas designated by the secretary of agriculture "as having severe and chronic erosion-related or water management-related problems." Contracts for such work, authorized to run for up to ten years, were to be based on specific conservation plans approved by the secretary and the local SCD. They might cover all or part of the farm or ranch as determined to meet the conservation problems, and they did not disqualify the operator from participating in other programs of the USDA. A program of matching grants for noncapital expenditures of local governmental units in formulating and implementing long-term conservation improvements was also authorized, with these grants also specifically stipulated to augment rather than to replace existing technical and financial assistance programs of the USDA.[22]

Meanwhile the Economic Research Service of the USDA explored the benefits of alternative measures. The analyst in a 1985 study estimated that with the available funding, from 21.4 million to 25.8 million acres should be retired if the primary consideration were profitability of operation—4.8 to 7 million acres of corn, 2 to 3 million acres of soybeans, and 13.9 to 15.8 million acres of wheat, dependent upon acreage productivity. "High erodibility" delineated 7.4 million acres of the corn land, 7.6 million acres of that in soybeans, but only 5.6 million acres of that in wheat. The program recommendation, nevertheless, called for withdrawing approximately 5.3 million acres from corn, 2.5 million acres from soybeans, and over 14 million acres from wheat. Of the wheat acreage to be retired, 3.7 million acres were classified as "nonerodible" and 4.8 million acres as merely "erodible." The basic consideration again centered on reduction of surpluses.[23]

Cost considerations further modified the recommendation. Annual rental costs would have amounted to as little as $2.14 per ton of reduced erosion, or $56.56 an acre, had erodibility been the sole criterion. Instead, the proposal tied erodibility to reduction of surpluses, by calling for long-term contracts to retire some 22 million acres targeted for erodibility out of 38 million acres estimated as surplus production. Under this combination the average annual government cost per idled acre would amount to $76 and the cost of erosion reduction would be increased to $4 a ton; but compared with the cost of $129 an acre, for annual diversion payments merely to reduce surplus, a saving of $53 would be effected.[24]

Under the Food Security Act of 1985, Congress embraced a dual approach, authorizing the customary payments according to marketing quotas and acreage limitations without targeting among wheat-producing farms but allowing the secretary of agriculture to make adjustments as deemed necessary. Whether or not general acreage limitations were applied to wheat, the secretary was authorized to enter into contracts calling for land-diversion payments based upon commitment of the acreage to conservation uses, "to assist in adjusting the total national acreage of wheat to desirable goals."

Under such arrangements limited use of the land for grazing was permitted at the request of state conservation committees. The legislation also, however, provided for negotiation of "multiyear set-aside contracts," applicable to base acreages in wheat, feed grains, upland cotton, and rice, for a period not to extend beyond the 1990 crops, which like the "conservation reserve" of the fifties prohibited use of the land even for stock-grazing "except in areas of a major disaster," as recommended by the secretary and decreed by the president. Farmers were required to provide and maintain vegetative cover on diverted and set-aside acreage, for which they were to be compensated "an appropriate share of the cost."[25]

A separate subtitle of the measure went much further, formally re-establishing a "conservation reserve" specifically applicable to "highly erodible cropland" as classified by SCS grades IV, VI, VII, and VIII or the soil-loss tolerance level established by the secretary of agriculture. That official was authorized to negotiate ten- to fifteen-year contracts to place in reserve from 40 to 45 million acres, scheduled gradually over the next five years. Although not considered "highly erodible," lands threatened with continued loss of productivity because of salinity were also eligible for inclusion. Under the contracts farm and ranch owners or operators were required to implement a plan approved by the local conservation district or the secretary of agriculture for converting the acreage from crop production to less intensive usage, such as pasture, permanent grass, trees, or shrubs, according to a stipulated plan. Maintenance of vegetative cover with an expected life-span of at least five years was a requisite. Meanwhile, harvesting, grazing, or other commercial utilization of the land was specifically prohibited unless authorized in response to an emergency. Annual rental and cost-sharing payments were to be made in accordance with contract bids on the basis of the amount found necessary to encourage participation in the program. The secretary of agriculture was given much flexibility in establishing the criteria for the bidding pool and selection process.[26]

In returning to the conservation-reserve program, congressional framers sought to alleviate the major problems identified with the earlier experience. To minimize disturbance of local communities, acreage withdrawals were restricted to no more than 25 percent of the cropland in any one county, unless the secretary of agriculture determined "that such action would not adversely affect the local economy." Contracts were to run not less than ten, nor more than fifteen, years; but to forestall return of the land to production, the conservation plan governing the contract might provide for permanent retirement of the cropland base and allotment history for the acreage. Moreover, the owner or operator of the tract or, indeed, of any cropland defined as "highly erodible" was barred from access to federal price support payments, crop insurance, disaster relief, loans for farm storage facilities, or credit through the

Farmers' Home Administration if he cropped the land after 1989 or two years after the SCS had completed a soil survey of the farm—unless he operated according to a conservation plan approved by the local SCD or the secretary of agriculture. The secretary was directed, "as soon as . . . practicable," to complete soil surveys on nonfederal lands, at least to the extent deemed necessary to determine their land-capability class for the purposes of the legislation.[27]

The focus suddenly assigned to conservation in this legislation seems to have caught administrators by surprise. Earlier in the decade Congress had extended the Great Plains Conservation Program through the eighties when Secretary of Agriculture Bergland had proposed its renewal merely through 1982. The planners of the Economic Research Service prior to passage of the Food Security Act had dismissed as exorbitantly expensive program options that projected erosion control independent of reduction in crop output. The SCS was not yet prepared to provide specific identification of "highly erodible" tracts. Data on wind erosion, generally, the primary concern in dry-land districts, had not been available when the National Resource Inventory report was prepared in 1977 and updated in 1982; they were first incorporated in tabular data in the 1989 *Appraisal*. Even then, the report did not call for more than 13 million acres, nationally, to be permanently converted to sod or trees. Critics cited need for further research on the effects of new technology in dry-land tillage, on the social and community consequences of land-retirement programs, and on the equitability of the cross-compliance provisions, most applicable to operators already in difficulty. As late as 1987, a program leader on soil and water conservation in the USDA extension service commented before a symposium on the Conservation Reserve Program (CRP) that the "job now is to try to catch up with the conservation planning required by the conservation compliance provision and hope to incorporate the CRP contracts into those plans."[28]

Computerized sampling continued to form the basis for analysis. Yet visual mapping for farm adjustment in Montana as early as 1940 had shown great variability in land classification, with pockets of the first three grades of farmland interspersed in predominantly grazing areas throughout the eastern two-thirds of the state. Soil survey in that state had progressed earlier and farther than elsewhere in the northern Plains but was only 55 percent completed in 1983. At that time a range specialist of the Montana Extension Service, addressing the Great Plains Agricultural Council in support of "sod-buster" legislation, estimated that 59 percent of the land in eastern Montana was suitable for intensive cultivation, 11 percent, grade IV; and the remaining 30 percent, grazing grades V–VIII. He acknowledged that the exact acreage in land capability classes was not yet known, and, in fact, his allotments for the state as a whole, based on preliminary findings of the RCA inventory, amounted to approximately 3.3 million acres less cropland in grades I–III, about 1.6 million acres less for the marginal grade IV, but over 4 million acres more in the grazing categories than were ultimately so defined by the RCA.[29]

The discrepancies emphasize what had been repeatedly urged for years, that detailed environmental study was still needed on a localized, even individualized, basis. H. H. Finnell's criticism of "generalized zones" had stressed in 1949 that consideration of the variability of conditions was prerequisite to land-use planning and conservation. This conceptualization had underlay the whole-farm management program so long promulgated by the SCS. It remained in the *Second RCA Appraisal* as a warning attached to the source analysis and in the USDA's *Update* for program development from 1988 through 1997, reiterating the continued need for better methods to estimate erosion and quantify its damages.[30]

The process by which farmers offered land for conservation reserve required such data immediately. Through a contractual bidding arrangement, they themselves were to initiate the action. The program, estimated to cost an average of $450 to $500 an acre over the ten-year contract, more than double the prevailing selling price of cropland in the western Plains, was expected to retire the acreage permanently. The cross-compliance provisions applicable to "sod-busting" or intensive cultivation of highly erodible tracts reinforced the commitment.[31]

Upon receipt of an application, ASCS personnel were to review the cropping history "on a field-by-field basis" to determine whether it met cropland and ownership eligibility. If that review was satisfactory, the SCS then determined whether the tract was considered highly erodible on the basis of available soil survey data and existing conservation measures. Unless that determination was conclusive, SCS technicians were to conduct on-site investigation. With such certification the acreage could be presented by bid proposing a rental rate and specific detail concerning the commodity base or bases to be retired. These bids must meet approval, in turn, by the county ASCS committee and the state ASCS office before they were forwarded to a regional center and to Washington.

The USDA was to review all bids and set maximum acceptable rental rates for "predesignated areas," that is, state and substate pools. State ASCS offices were to decide which counties fit the generalized production, as well as erosion, target criteria and, under those delimitations, to select bids within the assigned bid caps. No bid could exceed $50,000 annually to a single producer. The SCS was to visit the site to verify the erodibility characteristics and then to work out with the producer a plan detailing the areas to be planted with permanent vegetation and the conservation measures to be practiced on the accepted acreage. Finally, a written contract was to be prepared, signed, and reviewed by the local conservation district and county committee.

The RCA distinction between erodibility and the existence of erosion as a factor in current production appears to have had little significance in relation to the CRP. While index rates of erodibility were high for much of Montana, fewer than 1,000 acres of cropland in the state were projected to lose as much as 5 percent productivity over the next hundred years under 1982 management practices (EPIC model). The USDA, nevertheless, ruled somewhat more than 8 million acres eligible for the CRP. The erodibility index rates showed less than 2.3 million acres in North Dakota and about 1.7 million acres in South Dakota erodible at rates requiring intensive conservation. In North Dakota, too, fewer than 1,000 acres were estimated to be losing productivity under current management practices at greater than 5 percent over a hundred years, but over 5 million acres were so categorized in South Dakota. The USDA, however, qualified for the CRP almost 2.8 million acres in North Dakota and only about 2 million acres in South Dakota. Crop surpluses, rather than erodibility, seem to have been the determinant.[32]

The retirement of some 60 million acres of cropland in the northern Plains and an additional 25 to 30 million acres, each, in the southern Plains and mountain states, as projected in the *Second RCA Appraisal*—or even the 40 to 45 million acres nationally projected for the CRP—yet has a long way to go. By July 1988, 6.9 million acres had been enrolled in the northern Plains, 4.5 million acres in the southern Plains, and 5.6 million acres in the mountain states, as part of 28 million acres enrolled nationally. In the three states considered for this study, 6,228 contracts, covering 2,264,770 acres, were enrolled in Montana, 12,647 contracts, for 2,175,123 acres, in North Dakota; and 7,488 contracts, for 1,222,860 acres, in South Dakota. This amounted to only 28 percent of the acreage designated as eligible in Montana, but 78 percent of that in North Dakota and 60 percent of that in South Dakota. Annual rental costs averaged $37.45 an acre and federal cost-share payments, presumably a one-time charge to establish ground cover, averaged $21.12 an acre in Montana. In North Dakota the rental cost averaged $38.22 and the cost-share payments, $31.02; and in South Dakota, $39.96 and $28.97, respectively.[33]

The drought of the early eighties in the lower Plains may have generated more interest there. Ten counties in southeastern Colorado had reached the limit of 25 percent of cropland acreage enrolled by mid-1987. The spread of drought northward brought increased enrollments in the Dakotas and Montana during 1987 and, although somewhat diminished, early 1988. It set the course for renewal of the CRP and continuation of the regional focus on the Plains, generally, in the agricultural legislation of 1990.[34]

Monitoring of the program hinges upon the effectiveness of the local ASCS committees and SCD supervisors. The authority of these institutional forces to regulate land use for purposes of conservation under popular mandate has been long-standing. If used, it could have obviated the argument for federal sod-busting measures. In most cases, however, the proportion of popular vote necessary to permit such action has been high—under federal commodity-control legislation, a two-thirds vote by referendum, and under SCD enabling statutes, ranging from approximately two-thirds in Montana and North Dakota to three-fourths in South Dakota. Assertion of direct regulatory action has consequently been restrained. Study of the operation of the committees in relation to earlier set-aside programs showed them to be reluctant to issue decisions that would cause forfeiture of program payments and reluctant to brand their fellow farmers as "defrauders" of the government. Because committees viewed the program requirements as complex and too frequently changing and because they often questioned the appropriateness of the policies so regulated, they tended to be tolerant of evasion. Restrictions on land use adopted by the Cedar Creek Grazing District in North Dakota, the rural zoning action undertaken by the Corson County Land Planning Committee in South Dakota, and, more recently, enactment of a land-use ordinance by the Lewis and Clark Conservancy District in Montana were measures notable for their singularity, actions taken by a strongly entrenched constituency against what were viewed locally to be intrusive external elements. For the most part, consensus dictates the measure of control exercised.[35]

Criticism questioning the representative quality and the effectiveness of these authoritative bodies has not been absent. Membership tended to be comprised of the same individuals. They were generally closely aligned with local power structures and frequently themselves moved into political office on the basis of connections thus developed. Some analysts hailed the numerous public meetings held in connection with the RCA appraisal as an opening of the process, but warned that it would bring increased controversy to the fore. Others denied that the basic decisions had been effected or greatly influenced by popular voice.[36]

Whether congressional approval as an expression of public support will sustain the program remains to be seen when the variability of climatic and market conditions brings a reversal of recent trends. The contrast in conceptualization of the agricultural problem within the short period between the publication of the NALS in 1981 and the legislation of 1985 is indicative of the impact of the circumstantial conditions on policy formation. The disruption of communities and the discontent of dislocated farmers in the thirties and fifties compelled retreat from earlier proposals for large-scale land retirement that were far less costly to the nation. As the generalization of sod-busting and conservation compliance provisions extends beyond the remunerated contract periods, local monitoring committees and legislative spokesmen may find little supportive consensus.

Few have attempted to base environmental concern on economic grounds. Discounting the value of productivity loss in bushel yields is difficult when the loss by erosion over a hundred years is projected to be less than 1 percent, as for the northern Plains (defined as the Dakotas, Nebraska, and Kansas) or even the 1.8 percent of the mountain states. The issue, instead, has been presented as one of ethical concern to preserve the heritage of soil and water resources for generations much further in the future. Hyperbolic extremes of the consequent dilemma are juxtaposed—balanced, on the one hand, by the highly publicized proposal to convert large parts of the Plains into a huge "buffalo commons," initiated by a reinvented "1990s version of the 1930s Resettlement Administration," on the other, by the Worldwatch Institute's projections that between the years 1980 and 2000 the world cropland base will expand about 4 percent while population growth will increase by some 40 percent. The ethic of social responsibility in program development applicable to environmental policy cannot ignore the fact that approximately 50 percent of the world population is now dependent on American agricultural productivity.[37]

Dry-land agriculture has provided major contributions toward meeting that demand. In 1987 North Dakota and Montana ranked second and third, behind Kansas, in volume of American wheat production. Irrigation had little relevance to the effort. The amount of irrigated cropland in eastern Montana in 1987, after a decade during which precipitation reached the long-term average only three years, totaled about 795,000 acres out of over 7,847,000 acres of harvested cropland. In western South Dakota less than 92,000 out of 3,206,000 acres of harvested cropland were irrigated; and in western North Dakota, less than 43,000 out of 6,763,000 acres.[38]

Under dry-land technology, Hardy Webster Campbell's program for maintenance of a top-soil mulch coupled with alternating years of cropping after a year of summer-tilled fallow has been greatly modified but not eliminated. Found to lead to soil blowing in periods of drought, the recommended mulch became during the twenties a rough stubble and in some areas was combined with strip cropping. SCS demonstrations a decade later emphasized these tillage procedures, linked with the rotational cropping of Thomas Shaw's earlier programmatic emphasis and contoured terracing, adapted from the methodology of the central Plains. Neither of the latter two practices found widespread acceptance in dry-land farming to the north, although under AAA incentive payments, contouring was extended to some plowing and cultivation operations. Stubble mulching, strip cropping, and use of alternating fallow remained the methodological basis of the development.

With the reinstitution of acreage cropping quotas in 1955, pressure to increase production on the restricted allotments led to the addition of chemical fertilizers, fungicides, and insecticides for cropping in a region where they had not been widely used before. Nitrogen use, which in 1950 had amounted to only 22,000 tons in the mountain states, including Montana, and a similar amount in the northern Plains (defined as the Dakotas, Nebraska, and Kansas), by 1956 had increased to 73,000 tons and 93,000 tons in these regions, respectively, and by 1961, to 150,000 and 326,000 tons. Phosphate use increased from 39,000 tons to 107,000 tons and to 140,000 tons in the mountain states and from 44,000 tons to 91,000 tons to 163,000 tons in the northern Plains for the cited years. Although still a less sizeable element in wheat production costs than elsewhere in the United States, outside the central Plains, by 1987, 622,000 tons of nitrogen and 229,000 tons of phosphate were being applied in the mountain states, and 1,797,000 tons of nitrogen plus 487,000 tons of phosphate in the northern Plains. A large part of the increased wheat yields during the period—from 18.5 bushels to 32.2 bushels an acre in Montana, from 13.9 bushels to 29.5 bushels (a peak of 36.4 bushels in 1985) in North Dakota, and from 10 bushels to 30.2 bushels in South Dakota—must be attributed to use of these additives. Costly and environmentally hazardous, their application will probably be intensified as cropland acreage is contracted.[39]

Cultivated summer fallow in 1987 accounted for some 10.3 million acres, 27.7 percent of the total cropland acreage in the mountain states and 16.9 million acres, 19.4 percent of that in the northern Plains (including the Dakotas, Kansas, and Nebraska). But with growing use of herbicides during the sixties, the loss of soil moisture through weed growth no longer required extensive tillage of fallow. At the same time, developing research on the accumulation and retention of soil moisture during fallow indicated that little was added beyond the first few months of recovery from moisture depletion in cropping. Through the early seventies fallow with minimum tillage, use of contact and pre-emergence herbicides shortly after harvest, and perhaps another herbicide application with a single sweep before planting was under test. Before the end of the decade this "no-till" fallow and even "no-till" cropping were being introduced as recommended practice. Problems still remained: undisturbed wheat stubble required greater power, larger tractors, and specially designed seeding equipment to avoid clogging; heavy equipment, in turn, tended to break down contouring and terracing; and large-scale use of herbicides and insecticides added to the danger of ground-water pollution. But for farmers in a level region with minimal drainage and runoff these environmental concerns were in large measure counteracted by the better snow retention, lessened surface evaporation, and reduced wind erosion of undisturbed stubble. The number of tillage operations was reduced from seven to ten under fine-mulch fallowing to none or perhaps one with "no till" practice. Use of such methodology helps to account for the minimal erosion anticipated for the region.[40]

Emerging problems with salinity in dry-land operations gave added impetus to the shift away from summer-fallow practice. Long identified as a concern in irrigated farming, one which in the late forties had compelled abandonment of the proposed diversion canal from the Ft. Peck Reservoir for irrigation of northwestern North Dakota, shallow soils, frequently described as "retractive" and clay-like in the top layer with calcareous hard-pan near the surface, had been noted in Gieseker's early surveys along the High Line of

Montana. As Gieseker extended his surveys southward into the Judith country and eastward into the Musselshell district, he had found more areas of gravel-capping with limestone rock near the surface, particularly adjacent to the foothills and mountains. Dry-land benches of glacial drift and rock sedimentation interspersed in these districts afforded areas of rich farmland, but experience revealed that they required careful practice to forestall not only blowing but also rising salinity. The long period of increased rainfall from the forties into the seventies, husbanded under more widespread and efficient practice of summer tillage, built up deposits of mineral shale in poorly drained soils that gradually resulted in seedling kill or stunted plant growth. Saline seepage emerging downward from the affected areas often gave the first warning of a long-developing condition.

By 1971 over 80,000 acres in Montana were showing evidence of the problem. Research then initiated at a site near Sidney led to the conclusion that if summer fallow were practiced less often and more intensive cropping pursued, much of the water contributing to the seeps could be used. This would require an alternative means of water recharge between harvest and planting, alternative methods of weed control, minimization of plant diseases resulting from repeated small-grain cropping, and more understanding of required fertilization. At mid-decade workshops were being held from Montana south to Colorado for discussion of methods by which water stored in root zones might be better utilized for cropping. Research was directed to improved measurement of soil-moisture levels and analysis of the biological processes of water use in plant growth. Color infrared photography and thermal infrared imagery from aerial surveys and satellite monitoring have in recent years made mapping of soil moisture conditions possible for guidance in adjustment of dry-farming practices. Intensive cropping with small grains, grasses, and deep-rooted crops such as alfalfa and safflower in identified water recharge areas can, it has been found, reclaim as well as prevent seep development. No-till, continuous cropping has been specially authorized as conservation practice in areas of western North Dakota and northern Montana.[41]

The *Second RCA Appraisal* in 1989 estimated that there were nearly 2.73 million acres in North Dakota highly saturated with either sodium or saline salts and an additional 9.4 million acres slightly saline. Montana and South Dakota, each, had approximately 2 million acres highly saturated and some 7 million acres slightly affected. The report concluded that the land could be reclaimed within five years under intensive cropping in the recharge areas.[42]

Improved varieties of wheat and barley have also contributed to regional productivity. Acreage in other small grains, however, has shown little increase in the northern Plains. Corn acreage has declined throughout the region. Sorghum remains a minor crop, as are flaxseed, sunflower seeds, safflower, triticale, and soybeans. Acreage in alfalfa, primarily for hay, has largely supplanted the earlier efforts to raise such erosive cultivated crops as corn and sorghum in the region. Since 1950 the acreage in alfalfa has more than doubled in eastern Montana, has increased by nearly five times in western North Dakota and by eight times in western South Dakota; but it remains less than 29 percent of the harvested crop acreage in western South Dakota, 11 percent in western North Dakota, and 10 percent in eastern Montana.[43]

The basic structure of the cropping program has not greatly changed in eastern Montana since 1925. The acreage of harvested cropland, the acreage of harvested wheat land, and the acreage of total cropland have been increased at about the same ratios, 45 to 48 percent. Greater changes marked the agricultural activity in the Dakotas. The acreage of harvested wheat remained nearly the same in western North Dakota; but the amount of cropland harvested declined about 7 percent, while that of total cropland was increased 22 percent. As farm size was expanded, the increase represented greater use of summer fallow and tame hay pasturage. In western South Dakota the wheat acreage was increased 278 percent, while the amount of harvested cropland grew only slightly, 6 percent, and the acreage of total cropland declined 21 percent. The trend toward diversified cropping evident in that area prior to 1925 had ended.

Diversification within the regional economy remains in the form of a duality of wheat farming and cattle raising, primarily as combined operations. The number of cattle was increased to near record levels in dry-land districts of all three states by 1974. Despite considerable thinning of herds through the recent drought period, it still remains higher than at any census count prior to that year in the western Dakotas and 42 percent higher than in any census year prior to 1955 in eastern Montana.

This is not, and has not since 1910 in eastern Montana, since 1925 in western South Dakota, and since 1940 in western North Dakota been a region of farms smaller than the "stock-farming homestead" of 640 acres. There are relatively few very large, corporate enterprises, such as the Campbell Farming Corporation, in the region; but large-scale, mechanized, family operations, as projected by M. L. Wilson and Elmer Starch after the Fairway Farms experiment, have predominated. In 1987 farms in eastern Montana averaged 2,969 acres; those in western South Dakota, 2,727 acres; and those in western North Dakota, 1,343 acres. Size has been increased as operators incorporated larger holdings of pasture for livestock. In western North Dakota 59 percent of the land in farms is identified as cropland; but in eastern Montana only about one-third is cropland, as compared to 42 percent in 1924; and in western South Dakota less than one-quarter of the farm acreage is cropland, compared to 51 percent in 1924. Cropland averages 930 acres a farm in eastern Montana, 786 acres in western North Dakota, and 599 acres in western South Dakota.

As farm size has been expanded, the number of operators has been greatly reduced—from 37,000 in 1925 to 17,000 in 1987 in eastern Montana, from 31,000 to 15,000 in western North Dakota over the same period, and from 21,000 to 9,000 in western South Dakota. Contrary to popular belief, however, these decreases have stabilized over the past twenty years. Despite the drought and general agricultural depression of the past decade, the number of operators has increased somewhat in eastern Montana and western South Dakota, while remaining relatively unchanged in western North Dakota.[44]

The Wilson-Starch initiative of 1930 in advocacy of the Christgau proposal rested on the belief that large-scale wheat cropping in the northern Plains would be more economically viable than production among its competitors, at home and abroad. It cannot be said that agriculture in the region, or generally in the nation, has been prosperous through the recent period of low prices and drought. Excluding direct governmental payments, positive residual returns to management and risk, apart from allocations for owned inputs of operating capital, machinery, land, and labor, have been rare for producers of the commercial crops or livestock reported. Only the Northeast, with a small production of soft red winter wheats, accumulated such a profit from wheat cropping over the fifteen years from 1975 through 1989. The north-central region, with 9 percent of the crop, also soft red winter varieties, recorded the least loss, $31.03 per acre; and the Northwest ranked next, with a cumulative loss of $61.38 an acre, applicable to 12 percent of the total production, much of it irrigated white wheat, a soft variety of particular demand in the newly opened trade with China. The hard red winter wheats of the central and northern Plains, with about one-third of the national wheat production, ranked next, with losses of $134.37 and approximately $152.08 an acre, respectively. Hard red spring wheat in the northern Plains, 19 percent of the total crop, lost $223.83 an acre during the period, and the remaining regions of the nation, including the national average, whether as hard or soft wheats, did markedly poorer. A continuing price preference for hard red spring wheat denotes sustained demand for that production in the northern Plains, but increased emphasis upon winter wheat in Montana and South Dakota during the period reflects the cost advantages. Even such considerations have not counteracted the difficulties across the South. Nearly a quarter of the wheat cultivation in the United States has been carried on in areas of less profitability and higher crop cost than in the northern Plains despite the problems of the eighties. This has encouraged expanded production, not land retirement, in the region. Given the lack of economically remunerative alternatives, it is not likely to change in a competitive market.[45]

Misconceptions and myths have too frequently entered into the national land-use planning. There are still those who see the Plains as a "Great American Desert"; still those who, ignoring the ravages of drought without supplementary feed grains and tame hay for livestock, envisage production of range cattle as the only suitable land utilization in the region; still those who look to

TABLE 9.2. Residual Returns to Management and Risk on Wheat Cropping,
United States and Regions, 1975–1989 (dollars per seeded acre)

	1975	1976	1977	1978	1979	1980	1981
United States	18.72	– 2.67	–10.55	–12.49	9.03	– 8.81	–17.50
Central Plains (HRWW)	11.17	– 5.63	–12.51	5.56	20.75	– 0.41	–20.03
North Central (SRWW)	36.85	24.00	19.09	16.83	27.71	10.78	– 7.26
Northeast (SRWW)	81.05	48.97	34.32	29.65	28.71	15.65	2.37
Northern Plains (HRSW)	18.11	– 8.52	–10.98	–10.10	–13.09	–32.53	–25.17
(HRWW)	23.34	1.04	–10.34	7.79	– 4.10	–18.13	– 8.30
Northwest (White)	49.86	34.91	–12.61	–19.69	– 6.40	20.79	20.00
Southeast (SRWW)	1.72	18.93	– 1.73	1.37	10.90	1.64	–12.88
Southern Plains (HRWW)	9.69	–10.60	–12.01	– 7.52	23.55	–16.71	–20.91
Southwest (HRWW)	56.20	9.48	–28.42	– 0.03	45.89	8.30	–38.41

Sources: Computations based on Robert G. McElroy and Cole Gustafson, Costs of Pro-
ducing Major Crops, 1975–81; USDA, ERS, National Economics Division, ERS Staff Report no.
AGES850329 (Washington, D.C., 1985), pp. 34, 36–39, 41–43, 46, tables 24, 26–29, 31–33, 36;
USDA, ERS, Economic Indicators of the Farm Sector: Costs of Production, 1983, ECIFS 3-1
(Washington, D.C., 1984), pp. 56, 58–61, 63–65, 68, tables 21, 23–26, 28–30, 33; 1984, ECIFS
4-1 (Washington, D.C., 1985) pp. 51, 53–56, 58–60, 63, tables 21, 23–26, 28–30, 33; 1985,
ECIFS 5-1 (Washington, D.C., 1986), pp. 53, 55–58, 60–63, 65, tables 21, 23–26, 28–31, 33;
1986, ECIFS 6-1 (Washington, D.C., 1987), pp. 47, 49–52, 54–57, 59, tables 21, 23–26, 28–31,
33; 1987, ECIFS 7-3 (Washington, D.C., 1989), pp. 43–51, tables 20–28; 1988, ECIFS 8-4
(Washington, D.C., 1990), pp. 69–77, 83, tables 51–59, 70; 1989, ECIFS 9-5 (Washington, D.C.,
1991), pp. 40–48, tables 17–25.

1982	1983	1984	1985	1986	1987	1988	1989	Cumulative Total
−21.13	−12.11	−16.84	−29.26	−44.06	−35.95	−25.65	−34.10	−243.17
−14.63	− 1.70	− 8.17	− 7.72	−32.49	−24.02	− 9.66	−34.88	−134.37
−29.50	−19.62	−24.03	−19.47	−47.12	−23.49	2.36	1.84	− 31.03
—22.82	− 5.24	− 1.12	11.73	−11.98	−13.29	45.63	−20.53	223.10
−16.95 −13.57	−22.64 −14.23	−15.01 −28.59 }	−22.67	−38.13	−31.27	−48.55	−33.32	{ −223.83 −152.08
−9.24	16.80	− 8.36	−43.24	−53.32	−36.45	8.89	−23.32	− 61.38
−33.70	−31.32	−21.03	−48.23	−50.21	−38.56	1.10	−24.81	−226.81
−22.37	− 7.64	−20.36	−45.97	−57.50	−55.35	−41.86	−55.56	−341.12
−97.50	−92.96	−104.29	−56.76	−89.36	−98.47	− 7.23	−67.74	−561.30

Note: This series presents discrepancies. On the assumption that divergences in later compilations represent corrections, the latest data found have been used. The reporting for the northern Plains as a generalization of varieties since 1984 complicates comparison with the earlier period when spring (HRSW) and winter (HRWW) wheats were listed separately. In Montana, winter wheat has comprised from 45 to 67 percent of the production since 1975; in South Dakota, it has varied from 17 percent in 1979 to well over 50 percent in recent years. North Dakota and Minnesota primarily produce spring varieties. For purposes of computation from 1985 through 1989 the reported losses were divided between spring and winter varieties. The northern Plains as here defined cover the area generally identified as Hard Red Spring Wheat, from Minnesota to Montana inclusive and South Dakota. Soft wheats (SRWW or White) generally serve differing markets from the hard wheats of the Plains.

irrigation as the only basis for successful agriculture there; and still those who cherish the small, diversified homestead operation as the agrarian dream, regardless of the environment. Over the past half-century the myth of regional unity has established new distortions in popular perceptions. For residents of the Plains, generally, the sharing of drought experience has developed strong community loyalty, pride in coping ability, but defensiveness against external intrusion. In the public mind such a generalized identity has too often obliterated recognition of local and individual variations. Climate, terrain, soil, and vegetational characteristics are markedly different within the Plains and between localities even within individual states. Management requirements and managerial responses are diverse. These facets of dry-farming experience are not easily cast into broad land-capability classes for incorporation into bidding pools.

An early critic of the RCA survey, noting the contention among interested agencies and pressure groups during the drafting process, expressed alarm that it would be accepted as the product of scientific analysis when, in fact, the political role in the discussions, as described by a high-level participant, had been "frightening." Congressional representatives of producer constituents, having long fought to preserve existing quota and price-support arrangements, had established political connections upon which they drew for support. Politicization of proposals through the interaction of constituent interest and interagency rivalries was inevitable. Livestock and range-management interests were among the most active proponents of the program as long as it embodied state and local participation. Copies or summaries of the *1982 Final Program Report* under the RCA were sent to some 16 federal agencies and 128 national interest groups, including not only agricultural, civic, consumer, and environmental bodies but also "religious, minority, and youth organizations." The responsiveness to political pressures was evident in the decision to plan production adjustments under the assumption that little change would develop in marketing programs.[46]

Since the mid-twenties a powerful segment within the USDA has sought targeted land retirement as a more efficient approach. During the thirties the drought provided a justification for public action focused upon the northern Plains; during the fifties and eighties that rationalization was far less applicable to the region. The RCA proposals and their formalization in the CRP, offered as an environmental agenda, in effect shaped environmental and agricultural interests, alike, to fiscal considerations. Since analysis indicated that surplus domestic stocks could be brought under control with 45 million acres out of production and the commodity payments for annual crop diversions would be more costly than a program incorporating long-term land retirement, the conservation reserve was targeted to remove wheat- and corn-producing acreage. Far fewer contracts were negotiated in the midwestern, Delta, and southeastern states that produced the more heavily eroding soybean crop, of stronger market demand. The acre-cost of environmental conservation was thereby raised, while areas of major degradation were ignored. Indeed, if the experience of the earlier program of "conservation reserve" is repeated, soils of better quality will be overutilized to compensate for income reductions.

How fervently the Soil Conservation Service itself pressed for the CRP may be questioned. Even in the *Second RCA Appraisal* its recommendation for permanent conversion of soil "to sod or trees, or long rotations with several years of sod," is applied only to areas of very high susceptibility to erosion damage, areas with an erodibility index of fifteen or greater. Under the SCS recommendations of the Great Plains Program in the late fifties only 11 to 14 million acres had been estimated to require retirement. The SCS program, with its focus on whole-farm management, had rested upon the conception that farm plans should delineate and treat distinctively an intermingling of soil conditions. The generalized criticism of government programs that fostered production, as voiced by Held and Clawson, greatly disturbed so ardent a proponent of soil conservation as R. Neil Sampson, long associated with the National Association of Conservation Districts and in 1981 author of the volume *Farmland or Wasteland: A Time to Choose*. He saw the approach as a threat to funding of more balanced efforts for conservation. It was, he contended, a result of a "surplus syndrome" that was dying hard with a generation of agricultural policymakers who had spent their careers coping with low prices as a consequence of crop surpluses. It would die, he predicted, as demand for agricultural production increased while the effects of American technological breakthroughs declined.[47]

Those developments had in the postwar years elevated expectations of peoples around the world. Despite the frequently mistaken cultural assumptions of activities under President Truman's Point Four program, the Cold War rivalries of Public Law 480 funding, and the cutthroat trading of the struggle for export markets, generally, American productivity was of life-saving importance again and again, in every decade since 1940. While world food production has reached unprecedented levels, 20 percent of the population, according to the United Nations, continues to suffer from poverty and starvation. As population increases, the per capita acreage available for cultivation decreases. Cereal imports of developing countries rose from 20 million metric tons in 1969–1971 to 69 million metric tons in the years 1983–1985, and current projections are that they will reach 112 million metric tons by the year 2000. Ninety-five percent of the growth will come in developing nations, where urbanization is increasing far faster than in industrialized states. Income in the developing nations is also growing, at three to four times the rate in the latter, and with increased income comes growing demand for the specialized foods of wealthier societies. Estimated export demands on the United States were raised under RCA appraisals between the early eighties and 1989 to encompass higher requirements for corn, sorghum, soybeans, and wheat; but a projection by Resources for the Future as early as 1982 had called for more than double the food and fiber estimates of the intermediate scenario under the more recent RCA. In mid-1991 the United Nations raised its calculation of world population in the year 2025 by 38 million, about half a percentage point above its projections only two years earlier. "Periods of stress and pessimism [regarding the ability to meet needs] are occurring with increasing frequency," the authors of the *Second RCA Appraisal* noted.[48]

Planning for a lowering of production in the United States, which in 1987 provided over 43 percent of international exports of wheat and flour, almost twice as much as its nearest competitor, poses awesome responsibility in relation to world needs. Conceding that the volatility of export demand made such projection difficult, the authors of the *Second RCA Appraisal* concluded that under intermediate conditions higher yields through new developments in technology would offset need for increased acreage in cropland to meet export requirements during the period to 2030. For wheat acreage they anticipated a requirement of about 7 million acres less than were so utilized in 1982. Under high stress conditions an increase of approximately 30 million acres might be necessary. Since study had identified about 35 million acres in the United States as offering "high potential" for conversion to crop use and as much as 57.7 million acres of range and pasture in the Great Plains with "medium" to "high potential" for such utilization, they saw little prospective danger of shortages. In the temper of the times one proponent of the CRP criticized even such planning analysis as indicative of "some serious policy conflicts" within the USDA in commitment to conservation under the acreage reduction program.[49]

The analysis did raise question why such vastly larger land retirement should be initiated as an immediate agenda. Conceding that their projections of acreage requirements "implicitly" assumed continuation of the incentives over the next half century as they currently existed, "prices higher-than-market and set-aside acres on virtually every farm," the RCA analysts argued that if market-clearing prices and government-subsidized surpluses were abandoned, lower land prices "would serve to alter production toward 'land-using' technologies and away from 'land-saving' technologies." The forecast of large-scale operations was, indeed, the adjustment that the Wilson-Starch initiative had anticipated in the Fairway Farms proposal and that local adjustment patterns in dry-land areas had developed. That it had resulted or would result in the predicted environmental disaster was less obvious.[50]

The 1982 National Resources Inventory (NRI) listed less than 5 million acres of nonfederal cropland for the *whole state* of Montana under land capability grades IV through VIII, that is, only 62 percent of the acreage declared eligible for retirement as cropland under the CRP. Over 3.4 million acres of the NRI classification were in the marginal agricultural grade IV. Much of that acreage was already included in the nearly 1 million acres of *eastern* Montana cropland devoted to alfalfa and in 771,000 acres of cropland in "plowable pasture" reported for that area.[51] The RCA assessment projecting minimal loss of productivity under 1982 management practices over a hundred years does not indicate the prevalence of serious environmental neglect in the region. Where erosive practices are pursued, the sanctioning authority of monitoring local committees has been available, as it must operate under the current legislation, dependent upon popular agreement concerning the need for action. Dry-farming methodology has been adapting to environmental imperatives there for at least seventy years. The United States in seeking General Agreement on Tariffs and Trade has asked that the nations of the world move strongly toward competitive free enterprise. The Wilson-Starch proposal set the same challenge for domestic as well as international trade policy. Settlers freed to risk pursuit of dry-land development not only conducted the experimental process, they established viable interests—for themselves and their communities—that ought not now be sacrificed to hyperbole, myth, or politics.

NOTES

INTRODUCTION

1. The background provided in this and the following seven paragraphs is drawn from Mary W. M. Hargreaves, *Dry Farming in the Northern Great Plains, 1900–1925*, Harvard University Press, 1957), pp. 3–70, 83–125, 179–95, 204, 452, 464–73 passim.

2. Cited passages are from Jefferson Davis, "Report of the Secretary of War on the Several Railroad Explorations," USWD, *Reports of Explorations and Surveys to Ascertain the Most Practicable and Economical Route for a Railroad from the Mississippi River to the Pacific Ocean*, 33d Cong., 2d sess., *House Executive Document* no. 91 (Washington, D.C., 1855), 1:84–85; F. H. Newell, "Water Supply for Irrigation," USGS, *Thirteenth Annual Report, 1891–1892*, pt. 3, pp. 45–62, 69–73.

3. Quotation from *Campbell's 1902 Soil Culture Manual: Explains How the Rain Waters Are Stored and Conserved in the Soil; How Moisture Moves in the Soil by Capillary Attraction, Percolation, and Evaporation; and How These Conditions May be Regulated by Cultivation* (Holdrege, Nebr.: n.p., 1902), p. 92.

4. Computed from Hargreaves, *Dry Farming . . . 1900–1925*, tables 5 and 6 pp. 441, 468–69, and Table 1.1 in this volume. A small percentage of the acreage in alfalfa had been irrigated; irrigation of the other crops was negligible.

5. H. Ross Toole, *Twentieth Century Montana: A State of Extremes* (Norman, Okla.: University of Oklahoma Press, 1972), p. 26. Milburn Lincoln Wilson more conservatively placed abandonment in the "triangle" area at 49 percent by the spring of 1922. (Wilson, *Dry Farming in the North-Central Montana "Triangle,"* MSC, *Extension Service Bulletin* no. 66 [Bozeman, Mont., 1923], p. 123).

6. Wilson, *Dry Farming in the North-Central Montana "Triangle,"* pp. 28, 123–25.

7. E. A. Willson, H. C. Hoffsommer, and Alva H. Benton, *Rural Changes in Western North Dakota: Social and Economic Factors Involved in the Changes in Number of Farms and Movement of Settlers from Farms*, NDAES, *Bulletin* no. 214 (Fargo, N.D., 1928), pp. 50, 52, 57–58.

8. Hargreaves, *Dry Farming . . . 1900–1925*, pp. 499–502.

9. Ibid., pp. 426–31, 505–7; W. A. Schoenfeld and others, "The Wheat Situation; A Report to the President by the Secretary of Agriculture—The Price and Purchasing Power of Wheat," USDA, *Yearbook of Agriculture, 1923*, p. 119; Willson, Hoffsommer, and Benton, *Rural Changes*, table 6 p. 29; L. A. Reynoldson, "An Organization System for Farms in Northeastern Montana, Preliminary Report" (USDA, BAE, mimeographed report; Washington, 1925), pp. 1–2; L. A. Reynoldson, "An Organization System for Farms in Southeastern Montana, A Preliminary Report" (USDA, BAE, mimeographed report; Washington, 1925), pp. 1–2; L. A. Reynoldson, "The Progress of Farmers in Northeastern Montana for Two Years, Preliminary Report" (USDA, BAE, Division of Farm Management and Land Economics, mimeographed report; Washington, 1923), pp. 3–4.

10. Hargreaves, *Dry Farming . . . 1900–1925*, pp. 509–18; *Twelfth Census of the United States, Taken in the Year 1900*, vol. 5, *Agriculture*, pt. 1, table 19 pp. 287, 292, 296–97; *United States Census of Agriculture, 1925*, pt. 1, pp. 1010–17, 1066–75; pt. 3, pp. 90–97—county table 2.

11. Nils A. Olsen and others, "Farm Credit, Farm Insurance, and Farm Taxation," USDA, *Yearbook of Agriculture, 1924*, table 1 p. 188. No other region showed interest costing over $180.

12. *Federal Reserve Bulletin* 1 (June 1915):75, and 4 (April 1918): 312; Olsen and others, "Farm Credit," p. 194. Almost one-third of the state banks reporting to the Montana superintendent of banks in 1921 held less than the required legal reserves (Toole, *Twentieth Century Montana*, p. 81).

13. Olsen and others, "Farm Credit," p. 193; Gabriel Lundy, *Farm-Mortgage Experience in South Dakota, 1910–40, with Special Reference to Three Townships in Each of the Counties of Brookings, Clark, Haakon, Hyde, and Turner*, SDAES, *Bulletin* no. 370 (Brookings, S.D., 1943), table 6 pp. 28–29.

14. Lundy, *Farm-Mortgage Experience*, table 6 pp. 28–29; Gabriel Lundy, *Mortgage Loans on Farm Real Estate in Haakon County South Dakota 1910–1930*, SDAES, *Circular* no. 5 (Brookings, S.D., 1932), p. 45.

15. *Scobey Sentinel* (Mont.), April 9, 1920; Hargreaves, *Dry Farming . . . 1900–1925*, pp. 521–23. Unless otherwise indicated, the ensuing section on government credit programs is adapted from the latter source, pp. 523–31. See also Lundy, *Mortgage Loans*, pp. 37–46.

16. Montana Register of Lands, *Report, 1926–28*, p. 63; North Dakota Commissioner of Lands, *Report, 1926–28*, p. 5, and *1928–30*, p. 5; W. A. Hartman, *State Land-Settlement Problems and Policies in the United States*, USDA, *Technical Bulletin* no. 357 (Washington, D.C., 1933), pp. 59–60.

17. Willson, Hoffsommer, and Benton, *Rural Changes*, p. 55 table 25.

18. L. C. Gray and others, "The Utilization of Our Lands for Crops, Pasture, and Forests," USDA, *Yearbook of Agriculture, 1923*, p. 450; James H. Shideler, *Farm Crisis: 1919–1923* (Berkeley, Calif.: University of California Press, 1957), pp. 46–47. See also A. B. Genung, "Agriculture in the World War Period," *Yearbook of Agriculture 1940: Farmers in a Changing World*, pp. 280–84, 288–92, 294–95.

19. Gilbert C. Fite, "The Farmers' Dilemma, 1919–1929," in John Braeman and others, eds., *Change and Continuity in Twentieth-Century America: The 1920's* (Columbus, Ohio.: Ohio State University Press, 1968), pp. 69, 72; U.S. Bureau of the Census, *Historical Statistics of the United States, 1789–1945: A Supplement to the Statistical Abstract* (Washington, D.C.: U.S. Government Printing Office, 1949), series E 88-104, p. 99, and series E 181–95, p. 106.

20. Shideler, *Farm Crisis*, p. 58; Hargreaves, *Dry Farming . . . 1900–1925*, p. 497. See also M. R. Benedict, *Freight Rates and the South Dakota Farmer*, SDAES, *Bulletin* no. 269 (Brookings, S.D., 1932), pp. 10–11.

21. L. F. Gieseker, *Soils of Sheridan County: Preliminary Survey Report*, MAES, *Bulletin* no. 158 (Bozeman, Mont., 1923), p. 3; L. F. Gieseker, *Soils of Daniels County: Soil Reconnoissance of Montana Preliminary Report*, MAES, *Bulletin* no. 174 (Bozeman, Mont., 1925), p. 6; L. F. Gieseker, *Soils of Phillips County: Soil Reconnoissance of Montana Preliminary Survey*, MAES, *Bulletin* no. 199 (Bozeman, Mont., 1926), p. 6; MAES, *Agricultural Service from the Montana Experiment Station: Thirtieth Annual Report, July 1, 1922, to June 30, 1923* (Bozeman, Mont., 1924), p. 6; USDA, *Annual Report of the Secretary, 1923*, p. 16. Hereafter reports of the state agricultural experiment stations will be cited by the number and terminal year of the publication.

22. On the development of protein measurement, see Hargreaves, *Dry Farming . . . 1900–1925*, pp. 533–34, and pp. 26–28 in this volume.

23. South and west of the Missouri River, the terrain was considerably more broken, with extensive "Badlands" areas in both the Dakotas and mountain outcroppings in central and southern Montana. Soils in the trans-Missouri region of South Dakota were frequently a heavy clay, slick when wet, relatively impermeable, and difficult to cultivate. For more extended description, see the Gieseker series of soil surveys in the bulletins of MAES and Rex E. Willard and O. M. Fuller, *Type-of-Farming Areas: North Dakota*, NDAES, *Bulletin* no. 212 (Fargo, N.D., 1927), pp. 21–22.

24. MAES, *Thirtieth Annual Report, 1923*, p. 8; *Twenty-Eighth Annual Report, 1921*, p. 9, are both quoted.

25. Hargreaves, *Dry Farming . . . 1900–1925*, tables 5 and 6, pp. 441, 468–69, 471–72, and Table 1.1 in this volume.

26. Computed from data in Table 1.1 in this volume.

27. Hargreaves, *Dry Farming . . . 1900–1925*, pp. 37–38.

28. Hargreaves, *Dry Farming . . . 1900–1925*, p. 544, quoted.

29. David F. Houston, "Annual Report of the Secretary," USDA, *Yearbook of Agriculture, 1919*, p. 27.

30. Lloyd D. Teigen and Florence Singer, *Weather in U.S. Agriculture: Monthly Temperature and Precipitation by State and Farm Production Region, 1950–86*, USDA, ERS, *Statistical Bulletin* no. 765 (Washington, D.C., 1988), pp. 144–45, 176–77, 204–5; Jim Langley and Suchada Langley, *State-Level Wheat Statistics, 1949–88*, USDA, ERS, *Statistical Bulletin* no. 779 (Washington, D.C., 1989), table 38 p. 42, table 52 p. 56, and table 63 p. 67.

Weighted according to crop areas, precipitation averaged less than 15 inches annually in all three states in 1952 and 1976 and in Montana and North Dakota in 1961 and 1988. Wheat yields of less than fifteen bushels an acre coincided with the low precipitation throughout the region only in 1952 but accompanied low precipitation in North Dakota in 1961 and 1988, South Dakota in 1976, and Montana in 1961 and 1985.

Notably indicative of the adaptation of the cropping in variety and tillage technology to the regional conditions was the fact that in North Dakota in 1967 and 1976, in South Dakota in 1974, and in Montana fifteen years of the thirty-eight-year period, above reported, the yield of wheat averaged fifteen bushels an acre or more, usually much more, despite less than 15 inches of annual precipitation.

31. Computed from *1987 Census of Agriculture*, vol. 1, pt. 26, pp. 180–87, 204–11, pt. 34, pp. 180–87, 204–11, pt. 41, pp. 185–93, 212–20—county tables 5 and 7 in each pt., and Table 1.1 in this volume; Langley and Langley, *State-Level Wheat Statistics*, table 1 p. 5, table 38 p. 42, table 52 p. 56, and table 63 p. 67; USDA, ERS, *Economic Indicators of the Farm Sector: Costs of Production, 1987*, ECIFS 7-3 (Washington, D.C., 1989), table 20 p. 43, and table 24 p. 47.

32. USDA, *The Second RCA Appraisal: Soil, Water, and Related Resources on Nonfederal Land in the United States, Analysis of Conditions and Trends* (Washington, D.C., 1989), table 8 p. 41; C. Tim Osborn, Felix Llacuna, and Michael Linsenbigler, *The Conservation Reserve Program: Enrollment Statistics for 1987–88 and Signup Periods 1–7*, USDA, ERS, *Statistical Bulletin* no. 785 (Washington, D.C., 1989), table 1 pp. 124–27; W. A. Laycock, "History of Grassland Plowing and Grass Planting on the Great Plains," John E. Mitchell, ed., *Impacts of the Conservation Reserve Program in the Great Plains, Symposium Proceedings, September 16–18, 1987, Denver, Colorado*, USDA, Forest Service, *General Technical Report* RM-158 (Fort Collins, Colo.: Rocky Mt. Forest and Range Experiment Station, 1987), p. 7; eligibility computations based on data in Linda A. Joyce and Melvin D. Skold, "Implications of Changes in the Regional Ecology of the Great Plains," Mitchell, *Impacts*, table 1 p. 118.

Over 5 million acres of cropland in South Dakota are projected to lose productivity over a hundred years above the naturally renewable level, much of this erosion attributable to the emphasis upon corn cultivation in the eastern half of the state, but only 2 million acres were authorized for retirement under the Conservation Reserve Program. Approximately 2,790,000 acres were declared suitable for retirement in North Dakota. In Montana, 8,061,000 acres were so categorized, out of a total of only 17,829,766 acres of cropland in the state.

CHAPTER ONE: UNSTABLE RECOVERY

1. Computed from data cited in Table 1.1.

2. Computed from *Fourteenth Census of the United States Taken in the Year 1920*, vol. 6, pt. 1, pp. 636–40, 668–74, and pt. 3, pp. 116–20—county table 4; *United States Census of Agriculture: 1925*, pt. 1, pp. 1026–33, 1086–95, and pt. 3, pp. 105–11—county table 4; *1930 Census*, vol. 2, pt. 1, pp. 1100–5, 1158–63, and pt. 3, pp. 136–40—county table 5. When the price of wheat dropped to seasonal average lows of $ .67 a bushel in 1930 and $ .39 in 1931, a Montana agricultural economist wrote that operators with yields below fifteen bushels an acre would soon be forced out of production; but he noted that land yielding ten bushels an acre on summer fallow was classified, and presumably taxed, as farmland. E. A. Starch, *Economic Changes in Montana's Wheat Area*, MAES, *Bulletin* no. 295 (Bozeman, Mont., 1935), p. 12. See also *Lewistown Democrat-News* (Mont.), March 18, September 21, 1927.

3. *Lewistown Democrat-News*, July 30, and August 5, 1926, and January 8, 1927.

4. E. A. Willson, H. C. Hoffsommer, and Alva H. Benton, *Rural Change in Western North Dakota: Social and Economic Factors Involved in the Changes in Number of Farms and Movement of Settlers from Farms*, NDAES, *Bulletin* no. 214 (Fargo, N.D., 1928), p. 13; *Adams County Record* (Hettinger, N.D.), August 5, and December 30, 1926, and August 15, 1929; Rex E. Willard and O. M. Fuller, *Type of Farming Areas: North Dakota*, NDAES, *Bulletin* no. 212 (Fargo, N.D., 1927), p. 163 fig. 117.

5. Computed from *United States Census of Agriculture: 1925*, pt. 1, pp. 1026–33, 1086–95, and pt. 3, pp. 105–11—county table 4; USDA, *Yearbook, 1925*, table 5 p. 746, and *1931*, table 5 p. 587.

6. O. R. Mathews and Verner I. Clark, *Results of Field Crop, Shelterbelt, and Orchard Investigations at the United States Dry Land Field Station, Ardmore, S.D., 1911–32*, USDA, *Department Circular* no. 421 (Washington, D.C., 1937), pp. 10, 20; O. R. Mathews and John S. Cole, "Special Dry-Farming Problems," USDA, *Yearbook, 1938*, p. 680; Milburn Lincoln Wilson, *Dry Farming in the North-Central Montana "Triangle,"* MSC, *Extension Service Bulletin* no. 66 (Bozeman, Mont., 1923), p. 87; T. R. Stanton, *Oats in the Western Half of the United States*, USDA, *Farmers' Bulletin* no. 1611 (Washington, D.C., 1929), p. 4. Cf. also Leroy Moomaw, *Tillage and Rotation Experiments at Dickinson, Hettinger, and Williston, N.Dak.*, USDA, *Department Bulletin* no. 1293 (Washington, D.C., 1925), pp. 20–21; Joseph Gladden Hutton, *Thirty Years of Soil Fertility Investigations in South Dakota*, SDAES, *Bulletin* no. 325 (Brookings, S.D., 1938), p. 60; MAES, *Agricultural Service from the Montana Experiment Station, Thirtieth Annual Report, July 1, 1922, to June 30, 1923*, p. 51. Hereafter the experiment station reports will be cited by the number and fiscal year ending.

7. USDA, *Climate and Man, Yearbook of Agriculture, 1941*, pp. 711, 722, 734; P. Patton, *The Relationship of Weather to Crops in the Plains Region of Montana*, MAES, *Bulletin* no. 206 (Bozeman, Mont., 1927), pp. 5, 11, 22, 38–39, 42.

8. C. F. Marbut, "Soils of the United States," pt. 3, *Atlas of American Agriculture* (Washington, D.C.: U.S. Government Printing Office, 1936), pp. 16–17; Mary W. M. Hargreaves, *Dry Farming in the Northern Great Plains, 1900–1925*, Harvard Economic

Studies vol. 101 (Cambridge, Mass.: Harvard University Press, 1957), pp. 314–19; *Adams County Record,* September 10, 1925. A general description of the textural properties of surface soils in western South Dakota, with mapping of land-use classification as defined by the U.S. Geological Survey, was published as the basis for R. H. Rogers and F. E. Elliott, *Types of Farming in South Dakota,* SDAES, *Bulletin* no. 238 (Brookings, S.D., 1929), pp. 7–10. The survey of McKenzie County, North Dakota, was conducted in the summer of 1931 (*Williston Herald* [N.D.], July 16, 1931).

9. L. F. Gieseker, *Soils of Sheridan County: Preliminary Report,* MAES, *Bulletin* no. 158 (Bozeman, Mont., 1923), pp. 6–7.

10. See pp. 235–36, 262.

11. *Historical Statistics of the United States, 1789–1945,* prepared by the Bureau of the Census with the cooperation of the Social Science Research Council (Washington, D.C.: U.S. Government Printing Office, 1949), series E 181-95, p. 106; *Handbook of Facts about North Dakota Agriculture,* NDAES, *Bulletin* no. 382 (Fargo, N.D., 1922), p. 72 table 35.

12. Harold G. Halcrow and Jay G. Diamond, *Trends in Market Prices for Montana Farm and Ranch Products,* MAES, *Bulletin* no. 394 (Bozeman, Mont., 1941), p. 5 table 1; *Scobey Sentinel* (Mont.), August 26, September 9, 1921; *Lewistown Democrat-News,* April 19, 1928; USDA, *Yearbook, 1921,* p. 537 table 41; *Eight Years' Work by the Montana Grain Inspection Laboratory, July 1, 1923, to June 30, 1931,* MAES, *Bulletin* no. 270 (Bozeman, Mont., 1933), p. 10; George W. Morgan and M. A. Bell, *Wheat Experiments at the Northern Montana Branch Station,* MAES, *Bulletin* no. 197 (Bozeman, Mont., 1926), pp. 3, 5.

13. USDA, *Annual Report of the Secretary, 1922,* p. 21; *Hill County Journal* (Havre, Mont.), November 21, 1929; W. O. Whitcomb and E. J. Bell, Jr., *The Protein Test in Marketing Wheat,* MAES, *Bulletin* no. 189 (Bozeman, Mont., 1926), p. 26.

14. C. E. Mangels, T. E. Stoa, and William Guy, *Protein Survey of the North Dakota Wheat Crops of 1925 and 1926,* NDAES, *Bulletin* no. 208 (Fargo, N.D., 1927), p. 3; Whitcomb and Bell, *Protein Test,* p. 14; *Adams County Record,* August 13, 1925, and August 9, 1928; *Williston Herald,* May 21, August 27, and October 22, 1925; *Lewistown Democrat-News,* August 9, 1926, September 25, October 25, 1927, and October 9, 1929, citing *Wall Street Journal; Hill County Journal,* October 23, 1930. At the time the *Wall Street Journal* report was written, test weight was, in fact, again becoming the primary concern.

15. R. E. Post, *Farmers' Elevators in the Spring Wheat Area of South Dakota,* pt. 1, *Business Operations, 1921–22 to 1930–31,* SDAES, *Bulletin* no. 282 (Brookings, S.D., 1933), pp. 46–48.

16. Lowell D. Hill, *Grain Grades and Standards: Historical Issues Shaping the Future* (Urbana, Ill.: University of Illinois Press, 1990), pp. 136–37, 200.

17. E. J. Bell, Jr., *Marketing High Protein Wheat,* MAES, *Bulletin* no. 213 (Bozeman, Mont., 1928), pp. 7, 10–11; *Eight Years' Work by the Montana Grain Inspection Laboratory,* p. 6; Mangels, Stoa, and Guy, *Protein Survey . . . of 1925 and 1926,* pp. 9, 13; C. E. Mangels, T. E. Stoa, and R. C. Dynes, *Protein and Test Weight of the 1927 North Dakota Wheat Crop,* NDAES, *Bulletin* no. 213 (Fargo, N.D., 1927), p. 8; *Lewistown Democrat-News,* September 14, 1928; *Williston Herald,* October 22, 1925, November 3, 1927, and August 23, 1928.

18. *Lewistown Democrat-News,* August 4, 9, September 24, 27, October 9, 1929, and August 1, 1930; *Adams County Record,* August 21, 1930.

19. USDA, *Yearbook, 1925,* p. 1033 table 450, *1926,* p. 1036 table 348, and *1931,* p. 826 table 354; P. L. Slagsvold, *Readjusting Montana's Agriculture, II, Montana Farm Prices,* MAES, *Bulletin* no. 308 (Bozeman, Mont., 1936), p. 15 table 3.

20. USDA, *Annual Report of the Secretary, 1924,* p. 5 (quoted); *Historical Statistics,*

series E88-104, p. 99; L. A. Reynoldson, "The Progress of Farmers in Northeastern Montana for Two Years, Preliminary Report" (Washington, D.C.: USDA, BAE, Division of Farm Management and Land Economics, mimeographed report, 1923), p. 15; L. A. Reynoldson, "The Progress of Farmers Who Have Settled in Southeastern Montana . . . , Preliminary Report" (Washington, D.C.: USDA, BAE, Division of Farm Management and Land Economics, mimeographed report, 1924), p. 11; E. R. Johnson and C. G. Worsham, "Organization of Farms in Western South Dakota and Progress of Farmers Who Have Settled in the Area, A Preliminary Report" (Washington, D.C.: USDA, BAE and South Dakota State Department of Agriculture and State College, mimeographed report, 1924), p. 9; Rex E. Willard and O. M. Fuller, *Type-of-Farming Areas: North Dakota*, NDAES, *Bulletin* no. 212 (Fargo, N.D., 1927), p. 6; Slagsvold, *Readjusting Montana's Agriculture, II*, p. 15 table 3.

21. *Historical Statistics*, series 88-104, p. 99; R. R. Renne, *Montana Farm Bankruptcies*, MAES, *Bulletin* no. 360 (Bozeman, Mont., 1938), p. 46 fig. 21; computations from *United States Census of Agriculture: 1925, Reports for States*, pt. 1, pp. 1002–9, 1056–64; pt. 3, pp. 82–89—county table 1; *Fifteenth Census of the United States: 1930*, vol. 1, pt. 1, pp. 1084–87, 1136–41, pt. 3, pp. 118–22—county table 1; Willard and Fuller, *Type-of-Farming Areas: North Dakota*, p. 109 fig. 73; Reynoldson, "Progress of Farmers . . . in Southeastern Montana," pp. 16–17; Reynoldson, "Progress of Farmers in Northeastern Montana," pp. 16–17; Johnson and Worsham, "Organization of Farms in Western South Dakota," p. 15; H. E. Selby, *Statistics of Dry-Land Farming Areas in Montana*, MAES, *Bulletin* no. 185 (Bozeman, Mont., 1926), p. 2; T. S. Thorfinnson, "Through Depression on a Plains Farm," *Land Policy Review* 4 (January 1941): 17–18. The proportion of failed acreage in western counties of Montana was more than double that in the eastern counties in 1929. Cf. the picture reported in Robert G. Athearn, *High Country Empire: The High Plains and the Rockies* (Lincoln, Nebr.: University of Nebraska Press, 1960), p. 299.

22. O. E. Baker, "Government Research in Aid of Settlers and Farmers in the Northern Great Plains of the United States," W. L. G. Joerg, ed., *Pioneer Settlement*, American Geographical Society, *Special Publication* no. 14 (New York, N.Y.: American Geographical Society, 1932), pp. 63, 78–79. See also USDA, *Yearbook of Agriculture, 1927*, p. 1137 table 479. Baker's view was more optimistic than that held by his colleagues in BAE. See p. 107 in this book.

23. See, e.g., *Havre Plaindealer* (Mont.), April 2, 1921 (quoted); *Fergus County Argus* (Lewistown, Mont.), April 7, 1922; *Lewistown Democrat-News*, January 11, 15, February 2, and December 3, 1926; *Hardin Tribune* (Mont.), April 14, May 5, 1922; *Williston Herald*, October 7, 14, 1926; and South Dakota Department of Immigration, *South Dakota, Its Resources and Opportunities, a Description by Counties of the Western Half of the State* . . . (Pierre, S.D., 1921, p. 8.

24. South Dakota Commissioner of Immigration, *Sixth Biennial Report, 1922*, pp. 7–8; *Dickinson Recorder-Post* (N.D.), January 2, 1920; *Hardin Tribune*, June 11, 1920; Roy V. Scott, *Railroad Development Programs in the Twentieth Century*, The Henry A. Wallace Series on Agricultural History and Rural Studies (Ames, Iowa: Iowa State University Press, 1985), p. 82.

25. *Kadoka Press* (S.D.), March 4, July 15, 1921; *Williston Herald*, March 9, 16, 30, July 22, and October 12, 1922; *Fergus County Argus*, December 1, 1922; *Scobey Sentinel*, January 26, February 2, June 8, October 19, and November 16, 1923; Scott, *Railroad Development Programs*, pp. 82–84.

26. Quoted in *Hardin Tribune*, June 11, 1920.

27. *Hardin Tribune*, August 13, 1920; *Lewistown Democrat-News*, February 29, 1928; Stuart Mackenzie, "The Greatest Wheat Farmer in the World," *American Maga-*

zine 96 (October 1923): 37–39, 166, 168; Hiram M. Drache, *Beyond the Furrow, Some Keys to Successful Farming in the Twentieth Century* (Danville, Ill.: The Interstate Printers and Publishers, 1976), ch. 4, with bibliographic notes.

28. South Dakota Commissioner of Immigration, *Sixth Biennial Report, 1922,* pp. 6, 7; *Seventh Biennial Report, 1924,* pp. 8–10; *Kadoka Press,* April 29, 1921; *Adams County Record,* April 4, 1929; *Fergus County Argus,* July 20, 1923.

29. Hargreaves, *Dry Farming . . . 1900–1925,* pp. 238–39; *Fergus County Argus,* November 8, 1923; *Scobey Sentinel,* November 9, 1923; Montana Department of Agriculture, Labor, and Industry, *Biennial Report, July 1, 1923–November 30, 1924,* pp. 36–39, 42–43, 45.

30. W. A. Hartman, *State Land-Settlement Problems and Policies in the United States,* USDA, *Technical Bulletin* no. 357 (Washington, D.C., 1933), p. 48.

31. South Dakota Land Settlement Board, *Report, 1919–20,* pp. 3–4; South Dakota Commissioner of Immigration, *Seventh Biennial Report, 1924,* p. 12; Hartman, *State Land-Settlement Problems,* pp. 41–42; William A. Hartman, "State Policies in Regulating Land Settlement Activities," *Journal of Farm Economics* 13 (April 1931): 260–62.

32. It should be noted that upon initiating his work as immigration commissioner, Dr. Worst instructed his deputies to "Tell the truth about North Dakota. Don't bring any settlers into the state under a false impression." (*Dickinson Recorder-Post,* January 2, 1920 [quoting Worst].) The Atkinson letter was quoted in *Lewistown Democrat-News,* January 29, 1927.

33. *Lewistown Democrat-News,* February 28, March 7, 1926, July 2, 1927, and January 21, 1929; *Adams County Record,* May 26, 1927; *Hill County Democrat* (Havre, Mont.), January 1, 1929.

34. *Hill County Journal,* September 26, 1929, March 20, December 4, 1930; *Lewistown Democrat-News,* September 27, 1929, September 11, December 6, 1930, and January 31, 1931; *Adams County Record,* June 4, 1925, and March 13, 1930.

35. U.S. General Land Office, *Report, 1924,* p. 55, and *1931,* p. 57; *Hill County Democrat,* April 2, 30, 1939; *Williston Herald,* May 28, 1925, and August 25, 1926; *Lewistown Democrat-News,* June 27, 1937, and March 30, 1929.

36. Computations based on data cited in Table 1.1 in this volume. On sod-breaking in this period, see, e.g., *Lewistown Democrat-News,* May 13, 1927; R. J. Penn, W. F. Musbach, and W. C. Clark, *Possibilities of Rural Zoning in South Dakota, A Study in Corson County,* SDAES, *Bulletin* no. 345 (Brookings, S.D., 1940), p. 9.

37. *Lewistown Democrat-News,* December 20, 1927 (quoted); April 19, 23, and 26, 1928; *Adams County Record,* May 26, 1927; calculation on holdings under Northern Pacific sales is based on data in *Hill County Democrat,* April 2, 1929.

38. Computations based on data cited in Table 1.1.

39. On the adjustment in farm size, see pp. 57–61.

CHAPTER TWO: EMERGING ADJUSTMENTS

1. Montana Department of Agriculture, Labor, and Industry, *Montana,* Tourist ed. (Helena, Mont.: Independent Publishing Co., 1921), pp. 21, 23; Montana Legislative Assembly, *Journal of the Seventeenth Session . . . 1921, Senate,* pp. 386, 565, *House,* p. 640.

2. Milburn Lincoln Wilson, *Dry Farming in the North-Central Montana "Triangle,"* MSC, *Extension Service Bulletin* no. 66 (Bozeman, Mont., 1923), pp. 41, 98; *Fergus County Argus* (Lewistown, Mont.), January 23, 1920; *Hardin Tribune* (Mont.), February 23, March 2 and 30, 1923; L. F. Gieseker, *Soils of Blaine County, Soil Reconnoissance*

of Montana, Preliminary Report, MAES, *Bulletin* no. 228 (Bozeman, Mont., 1930), p. 58; Montana Department of Agriculture, Labor, and Industry, *Biennial Report July 1, 1923–November 30, 1924*, p. 36.

3. U.S. Bureau of Reclamation, *Nineteenth Annual Report, 1919–20*, p. 301; *Twenty-Fourth Annual Report . . . for the Fiscal Year Ended June 30, 1925*, p. 87; *Twenty-Fifth Annual Report . . . for the Fiscal Year Ended June 30, 1926*, p. 8; *Williston Herald* (N.D.), November 5, December 17, 1925, January 14, February 4, 1926, and April 28, 1927 (quoted).

4. *Dickinson Recorder-Post* (N.D.), May 5 and June 9, 1922; *Adams County Record* (Hettinger, N.D.), September 24, 1925; U.S. Bureau of Reclamation, *Twenty-Fifth Annual Report . . . for the Fiscal Year Ended June 30, 1926*, p. 90; F. C. Youngblutt (Supt. of Belle Fourche Project) to Elwood Mead, June 3, 1930, Denver Records Center (DRC), RG 115, General Correspondence File (Straights), 42 Lands General, FRC 220868, p. 02549.

5. Elwood Mead, quoted in *Williston Herald*, July 2, 1925; 43 *U.S. Stat.* 702 (December 5, 1924); 44 *U.S. Stat.* 636-50 (May 25, 1926); Elwood Mead, speech broadcast February 5, 1929, released to press February 6, 1929, file copy in DRC, RG 115, Office of Chief Engineer, General Correspondence File (Straights), 42 Lands General, FRC 220868, p. 02786.

6. For the reports, see DRC, RG 115, Office of the Chief Engineer, General Correspondence File (Straights), 42-A3, FRC 220870—that on the Milk River project dated August 21, 1924, pp. 08547-53 (quotation on p. 08553); that on the Huntley project dated August 30, 1924, pp. 08514–19; that on the Sun River project dated August 31, 1924, pp. 08477–84; that on the Williston project dated September 8, 1924, pp. 08370–79; and that on the Belle Fourche project dated August 27, 1924, pp. 08505–11. The summarizing statistics are given in Sherman E. Johnson, "Irrigation Policies and Programs in the Northern Great Plains Region," *Journal of Farm Economics* 18 (August 1936): 544. See also MAES, *Some Outstanding Accomplishments . . . , Thirty-Fourth Annual Report, July 1, 1926, to June 30, 1927*, p. 5 (quoted); Marvin P. Riley, W. F. Kumlien, and Duane Tucker, *Fifty Years Experience on the Belle Fourche Irrigation Project*, SDAES, *Bulletin* no. 450 (Brookings, S.D., 1955), pp. 14–17. Henceforth experiment station reports will be cited by the number and year of fiscal ending.

7. William M. Jardine, "Report of the Secretary," in USDA, *Yearbook, 1927*, pp. 26-27; John W. Haw, "Federal Reclamation Is a Sound National Policy," file copy in DRC, RG 115, Office of Chief Engineer, General Correspondence File (Straights), 42 Lands General, FRC 220868, pp. 02759–62; Elwood Mead to William H. Adams, November 5, 1928, DRC, RG 115, Office of Chief Engineer, General Correspondence File (Straights), 42 Lands General, FRC 220868, p. 02829.

8. Mead to Adams, November 5, 1928; P. W. Dent to Roy R. Gill, March 8, 1928, quoting *Chicago Tribune* editorial, DRC, RG 115, Office of Chief Engineer, General Correspondence File (Straights), 42 Lands General, FRC 220868, p. 02625; Mead to Administrative Staff, August 8, 1928, and enclosed report of the Lippincott committee of the American Society of Civil Engineers, DRC, RG 115, Office of Chief Engineer, General Correspondence File (Straights), 42 Lands General, FRC 220868, pp. 02888–89, 02902–904 (quotation on p. 02903).

9. George C. Kreutzer to Mrs. Jack Hoffman, June 15, 1927; Kreutzer? to R. E. Kelly, June 10, 1927; Kreutzer to Elwood Mead, April 16, 1927; Mead to J. W. Haw, November 7, 1927; R. F. Walter to Elwood Mead, September 25, 1928—all DRC, RG 115, Office of Chief Engineer, General Correspondence File (Straights), 42 Lands General, FRC 220868, pp. 03100–1, 03108, 03136–38, 03005–6, 02911-B, 02851; *Williston Herald*, August 5, 1926.

10. Elwood Mead to Secretary of the Interior, May 24, 1928, DRC, RG 115, 42 Lands General, FRC 220868, pp. 02909–22 (quotation on p. 02922).

11. Calculated from data in *Fifteenth Census of the United States: 1930, Irrigation of Agricultural Lands*, county table, pp. 147–51, 188, 211; *Agriculture*, vol. 2, pt. 1, pp. 1084–87, 1136–41; pt. 3, pp. 118–22—county table 1. The "irrigable" acreage was limited to that in enterprises.

12. Publicity bulletins and reports of the state immigration agencies, the railway lines, and the state bankers' associations reflect the promotional aspect. See also Roy V. Scott, *Railroad Development Programs in the Twentieth Century*, The Henry A. Wallace Series on Agricultural History and Rural Studies (Ames, Iowa: Iowa State University Press, 1985), pp. 90, 95; *Dickinson Recorder-Post*, February 10, 1922, February 16, May 11, 1923; *Williston Herald*, March 18, April 1, 1920, April 20, 1922, August 19, 1926, January 19, 26, April 26, May 17, 1928, January 7, 28, October 26, 1929, and May 13, 1930; *Adams County Record*, July 30, September 24, 1925, May 20, 1926, March 22, August 9, 1928, September 26, 1929, and July 24, 1930; *Hardin Tribune* (Mont.), May 14, 1920, June 9, 1922, and February 9, March 30, 1923; *Wibaux Pioneer Gazette* (Mont.), March 11, April 1, 1920; *Scobey Sentinel* (Mont.), March 30, April 13, 1923; *Hill County Democrat* (Havre, Mont.), August 8, September 26, 1929; *Fergus County Argus* (Lewistown, Mont.), November 26, 1920, March 11, 1921, January 27 and passim through June and July, 1922, and April 13, November 29, 1923; *Lewistown Democrat-News* (Mont.), January 8, 15, 1926, March 4, 1928, April 22, May 26, October 13, 1929, January 19, June 7, November 22, 1930.

13. *Lewistown Democrat-News*, April 22, 1928, reprinting *Minneapolis Tribune*. See also *Scobey Sentinel*, October 19, 1923; *Williston Herald*, June 18, 1925; *Adams County Record*, August 9, 1928; and Mary W. M. Hargreaves, *Dry Farming in the Northern Great Plains, 1900–1925* Harvard Economic Studies vol. 101 (Cambridge, Mass.: Harvard University Press, 1957), pp. 182–220, 251–53.

On contemporary attention to diversification as a theme of national agricultural policy, see *Agricultural Review* 16 (November 1923): 18 and 18 (December 1925): 3; James H. Shideler, *Farm Crisis, 1919–1923* (Berkeley, Calif.: University of California Press, 1957), pp. 82–83, 207; William D. Rowley, *M. L. Wilson and the Campaign for the Domestic Allotment* (Lincoln, Nebr.: University of Nebraska Press, 1970), pp. 16–17, 37, 44, 53.

14. MAES, *Twenty-Ninth Annual Report, 1922*, p. 8, *Thirtieth Annual Report, 1923*, pp. 5–6, 9, and *Thirty-Seventh Annual Report, 1930*, pp. 24–25; Wilson, *Dry Farming in the North-Central Montana "Triangle,"* pp. 41, 46–59, 61; Hargreaves, *Dry Farming . . . 1900–1925*, p. 151.

15. O. R. Mathews, *Dry Farming in Western South Dakota*, USDA, *Farmers' Bulletin* no. 1163 (Washington, D.C., 1920), pp. 3, 6–9, 15; O. R. Mathews and Verner I. Clark, *Results of Field Crop, Shelterbelt, and Orchard Investigations at . . . Ardmore, S.D., 1911–32*, USDA, *Department Circular* no. 421 (Washington, D.C., 1937), p. 21; R. A. Oakley and H. L. Westover, *Forage Crops in Relation to the Agriculture of the Semi-Arid Portion of the Northern Great Plains*, USDA, *Department Bulletin* no. 1244 (Washington, D.C., 1924), pp. 1, 13, 16–19, 50–52; Hargreaves, *Dry Farming . . . 1900–1925*, p. 468 table 6 and pp. 472–74; and Table 1.1 in this volume.

16. Wilson, *Dry Farming in the North-Central Montana "Triangle,"* pp. 60–65; Oakley and Westover, *Forage Crops . . . of the Northern Great Plains*, p. 15; Mathews, *Dry Farming in Western South Dakota*, p. 15; Rex E. Willard and O. M. Fuller, *Type-of-Farming Areas: North Dakota*, NDAES, *Bulletin* no. 212 (Fargo, N.D., 1927), pp. 195, 196 figs. 141, 142; *Lewistown Democrat-News*, August 16 and September 11, 1930.

17. Except as noted, production data in this and the following four paragraphs are based on Table 1.1.

18. Niel W. Johnson and M. H. Saunderson, *Types of Farming in Montana, Part I, Physical Environment and Economic Factors Affecting Montana Agriculture*, MAES, *Bulletin* no. 328 (Bozeman, Mont., 1936), p. 46; Oakley and Westover, *Forage Crops of the Northern Great Plains*, pp. 15, 25; Mathews, *Dry Farming in Western South Dakota*, pp. 9–11, 14; Samuel Garver, *Alfalfa in South Dakota, Twenty-One Years Research at the Redfield Station*, SDAES, *Bulletin* no. 383 (Brookings, S.D., 1946), p. 4; A. N. Hume, Edgar Jay and Clifford Franzke, *Twenty-One Years of Yields from Cottonwood Experiment Farm*, SDAES, *Bulletin* no. 312 (Brookings, S.D., 1937), p. 21; G. W. Morgan and A. E. Seamans, *Dry Farming in the Plains Area of Montana*, MAES, *Circular* no. 89 (Bozeman, Mont., 1920), pp. 19–20; Virgil D. Gilman, *Types of Farming in Southeastern Montana, Based on Studies Made in 1926 and 1929*, MAES, *Bulletin* no. 287 (Bozeman, Mont., 1934), pp. 30–31, 37–38, 41–42, 54, 57–58; *Lewistown Democrat-News*, January 19, 1930; *Williston Herald*, September 2, 1926 and January 26, 1928. See also Willard and Fuller, *Type-of-Farming Areas: North Dakota*, p. 103 fig. 69.

The Dickinson branch experiment station, which had been testing crested wheat grass since 1920, endorsed it enthusiastically in the spring of 1927, but the acreage cultivated was slight (*Adams County Record*, April 7, 1927).

19. *Lewiston Democrat-News*, September 5, November 27, December 9, 1930 and January 4, 1931; Theodore E. Stoa, *Varietal Trials with Oats in North Dakota*, NDAES, *Bulletin* no. 164 (Fargo, N.D., 1922), p. 26; Ralph W. May, *Oats and Barley in Central Montana*, MAES, *Bulletin* no. 209 (Bozeman, Mont., 1927), pp. 2, 3, 11, 17; Mathews and Clark, *Results of Field Crop . . . Investigations at . . . Ardmore, S.D.*, pp. 15–16, 31–32; Oscar R. Mathews, *Summer Fallow at Ardmore, S. Dak.*, USDA, *Department Circular* no. 213 (Washington, D.C., 1932), p. 7; Morgan and Seamans, *Dry Farming in the Plains Area of Montana*, p. 12; MAES, *Thirty-Fifth Annual Report, 1928*, p. 15.

20. Paul W. Gates, *History of Public Land Law Development* (Washington, D.C.: U.S. Government Printing Office, 1968), pp. 608–10; Gilman, *Types of Farming in Southeastern Montana*, p. 59; MAES, *Thirty-Seventh Annual Report, 1930*, pp. 25 (quoted), 32.

21. Roland R. Renne, *The Flaxseed Market and the Tariff*, MAES, *Bulletin* no. 272 (Bozeman, Mont., 1933), pp. 6, 45, 56–57; A. C. Dillman, "Improvement in Flax," USDA, *Yearbook, 1936*, pp. 752, 758; Willard and Fuller, *Type-of-Farming Areas: North Dakota*, p. 117 fig. 78 and pp. 145–60, figs. 101–15 passim; computations from data in Table 1.1.

22. USDA, *Yearbook, 1923*, p. 602 table 3, *1925*, p. 744 table 3, *1927*, p. 740 table 2 and *1931*, p. 584 table 2; Willard and Fuller, *Type-of-Farming Areas: North Dakota*, pp. 13, 98 fig. 65; W. M. Jardine, "Report of the Secretary of Agriculture," USDA, *Yearbook, 1927*, pp. 11 (quoted), 13–14; MAES, *Thirty-Seventh Annual Report, 1930*, p. 32 (quoted).

23. Ralph W. May, *Wheats in Central Montana*, MAES, *Bulletin* no. 203 (Bozeman, Mont., 1927), pp. 4–6, 9, 18; MAES, *Thirty-Sixth Annual Report . . . 1929*, p. 27. On the development of protein testing, see pp. 26–28. On the role of plant introduction and varietal development as an aspect of the early dry-farming movement, see Hargreaves, *Dry Farming . . . 1900–1925*, pp. 301–9.

24. *Dickinson Recorder-Post*, May 11, 1923; T. E. Stoa, R. W. Smith, and O. E. Mangels, *Spring Wheat Varieties for North Dakota*, NDAES, *Bulletin* no. 209 (Fargo, N.D., 1927), pp. 2, 22–23; May, *Wheats in Central Montana*, pp. 19, 21, 24; K. H. Klages, *Spring Wheat Varieties for South Dakota*, SDAES, *Bulletin* no. 268 (Brookings, S.D., 1931), pp. 10–12; Carleton R. Ball, "The History of American Wheat Improvement," *Agricultural History* 4 (April 1930): 30. Nodak was distributed in 1923.

25. *Hardin Tribune*, July 15, 1921; Ball, "History of American Wheat Improve-

ment," pp. 68–69; Stoa, Smith, and Mangels, *Spring Wheat Varieties for North Dakota*, p. 2; May, *Wheats in Central Montana*, p. 23; Mathews and Clark, *Results of Field Crop . . . Investigations at . . . Ardmore, S.D.*, p. 29; USDA, *Annual Report of the Secretary, 1924*, p. 77.

26. Ball, "History of American Wheat Improvement," p. 69; Stoa, Smith, and Mangels, *Spring Wheat Varieties for North Dakota*, p. 2; May, *Wheats in Central Montana*, pp. 19–20, 22; M. A. Bell, *The Effect of Tillage Methods, Crop Sequence, and Date of Seeding upon the Yield and Quality of Cereals and Other Crops Grown under Dry-Land Conditions in North-Central Montana*, MAES, *Bulletin* no. 336 (Bozeman, Mont., 1937), p. 76; Klages, *Spring Wheat Varieties for South Dakota*, pp. 19, 30, 42; *Adams County Record*, February 23, 1928, January 17, 1929, and July 24, 1930; *Lewistown Democrat-News*, March 16, November 12, 1927, November 18, 1928, January 18 and February 21, 1929; *Hill County Democrat*, January 18, 1929.

27. Hargreaves, *Dry Farming . . . 1900–1925*, pp. 85–90.

28. Morgan and Seamans, *Dry Farming in the Plains Area of Montana*, pp. 4, 5, 7; May, *Wheats in Central Montana*, pp. 15–16, 27–28; A. Osenbrug, *Cultural Methods for Winter Wheat and Spring Wheat in the Judith Basin*, MAES, *Bulletin* no. 205 (Bozeman, Mont., 1927), pp. 2, 4; MAES, *Twenty-Eighth Annual Report, 1921*, p. 32; *Thirty-Fifth Annual Report, 1928*, p. 49; Mathews and Clark, *Results of Field Crop . . . Investigations at . . . Ardmore, S.D.*, p. 9; Mathews, *Summer Fallow at Ardmore, S. Dak.*, p. 3; Mathews, *Dry Farming in Western South Dakota*, p. 13; SDAES, *Report, 1920*, p. 19; *Fergus County Argus*, November 25, 1921; *Hardin Tribune*, May 4, 1923.

29. *Hardin Tribune*, November 11, 1921, and February 17, April 14, 1922; *Scobey Sentinel*, September 30 [should be October 7], November 11, 1921, and June 8, November [should be October] 19, 1923; *Fergus County Argus*, November 25, 1921, June 23, 1922; *Williston Herald*, April 14, 1921, and December 7, 1922; Montana Bankers' Association, *Proceedings of the Eighteenth Annual Convention . . . at Helena, August Fifth and Sixth, Nineteen Twenty-One*, pp. 34, 40, 42.

30. Quotations from *Hardin Tribune*, May 26, 1922; *Scobey Sentinel*, November [should be October] 19 and November 16, 1923.

31. Osenbrug, *Cultural Methods for Winter Wheat and Spring Wheat in the Judith Basin*, pp. 2, 3; MAES, *Twenty-Eighth Annual Report, 1921*, p. 81; Edmund Burke and Reuben M. Pinckney, *The Influence of Fallow on Yield and Protein Content of Wheat*, MAES, *Bulletin* no. 222 (Bozeman, Mont., 1929), passim.

32. *Lewistown Democrat-News*, August 11, 1927, May 31, 1928; computations from *Fifteenth Census of the United States: 1930*, vol. 2, pt. 1, pp. 1084–87, 1136–41, and pt. 3, pp. 82–89—county table 1. Cf. also Willard and Fuller, *Type-of-Farming Areas: North Dakota*, p. 111 fig. 74.

The practice of alternating fallow was probably most common in central Montana. A three-year rotation of successive grain crops on fallow was prevalent in the northeastern counties and in combination wheat-livestock farming in the southeastern district of that state. On specialized wheat-growing farms of southeastern Montana only about 25 percent had as much as one-half of the tillable land in summer fallow and less than one-half had as much as one-third (L. A. Reynoldson, "An Organization System for Farms in Northeastern Montana, A Preliminary Report" Washington, D.C.: USDA, BAE, mimeographed report, 1925, p. 8; and Gilman, *Types of Farming in Southeastern Montana*, pp. 44, 46, 53). The rarity of the practice in the wheat district of northwestern North Dakota was indicated in an article by the Williams county agent, who in 1927 urged its use in a two- or three-year rotation. (*Williston Herald*, February 24, 1927).

33. *Scobey Sentinel*, November 16, 1923 (quoted); Haney, "Diversified Farming as Safe Farming," Montana Bankers' Association, *Proceedings, 1922*, pp. 83–85.

34. *Fergus County Argus*, March 11, 1921, and March 24, 1922; *Lewistown Democrat-News*, April 19, August 27, 1926; *Hardin Tribune*, May 4, 1923; Wilson, *Dry Farming in the North-Central Montana "Triangle,"* p. 72; Osenbrug, *Cultural Methods for Winter Wheat and Spring Wheat in the Judith Basin*, p. 4; Morgan and Seamans, *Dry Farming in the Plains Area of Montana*, p. 8; Ralph W. May and Clyde McKee, *Furrow Drill for Sowing Winter Wheat in Central Montana*, MAES, *Bulletin* no. 177 (Bozeman, Mont., 1925), passim; May, *Wheats in Central Montana*, pp. 12–13; MAES, *Twenty-Eighth Annual Report, 1921*, p. 83; *Thirty-Fourth Annual Report, 1927*, p. 24.

35. Wilson, *Dry Farming in the North-Central Montana "Triangle,"* pp. 81–82; E. A. Starch, *Economic Changes in Montana's Wheat Area*, MAES, *Bulletin* no. 295 (Bozeman, Mont., 1935), p. 43. On early stripping, see also recollections of Harry Laubach, "Montanans at Work," MHS, OH 234.

36. Starch, *Economic Changes in Montana's Wheat Area*, pp. 35–36; Mathews, *Summer Fallow at Ardmore, S. Dak.*, p. 3; Wilson, *Dry Farming in the North-Central Montana "Triangle,"* p. 75; MAES, *Thirty-Fourth Annual Report, 1927*, p. 25; *Thirty-Eighth Annual Report, 1931*, p. 86; *Hill County Journal* (Havre, Mont.), May 15, 1930; *Williston Daily Herald*, February 27, 1942.

37. A. E. Seamans, *Fallow Experiments in South Central Montana: Results from the Experimental Dry Farm, Huntley, Montana*, MAES, *Bulletin* no. 142 (Bozeman, Mont., 1921), p. 24; Osenbrug, *Cultural Methods for Winter Wheat and Spring Wheat in the Judith Basin*, p. 3; May, *Oats and Barley in Central Montana*, p. 2; MAES, *Thirty-Eighth Annual Report, 1931*, pp. 79–80; Mathews, *Dry Farming in Western South Dakota*, pp. 12, 14; Mathews and Clark, *Results of Field Crop . . . Investigations at . . . Ardmore, S.D.*, pp. 18–20; Leroy Moomaw, *Tillage and Rotation Experiments at Dickinson, Hettinger, and Williston, N. Dak.*, USDA, *Department Bulletin* no. 1293 (Washington, D.C., 1925), pp. 15–16; T. R. Stanton, *Oats in the Western Half of the United States*, USDA, *Farmers' Bulletin* no. 1611 (Washington, D.C., 1929), pp. 1, 3.

38. Wilson, *Dry Farming in the North-Central Montana "Triangle,"* pp. 82–83; L. P. Reitz, *Crop Regions in Montana as Related to Environmental Factors*, MAES, *Bulletin* no. 340 (Bozeman, Mont., 1937), p. 69; H. E. Selby, *Big-Scale Corn Raising in Montana*, MAES, *Bulletin* no. 171 (Bozeman, Mont., 1925), p. 10; MAES, *Twenty-Ninth Annual Report, 1922*, pp. 32–33; *Thirty-Eighth Annual Report, 1931*, pp. 63–64; Mathews, *Summer Fallow at Ardmore, S. Dak.*, pp. 1, 13.

39. Selby, *Big-Scale Corn Raising*, pp. 4, 10, 32–35, 38; *Adams County Record*, June 2, 16, July 28, and August 4, 1927; *Lewistown Democrat-News*, March 24, 1927 (quoted); MAES, *Thirty-Eighth Annual Report, 1931*, p. 63. On the introduction of trench silos as a "new form of winter storage for feed" in northwestern South Dakota, see *Adams County Record*, February 5, 1925.

40. Starch, *Economic Changes in Montana's Wheat Area*, p. 36; W. M. Hurst and L. M. Church, *Power and Machinery in Agriculture*, USDA, *Miscellaneous Publication* no. 157 (Washington, D.C., 1933), p. 29; Moomaw, *Tillage and Rotation Experiments at Dickinson, Hettinger, and Williston, N. Dak.*, p. 14; F. A. Lyman, "What's New in Farm Machinery? or Farm Machinery Up to Date," MS copy, p. 6, in Farm Equipment Institute, Chicago, Ill. (article prepared for *The Farmer*, December 3, 1927). Cf. Mathew Paul Bonnifield, *The Dust Bowl: Men, Dirt, and Depression* (Albuquerque, N.M.: University of New Mexico Press, 1979), pp. 51–52.

41. E. A. Starch, *Farm Organization as Affected by Mechanization*, MAES, *Bulletin* no. 278 (Bozeman, Mont., 1933), p. 66; MAES, *Thirty-Eighth Annual Report, 1931*, p. 61.

42. May, *Wheats in Central Montana*, pp. 26–27; Starch, *Farm Organization as Af-*

fected by Mechanization, pp. 10–13; Wilson, *Dry Farming in the North-Central Montana "Triangle,"* pp. 91, 93–94; MAES, *Twenty-Ninth Annual Report, 1922*, p. 9; *Thirtieth Annual Report, 1923*, p. 32; *Hardin Tribune*, March 30, 1923; *Scobey Sentinel*, March 30, 1923; L. A. Reynoldson, "Organization Systems for Farms in Southwestern Montana, A Preliminary Report" (Washington, D.C.: USDA, BAE, mimeographed report, 1925), p. 14.

43. MAES, *Thirty-Sixth Annual Report, 1929*, p. 11; Hurst and Church, *Power and Machinery in Agriculture*, pp. 2–3, 22–23; H. E. Selby, *The Gas Tractor in Montana*, MAES, *Bulletin* no. 151 (Bozeman, Mont., 1922), p. 3; Johnson and Saunderson, *Types of Farming in Montana, Part I, Physical Environment and Economic Factors*, pp. 33–35; O. E. Baker, "Government Research in Aid of Settlers and Farmers in the Northern Great Plains of the United States," W. L. G. Joerg, ed., *Pioneer Settlement*, American Geographical Society, *Special Publication* no. 14 (New York, N.Y.: American Geographical Society, 1932), p. 78 n. 28. See also Starch, *Farm Organization as Affected by Mechanization*, pp. 29–32.

44. Starch, *Farm Organization as Affected by Mechanization*, p. 47; Hurst and Church, *Power and Machinery in Agriculture*, pp. 22, 28–30; M. L. Wilson, "Mechanization, Management, and the Competitive Position of Agriculture," *Agricultural Engineering* 13 (January 1932): 4; R. S. Washburn, "Tractor Farming in Dry Regions Has Advantages," USDA, *Yearbook, 1926*, pp. 735–36; Lyman, "What's New in Farm Machinery," pp. 4, 7; Harry G. Davis, "A Special Study of the Tractor Industry in the United States," (Chicago, Ill.: Research Dept., Farm Equipment Institute, mimeographed, 1937), p. 7; Tractioneer, "The General Purpose Tractor," *The American Thresherman*, February 1931, pp. 6–7; Frank J. Zink, "Machinery Developments in 1932," radio talk, January 19, 1933, copy in FEI; C. D. Kinsman, *An Appraisal of Power Used on Farms in the United States*, USDA, *Department Bulletin* no. 1348 (Washington, D.C., 1925), pp. 49–50; P. J. Olson, H. L. Walster, and T. H. Hopper, *Corn for North Dakota*, NDAES, *Bulletin* no. 207 (Fargo, N.D., 1927), pp. 70–71; *Lewistown Democrat-News*, July 29, 1933, and June 7, 1935.

45. MAES, *Thirty-Eighth Annual Report, 1931*, pp. 35–36; Starch, *Economic Changes in Montana's Wheat Area*, p. 49; H. E. Murdock, *Tractor Hitches*, MAES, *Bulletin* no. 229 (Bozeman, Mont., 1930), p. 3; Starch, *Farm Organization as Affected by Mechanization*, pp. 28–32, 72, 74, 101; Stuart Mackenzie, "The Greatest Wheat Farmer in the World," *American Magazine* 96 (October 1923): 39; Hiram M. Drache, *Beyond the Furrow, Some Keys to Successful Farming in the Twentieth Century* (Danville, Ill.: The Interstate Printers & Publishers, 1976), p. 113.

46. L. C. Aicher, *Growing Grain on Southern Idaho Dry Farms*, USDA, *Farmers' Bulletin* no. 769 (Washington, D.C., 1916), pp. 10–11; A. E. Starch and R. M. Merrill, *The Combined Harvester-Thresher in Montana*, MAES, *Bulletin* no. 230 (Bozeman, Mont., 1930), pp. 7, 9; L. A. Reynoldson and others, *The Combined Harvester-Thresher in the Great Plains*, USDA, *Technical Bulletin* no. 70 (Washington, D.C., 1928), pp. 11–14, 30, 35–36.

47. Reynoldson and others, *The Combined Harvester-Thresher*, pp. 21–23, 32, 37; Starch, *Farm Organization as Affected by Mechanization*, p. 14; Alva H. Benton and others, *The Combined Harvester-Thresher in North Dakota*, NDAES, *Bulletin* no. 225 (Fargo, N.D., 1929), pp. 25–26; *Hill County Democrat*, March 29, 1929.

48. *Lewistown Democrat-News*, September 7, 1927, and August 14, 1928; *Williston Herald*, July 12, 1928, and November 26, 1929; Tudor J. Charles, Jr., "Improved Equipment for Dakota Farms," MS dated November 22, 1929, FEI, p. 2; Thomas D. Isern, "Adoption of the Combine on the Northern Plains," *South Dakota History* 2 (Spring 1980): 114–15, map.

49. Reynoldson and others, *The Combined Harvester-Thresher*, pp. 45, 54; Isern,

"Adoption of the Combine," pp. 110–11; Benton and others, *The Combined Harvester-Thresher in North Dakota*, pp. 5–7, 14, 27, 48–49; Lyman, "What's New in Farm Machinery," pp. 2–3; *Lewistown Democrat-News*, August 14, 1928.

50. Thomas D. Isern, *Custom Combining on the Great Plains: A History* (Norman, Okla.: University of Oklahoma Press, 1981), pp. 113–14: Starch, *Farm Organization as Affected by Mechanization*, p. 40. As late as 1950, 68 percent of the wheat acreage in North Dakota was still combined from windrow (*Williston Daily Herald*, April 25, 1952, citing BAE data).

51. Reynoldson and others, *The Combined Harvester-Thresher*, pp. 9–11, 24; Starch and Merrill, *The Combined Harvester-Thresher in Montana*, p. 22; Benton and others, *The Combined Harvester-Thresher in North Dakota*, p. 26; Charles, "Improved Equipment for Dakota Farms," pp. 2, 6; USDA, *Annual Report of the Secretary, 1924*, p. 39.

52. *Fourteenth Census of the United States, Taken in the Year 1920*, vol. 6, pt. 1, p. 50 table 61; *Fifteenth Census of the United States: 1930*, vol. 2, pt. 1, pp. 1127, 1194, and pt. 3, p. 165—county table 12; Kinsman, *An Appraisal of Power Used on Farms*, p. 53; Starch, *Farm Organization as Affected by Mechanization*, p. 15.

53. Frank T. Hady, *Motor Truck Transportation in Western South Dakota*, SDAES, *Circular* no. 11 (Brookings, S.D., 1933), pp. 5, 6, 13, 15–16, 23.

54. *Fourteenth Census of the United States: 1920*, vol. 6, pt. 1, p. 50 table 61; *Fifteenth Census of the United States: 1930*, vol. 2, pt. 1, pp. 1127, 1194, and pt. 3, p. 165—county table 12; *Lewistown Democrat-News*, November 8, 1927, and April 7, 1930.

55. Hargreaves, *Dry Farming . . . 1900–1925*, p. 494; *Adams County Record*, February 26, November 12, 1925, August 22, 1929, and March 13, 1930; *Hill County Journal*, July 18, 1929, and July 7, 1932.

56. Computations based on data in Table 1.1. See also Starch, *Economic Changes in Montana's Wheat Area*, p. 28.

"Suitcase" farming has been defined as that conducted by non-residents who lived outside the county; "sidewalk" farming, as that by non-residents who lived in adjacent towns. The latter was common at that time in the northern Plains. The former also seems indicated by references in *Lewistown Democrat-News*, June 27, 1929; *Adams County Record*, September 15, 1927; *Hettinger County Herald*, in *Adams County Record*, March 24, 1927.

57. M. L. Wilson, "Research Studies in the Economics of Large Scale Farming in Montana," *Agricultural Engineering* 10 (January 1929): 3; Starch, *Farm Organization as Affected by Mechanization*, p. 10; E. A. Willson and others, *Rural Changes in Western North Dakota: Social and Economic Factors*, NDAES, *Bulletin* no. 214 (Fargo, N.D., 1928), pp. 5, 6, 41, 45–46; Willard and Fuller, *Type-of-Farming Areas: North Dakota*, p. 10; Rex E. Willard, *Some Farming Changes in Southwestern North Dakota, 1922 to 1925*, NDAES, *Bulletin* no. 201 (Fargo, N.D., 1926), pp. 5, 13.

58. Baldur H. Kristjanson and C. J. Heltemes, *Handbook of Facts about North Dakota Agriculture*, NDAES, *Bulletin* no. 382, rev. ed. (Fargo, N.D., 1952), p. 76 tables 38 and 39; Gilman, *Types of Farming in Southeastern Montana*, pp. 44, 52; Starch, *Economic Changes in Montana*, p. 30; Jessie E. Richardson, *The Quality of Living in Montana Farm Homes*, MAES, *Bulletin* no. 260 (Bozeman, Mont., 1932), p. 8; *Williston Herald*, November 24, 1927, and May 17, 1928; *Adams County Record*, March 24, 1927, and November 28, 1929; *Lewistown Democrat-News*, October 18, November 24, 1927, October 2, December 7, 1928, and May 3, September 10, October 23, 1929.

59. For sources on data in this and the following paragraph, see Hargreaves, *Dry Farming . . . 1900–1925*, pp. 321–23; Rex E. Willard, *Cost of Producing Crops in North Dakota*, NDAES, *Bulletin* no. 199 (Fargo, N.D., 1926); Willard, *Some Farming Changes in Southwestern North Dakota*.

60. Wilson, "Research Studies in the Economics of Large Scale Farming," pp. 3,

11–12 (quotation, p. 12); MAES, *Thirty-Eighth Annual Report, 1931,* pp. 61, 65; E. A. Starch, *Readjusting Montana's Agriculture, VII, Montana's Dry-Land Agriculture,* MAES, *Bulletin* no. 318 (Bozeman, Mont., 1936), pp. 14–15, 18–19.

61. For biographical detail on Wilson, see *Hill County Democrat,* July 5, 1934; Rowley, *M. L. Wilson,* passim.

62. M. L. Wilson, "The Fairway Farms Project," *Journal of Land and Public Utility Economics* 2 (April 1926): 156–62, 168–69.

63. Wilson, "Research Studies in the Economics of Large Scale Farming," pp. 9–10 (quotation, p. 10); Baker, "Government Research in Aid of Settlers," p. 73 n. 17.

64. *Lewistown Democrat-News,* February 24, 1927; and *Hill County Journal,* November 7, 1929.

65. *Lewistown Democrat-News,* February 17 (quoted), March 3, 4, 7, 18, 24, October 25, November 8, 1927, and January 3, 1928.

66. *Lewistown Democrat-News,* November 20, 1927, and January 11, October 13, 1929; *The Country Gentleman,* XCIV (April 1929): 28 (quoted).

67. *Lewistown Democrat-News,* January 3, February 22, 28, March 3, 4, 1928, February 23, March 1, 1928, and February 25, 27, September 25, October 10, 20, 26, 1929; *Williston Herald,* March 7, 14, 1929, March 21, 1930; *Adams County Record,* March 1, 1928, November 7, 1929, and February 6, 20, 1930.

Farm cost surveys on very limited sampling, conducted by the North Dakota Extension Service in a number of counties at the end of the decade, also showed that the more successful farmers operated larger than average size holdings, used more summer fallow, and gave greater attention to incorporation of corn, legumes, and livestock. While noting the availability of ample equipment on these operations, they did not stress the development of mechanization as a factor in the profitability (*Williston Herald,* June 18, 1931; *Adams County Record,* September 10, 1931).

68. *Hill County Democrat,* February 1, 18 (quoted), 19, and April 2, 1929. The train was run again in 1930, on the tracks of the Northern Pacific Railway, once more with Wilson as principal speaker (*Lewistown Democrat-News,* July 4, 1930).

69. Rowley, *M. L. Wilson,* pp. 54–55; above, p. 46. On the surveys of "successful" farmers, see Starch, *Economic Changes in Montana's Wheat Area,* p. 29; Gilman, *Types of Farming in Southeastern Montana,* p. 23. The experimental conclusions derived from the Fairway Farms project were reported in Starch, *Farm Organization as Affected by Mechanization.* See also Russell Lord, *The Wallaces of Iowa* (Boston, Mass.: Houghton Mifflin Co., 1947), p. 303, quoting interview with Wilson.

70. Rowley, *M. L. Wilson,* pp. 49, 63.

71. E. A. Starch to Victor Christgau, December 4, 1929, quoted in Rowley, *M. L. Wilson,* p. 53; *Congressional Record,* 71st Cong., 2d sess., pp. 12119–22 (quotation on p. 12120).

72. *Congressional Record,* 71st Cong., 2d sess., p. 12121 (quoted). On the broader debate over land utilization during the decade, see Albert Z. Guttenberg, "The Land Utilization Movement of the 1920s," *Agricultural History* 50 (July 1976): 447–90; and chaps. 4 and 5 in this volume.

The interpretation here given to Wilson's identification with the Christgau proposal differs considerably from the emphasis upon land planning which is presented in Richard S. Kirkendall, *Social Scientists and Farm Politics in the Age of Roosevelt* (Columbia, Mo.: University of Missouri Press, 1966), pp. 21–29. The distinction, I believe, rests on the timing of the initiation of the Christgau proposal. Conditions changed radically in relation to the regional agriculture, including the Fairway Farms experimental program, between 1929 and 1931.

73. M. L. Wilson to John D. Black, April 11, 1928, quoted in Rowley, *M. L. Wilson,* p. 40. See also Wilson's remarks on the "Low Cost Wheat Train," quoted in *Hill*

County Democrat, February 18, 1929; but cf. also his remarkably prophetic view of the agricultural revolution to come from increased mechanization and improved communication, quoted in Lord, *Wallaces,* pp. 303–4.

CHAPTER THREE: A DECADE OF DISASTER

1. *See,* in particular, *Annual and Seasonal Precipitation at Six Representative Locations in Montana,* MAES, *Bulletin* no. 447 (Bozeman, Mont., 1947); *Lewistown Democrat-News* (Mont.), August 2, 1934, July 5, 13, 1937, January 31, 1938 (quoted), and April 11, 1948; Neil W. Johnson, *Farm Adjustments in Montana, Study of Area VII: Its Past, Present and Future,* MAES, *Bulletin* no. 367 (Bozeman, Mont., 1939), pp. 10, 14; *Williston Daily Herald* (N.D.), June 2, 7, 18, 1937; *Adams County Record* (Hettinger, N.D.), July 14, 1954, reprinting *Mott Pioneer Press* (N.D.); M. B. Johnson, *Range Cattle Production in Western North Dakota,* NDAES, *Bulletin* no. 347 (Fargo, N.D., 1947), p. 17; J. T. Sarvis, *Grazing Investigations on the Northern Great Plains,* NDAES, *Bulletin* no. 308 (Fargo, N.D., 1941), p. 33; Raymond B. Hile, *Farm Size as a Guide to Planning in the Tri-County Soil Conservation District,* USDA, BAE, *Farm Management Report* no. 16 (Washington, 1940), p. 8 fig. 3; W. F. Kumlien, *A Graphic Summary of the Relief Situation in South Dakota (1930–1935),* SDAES, *Bulletin* no. 310 (Brookings, S.D., 1937), pp. 31, 37 fig. 34; S. P. Swenson, *Spring Wheat Varieties in South Dakota,* SDAES, *Bulletin* no. 342 (Brookings, S.D., 1940), p. 33; Paul H. Landis, *Rural Relief in South Dakota, with Special Attention to Rural Relief Families under the New Deal Relief Program,* SDAES, *Bulletin* no. 289 (Brookings, S.D., 1934), pp. 7–8. The Dakotas were again designated officially as a drought area in 1939, but the relief load declined sharply in North Dakota that autumn. *Williston Daily Herald,* November 1, 22, 1939.

2. *Williston Herald,* April 26, 1933 (quoted), May 10, 1934, April 12, 1935, and August 31, 1937; *Adams County Record,* May 10, 1934; *Lewistown Democrat-News,* May 9, 17, June 3, 1934; P. G. Peck and M. C. Forster, *Six Rural Problem Areas: Relief—Resources—Rehabilitation,* FERA, Division of Research, Statistics, and Finance, *Research Monograph* no. I (Washington, 1935), p. 137; W. F. Kumlien, Robert L. McNamara, Zetta E. Bankert, *Rural Population Mobility in South Dakota, 1928–1935,* SDAES, *Bulletin* no. 315 (Brookings, S.D., 1938), p. 15 fig. 9, and pp. 263, 264, and 355n19 in this volume.

3. *Hill County Journal* (Havre, Mont.), January 2, 1930, January 1, 1931, January 7, 1932, and August 13, September 14, 1933; *Adams County Record,* July 17, 1930, and August 6, 1936; *Lewistown Democrat-News,* June 18, 21, 1933, and August 11, 1934; *Williston Daily Herald,* June 14, July 2, 3, 8, 9, 10, 1935; Francis D. Cronin and Howard W. Beers, *Areas of Intense Drought Distress, 1930–1936,* WPA, Division of Social Research, *Research Bulletin,* ser. 5, no. 1 (Washington, D.C., mimeographed, 1937), pp. 7–8.

4. Sherman E. Johnson, *From the St. Croix to the Potomac—Reflections of a Bureaucrat* (Bozeman, Mont.: Big Sky Books, 1974), pp. 89–90; Johnson, *Range Cattle Production,* p. 17; Kumlien, *Graphic Summary of the Relief Situation,* pp. 42–43 figs. 36 D, E, F; *Lewistown Democrat-News,* July 28, August 9, 1931, June 15, 21, July 31, 1933, August 26, 1934, July 18, August 1, 8, 1935, July 17, 18, 1936, June 10, 1937, August 11, September 10, 1938, and May 13, June 30, July 20, 1939; *Hill County Journal,* July 4, 1937; *Williston Daily Herald,* June 22, 1933, July 28, August 1, 30, 1938, and May 15, 1939; *Adams County Record,* April 13, 1933, January 18, 1934, March 4, 1937, July 1, 28, 1938, and February 29, 1940. See also Sarvis, *Grazing Investigations,* p. 33.

5. L. W. Schaffner, *Economics of Grain Farming in Renville County, North Dakota,* NDAES, *Bulletin* no. 367 (Fargo, N.D., 1951), p. 7 table 2; Lloyd E. Jones and Olav Rogeness, *Farmers Face Farm-Management Facts in Ward County (North Dakota),*

USDA, BAE, *Farm Management Report* no. 15 (Washington, D.C., 1941), p. 36 table 15; T. S. Thorfinnson, "Through Depression on a Plains Farm," *Land Policy Review* 4 (January 1941): 17; Marion Clawson, M. H. Saunderson, and Neil W. Johnson, *Farm Adjustments in Montana, Study of Area IV: Its Past, Present, and Future*, MAES, *Bulletin* no. 377 (Bozeman, Mont., 1940), p. 17 table 3; Johnson, *Farm Adjustments in Montana, Study of Area VII*, p. 14; *Lewistown Democrat-News*, August 6, 1938, September 16, 1939, and September 12, 1940. On yields in the North Dakota "Slope" counties, see National Resources Planning Board, Land Committee, Subcommittee on Tax Delinquency, *Tax Delinquency and Rural Land-Use Adjustment, Technical Paper* no. 8 (Washington, D.C., 1942), table 37.

6. *Lewistown Democrat-News*, January 1, 1931, July 28, October 30, November 1, 2, 1932, and November 10, 1933; P. L. Slagsvold, *Readjusting Montana's Agriculture, II, Montana Farm Prices*, MAES, *Bulletin* no. 308 (Bozeman, Mont., 1936), p. 4. See also Gabriel Lundy, *Graphic Views of Changes in South Dakota Agriculture*, SDAES, *Circular* no. 78 (Brookings, S.D., 1949), p. 26 fig. 28. Prices received by North Dakota farmers were five to ten cents a bushel higher.

7. Slagsvold, *Readjusting Montana's Agriculture, II, Montana Farm Prices*, p. 7 fig. 4; Lundy, *Graphic Views*, pp. 26–28 figs. 28–30; T. Hillard Cox and L. M. Brown, *South Dakota Farm Prices, 1890–1937*, SDAES, *Bulletin* no. 317 (Brookings, S.D., 1938), p. 9 chart II; Baldur H. Kristjanson and C. J. Heltemes, *Handbook of Facts about North Dakota Agriculture*, NDAES, *Bulletin* no. 357 (Fargo, N.D., 1950), p. 67 table 27; *Historical Statistics of the United States, 1789–1945, A Supplement to the Statistical Abstract of the United States* (Washington, D.C.: U.S. Department of Commerce, Bureau of the Census, 1949), p. 102, ser. E 136–51.

8. MAES, *Forty-Third Annual Report, July 1, 1935, to June 30, 1936*, p. 8; R. S. Kifer and H. L. Stewart, *Farming Hazards in the Drought Area*, WPA, Division of Social Research, *Research Monograph* no. XVI (Washington, D.C., United States Government Printing Office, 1938), pp. 2–3 table 1.

9. Thorfinnson, "Through Depression on a Plains Farm," pp. 17–20.

10. E. A. Starch, *Farm Organization as Affected by Mechanization*, MAES, *Bulletin* no. 278 (Bozeman, Mont., 1933), p. 20; *Lewistown Democrat-News*, January 4, 1937; computations from *Fifteenth Census of the United States: 1930, Agriculture*, vol. 2, pt. 1, pp. 1124–26, 1190–93, and pt. 3, pp. 162–64—county table 11.

11. *Lewistown Democrat-News*, August 15, 16, September 14, 1931; *Hill County Journal*, April 16, July 23, 1931, and February 10, 1932; *Williston Herald*, July 9, August 6, 1931; Isaiah Bowman, *The Pioneer Fringe* (New York: American Geographical Society, 1931), p. 113; *United States Census of Agriculture: 1935*, vol. 1, pt. 1, pp. 290–94, 306–11, and pt. 3, pp. 812–16—county table 1; computed to include cropland harvested, failed, idle or fallow, and plowable pasture.

Relief payments in South Dakota were begun somewhat later than elsewhere in the region and remained relatively low until the end of 1933. Thereafter they were extended to a larger percentage of the population than anywhere else in the nation. The area of greatest need, however, was the central counties, along the Missouri River, extending from Jackson diagonally northwestward to Perkins and eastward almost to the Minnesota border (Kumlien, *Graphic Summary of the Relief Situation in South Dakota*, cover map, p. 26 figs. 30, 31).

12. Gabriel Lundy, *Farm-Mortgage Experience in South Dakota, 1910–40, with Special Reference to Three Townships in Each of the Counties of Brookings, Clark, Haakon, Hyde, and Turner*, SDAES, *Bulletin* no. 370 (Brookings, S.D., 1943), p. 12; Roland R. Renne, *Readjusting Montana's Agriculture: VIII, Tax Delinquency and Mortgage Foreclosures*, MAES, *Bulletin* no. 319 (Bozeman, Mont., 1936), pp. 1, 22; R. R. Renne, *Montana Farm Foreclosures: Number, Characteristics, and Causes of Farm Mortgage Foreclosures over a*

Seventy-Year Period, MAES, *Bulletin* no. 368 (Bozeman, Mont., 1939), pp. 17 fig. 6, and 57 table 2; R. R. Renne, *Montana Farm Bankruptcies: A Study of the Number, Character-istics, and Causes of Farm Bankruptcies over a Forty-Year Period,* MAES, *Bulletin* no. 360 (Bozeman, Mont., 1938), pp. 26, 27; Baldur H. Kristjanson and C. J. Heltemes, *Hand-book of Facts about North Dakota Agriculture, Revised,* NDAES, *Bulletin* no. 382 (Fargo, N.D., 1952), p. 84 table 43; Rainer Schickele and Reuben Engelking, *Land Values and the Land Market in North Dakota,* NDAES, *Bulletin* no. 353 (Fargo, N.D., 1949), p. 41 table 13. The delinquency rate had declined greatly in South Dakota during the latter half of the twenties and stood at only 3.9 on federal land bank loans in 1930. Lundy, *Farm Mortgage Experience in South Dakota,* pp. 8–9. See also Henry A. Steele, *Farm Mortgage Foreclosures in South Dakota, 1921–1932,* SDAES, *Circular* no. 17 (Brookings, S.D., 1934), p. 7 fig 4. This suggests that the greater gravity of the drought experience in this section of the region prior to World War I had led to a stricter credit policy in South Dakota.

13. 40 *U.S. Stat.* 274, 494 (August 10, 1917; March 28, 1918); 41 *U.S. Stat.* 1347 (March 3, 1921); 42 *U.S. Stat.* 467, 1322 (March 20, 1922; February 26, 1923); M. L. Wilson, *Dry Farming in the North-Central Montana "Triangle,"* MSC, *Extension Service Bulletin* no. 66 (Bozeman, Mont., 1923), pp. 18 fig. 9, 20 fig. 10, 21 fig. 11, and 127; USDA, *Annual Report of the Secretary, 1922,* p. 52 fig. 18; *Dickinson Recorder-Post* (N.D.), March 8, 1921; *Williston Daily Herald,* July 19, 1937; National Resources Board, *A Report on National Planning and Public Works in Relation to Natural Resources and Including Land Use* (Washington, D.C.: U.S. Government Printing Office, 1934), p. 179; National Resources Board, Land Planning Committee, *Maladjustments in Land Use in the United States, Report on Land Planning,* pt. 6 (Washington, D.C.: U.S. Govern-ment Printing Office, 1935), pp. 33–35; John J. Haggerty, *Public Finance Aspects of the Milk River Land Acquisition Project (LA-MT-2), Phillips County, Montana,* USDA, Reset-tlement Administration, Land Utilization Division, *Land Use Planning Publication* no. 18-a (Washington, D.C.: mimeographed, 1937), p. 4; Kumlien, *Graphic Summary of the Relief Situation,* p. 31; Norman J. Wall, *Federal Seed-Loan Financing and Its Relation to Agricultural Rehabilitation and Land Use,* USDA, *Technical Bulletin* no. 539 (Washing-ton, D.C., 1936), pp. 45–47 tables 14, 15 et passim.

14. Montana Legislative Assembly, *Laws . . . Extraordinary Session of the Fifteenth, 1918,* ch. 19, pp. 43–57 (February 23, 1918); *Extraordinary Session of the Sixteenth, 1919,* ch. 8, pp. 19–30 (August 11, 1919); North Dakota Legislative Assembly, *Laws Passed at the Special Session of the Fifteenth, 1918,* ch. 13, pp. 15–24 (January 30, 1918); Wilson, *Dry Farming in the North-Central Montana "Triangle,"* p. 127; North Dakota Commis-sioner of Agriculture, *Fifteenth Biennial Report . . . Ending June 30, 1918,* p. 13; *Sixteenth Biennial Report . . . Ending June 30, 1920,* p. 11; Fred R. Taylor, C. J. Heltemes, and R. F. Engelking, *North Dakota Agricultural Statistics,* NDAES, *Bulletin* no. 408 (Fargo, N.D., 1957), p. 93 table 86; Roland R. Renne and Bushrod W. Allin, *Montana Farm Taxes,* MAES, *Bulletin* no. 286 (Bozeman, Mont., 1934), pp. 12, 21; Renne, *Readjusting Mon-tana's Agriculture, VIII, Tax Delinquency,* pp. 6, 12; *Williston Daily Herald,* May 21 and June 29, 1942.

Farm tax rates per acre were approximately the same in Montana and South Dakota and were lower than those in North Dakota (Renne and Allin, *Montana Farm Taxes,* pp. 22, 23).

15. Kristjanson and Heltemes, *Handbook of Facts about North Dakota Agriculture, Revised,* p. 85; Renne, *Montana Farm Foreclosures,* p. 57 table II; Kifer and Stewart, *Farming Hazards in the Drought Area,* pp. 4, 5, 32–34; Schickele and Engelking, *Land Values in North Dakota,* pp. 41 table 13, and 42; Kristjanson and Heltemes, *Handbook of Facts about North Dakota Agriculture,* p. 86 table 40; *Adams County Record,* September 15, 1932.

16. Renne, *Readjusting Montana's Agriculture, VIII, Tax Delinquency,* p. 4 fig. 1; *Lewistown Democrat-News,* January 4, 1937, and September 27, 1940; NRPB, Land Committee, Subcommittee on Tax Delinquency, *Tax Delinquency and Rural Land-Use Adjustment,* p. 127 and table 34, but cf. p. 118 fig. 5. See also R. R. Renne and H. H. Lord, *Montana Farm Taxes: The Significance of Farm Taxes to Montana Agriculture,* MExtS, *Circular* no. 94 (Bozeman, Mont., 1938), p. 7.

17. Kifer and Stewart, *Farming Hazards in the Drought Area,* p. 36; *Williston Daily Herald,* January 13, 1937, and August 15, 1940; NRPB, Land Committee, Subcommittee on Tax Delinquency, *Tax Delinquency and Rural Land-Use Adjustment,* pp. 145, 152, 155, 160, 163–67 and tables 40, 42, 44, 45, 47, 48; Herbert S. Schell, "Adjustment Problems in South Dakota," *Agricultural History* 14 (April 1940): 71; R. B. Westbrook, *Tax Delinquency and County Ownership of Land in South Dakota,* SDAES, *Bulletin* no. 322 (Brookings, S.D., 1938), p. 39; Hile, *Farm Size as a Guide to Planning,* p. 9.

18. *Williston Daily Herald,* September 27, 1939, and April 4, 1941; Morris H. Taylor and Raymond J. Penn, *Management of Public Land in North Dakota,* NDAES, *Bulletin* no. 312 (Fargo, N.D., 1942), p. 10; Schickele and Engelking, *Land Values in North Dakota,* pp. 41–42; *Sixteenth Census of the United States: 1940, Agriculture,* vol. 1, pt. 2, pp. 414–18, 508–13, and pt. 6, pp. 50–54—county table 8. On the legislation for debt moratoria, see p. 85.

19. Computed from *Fifteenth Census of the United States: 1930, Population,* vol. 3, pt. 2, pp. 22–25, 419–22, 833–37—county table 13; *Sixteenth Census of the United States: 1940, Agriculture,* vol. 1, pt. 2, pp. 382–86, 468–73, and pt. 6, pp. 16–20—county table 1; *Population,* vol. 2, pt. 4, pp. 531–34, pt. 5, pp. 478–81, and pt. 6, pp. 496–500 table 27; Neil W. Johnson and M. H. Saunderson, *Types of Farming in Montana: Part I, Physical Environment and Economic Factors Affecting Montana Agriculture,* MAES, *Bulletin* no. 328 (Bozeman, Mont., 1936), p. 37; Kifer and Stewart, *Farming Hazards in the Drought Area,* p. xx.

Cropland as defined in the census includes farm land that is idle, in fallow, failed under cropping, and "plowable pasture," that is, land that was in pasture but that had been cropped within the past five years.

On the increasing rate of tenancy, see Cap E. Miller and Willard O. Brown, *Farm Tenancy and Rental Contracts in North Dakota,* NDAES, *Bulletin* no. 289 (Fargo, N.D., 1937), pp. 4 table 1, 5, and 7–8 figs. 4 and 5; Renne and Lord, *Montana Farm Taxes,* p. 7. On population movement, see Carl F. Kraenzel, *Farm Population Mobility in Selected Montana Communities,* MAES, *Bulletin* no. 371 (Bozeman, Mont., 1939), pp. 10–14, 20, 30; Kumlien, McNamara, and Bankert, *Rural Population Mobility in South Dakota,* pp. 14–15; W. F. Kumlien, *Basic Trends of Social Change in South Dakota: I, Population Tendencies,* SDAES, *Bulletin* no. 327 (Brookings, S.D., 1939), p. 15 figs. 9 and 10; John P. Johansen, *The Influence of Migration upon South Dakota's Population, 1930–50,* SDAES, *Bulletin* no. 431 (Brookings, S.D., 1953), p. 29; Peck and Forster, *Six Rural Problem Areas,* pp. 67–68 fig. 10; Walter L. Slocum, *Migrants from Rural South Dakota Families, Their Geographical and Occupational Distribution,* SDAES, *Bulletin* no. 359 (Brookings, S.D., 1942).

20. See reminiscences by Laura Gessaman, Selmer Helland, Hazel Klozbuecher, Ted Worrall, and Mary Frank Zanto, MHS, OH; Avery Bates, "My Life: Farming and the Depression," mimeographed text supplied by Grace Bates, Manhattan, Mont.; Harold G. Halcrow, *Failures among Montana's Dry-Land Farmers,* MAES, *Mimeographed Circular* no. 19 (Bozeman, Mont., 1939), pp. 1 (quoted)–3, 5–9; *Lewistown Democrat-News,* April 17, 1935.

21. *Lewistown Democrat-News,* January 10, 1930 (quoted), December 21, 1931, September 1, 1937, and February 8, 1938; *Williston Herald,* April 14, 1932, and May 23, 1935; *Adams County Record,* March 7, 1935; Peck and Forster, *Six Rural Problem Areas,*

pp. 29, 51; Berta Asch and A. R. Mangus, *Farmers on Relief and Rehabilitation*, U.S. WPA, Division of Social Research, *Research Monograph* no. VIII (Washington, D.C.: U.S. Government Printing Office, 1973), p. 76; H. L. Stewart, *Natural and Economic Factors Affecting Rural Rehabilitation Problems in Northwestern North Dakota and Northeastern Montana (as Typified by Divide County, North Dakota)*, Resettlement Administration, *Research Bulletin* no. K-7 (Washington, D.C.: mimeographed, 1936), p. 16; H. L. Stewart, *Natural and Economic Factors Affecting Rural Rehabilitation Problems in Southwestern North Dakota (as Typified by Hettinger County)*, Resettlement Administration, *Research Bulletin* no. K-4 (Washington, D.C.: mimeographed, 1936), p. 16; Kumlien, *Graphic Summary of the Relief Situation*, pp. 24 fig. 27, 32, and 47 fig. 44; Cronin and Beers, *Areas of Intense Drought Distress*, pp. 26 fig. 6, 43–45, and 51; Lorena Hickok to Harry L. Hopkins, November 1 and 3, 1933, Richard Lowitt and Maurine Beasley, eds., *One Third of a Nation: Lorena Hickok Reports on the Great Depression* (Urbana, Ill.: University of Illinois Press, 1981), pp. 62–64, 69.

22. Hickok to Hopkins, October 30 and November 1, 3, 9, 18, 1933, Lowitt and Beasley, eds., *One Third of a Nation*, pp. 56 (quoted), 62–63, 68 (quoted), 69, 83–84, 93–95.

23. Waller Wynne, Jr., and Gordon Blackwell, *Survey of Rural Relief Cases Closed for Administrative Reasons in South Dakota*, WPA, Division of Social Research, *Research Bulletin*, ser. 2, no. 12 (Washington, D.C.: mimeographed, 1936), pp. 2–5, 9.

24. Kifer and Stewart, *Farming Hazards in the Drought Area*, p. 124 table 2; Stewart, *Natural and Economic Factors Affecting Rural Rehabilitation . . . Hettinger County*, p. 16; Peck and Forster, *Six Rural Problem Areas*, pp. 8, 48, 82, 125 table 17, 127 table 19, and 128 table 20-B; Kumlien, *Graphic Summary of the Relief Situation*, pp. 27 and 60 table 19; Wynne and Blackwell, *Survey of Rural Relief Cases*, pp. 2–3; Hickok to Hopkins, October 30, 1933, Lowitt and Beasley, eds., *One Third of a Nation*, p. 56.

25. Mary W. M. Hargreaves, *Dry Farming in the Northern Great Plains, 1900–1925*, Harvard Economic Studies, vol. 101 (Cambridge, Mass.: Harvard University Press, 1957), p. 531; *Lewistown Democrat-News*, October 15, 1935; *Williston Herald*, January 21, 1932.

26. Theodore Saloutos and John D. Hicks, *Agricultural Discontent in the Middle West, 1900–1939* (Madison, Wisc.: University of Wisconsin Press, 1951), pp. 441–43; Lowell K. Dyson, *Red Harvest: The Communist Party and American Farmers* (Lincoln, Nebr.: University of Nebraska Press, 1982), pp. 71–73.

27. *Williston Herald*, August 4, 1932; Larry Remele, "The North Dakota Farm Strike of 1932," *North Dakota History* 41 (Fall 1974): 6, 8.

28. Quoted passages from Remele, "North Dakota Farm Strike," p. 6, citing "most daily North Dakota newspapers for August 1, 1932."

29. *Williston Herald*, August 11, 25, September 22, 29, October 6, 13, 1932, and January 12, 1933; *Adams County Record*, October 6, 1932.

30. John E. Miller, "Restrained, Respectable Radicals: The South Dakota Farm Holiday," *Agricultural History* 59 (July 1985): 429–47 passim; John L. Shover, *Cornbelt Rebellion: The Farmers' Holiday Association* (Urbana, Ill.: The University of Illinois Press, 1965), pp. 39–40, 93; Herbert S. Schell, *History of South Dakota* (Lincoln, Nebr.: University of Nebraska Press, 1961), pp. 287–88. On the timing of the extension of the AAA program to corn and hog production in relation to the FHA strike call, see Van L. Perkins, *Crisis in Agriculture: The Agricultural Adjustment Administration and the New Deal, 1933*, University of California Publications in History, vol. 81 (Berkeley, Calif.: University of California Press, 1969), pp. 176–79.

31. *Lewistown Democrat-News*, September 2, October 17, 1932, and March 21, November 28, December 5, 1933.

32. Dyson, *Red Harvest*, pp. 31–32, 45, 47–48, 69–70, 120–28, 133; Charles Vindex,

"The Agonizing Years: Radical Rule in Montana," *Montana: Magazine of Western History* 18 (January 1968): 6–9, 13–15; William C. Pratt, "Rethinking the Farm Revolt of the 1930s," *Great Plains Quarterly* 8 (Summer 1988): 138–39. See also Allan J. Mathews, "Agrarian Radicals: The United Farmers League of South Dakota," *South Dakota History* 3 (Fall 1973): 408–21.

33. Dyson, *Red Harvest*, pp. 73–81 passim; *Lewistown Democrat-News*, January 1, 5, 1933.

34. *Williston Herald*, May 11, 1933; Saloutos and Hicks, *Agricultural Discontent*, p. 448; Arthur M. Schlesinger, Jr., *The Age of Roosevelt: The Coming of the New Deal* (Boston, Mass.: Houghton Mifflin Co., 1959), pp. 64–65. See also Perkins, *Crisis in Agriculture*, pp. 168–71, 174–79.

35. *Adams County Record*, October 19, 1933; *Williston Herald*, October 26 (quoted), December 7, 1933, and January 18, 1934.

36. *Williston Herald*, November 5, 1931, and January 21, 28, February 2, 1932; and *Adams County Record*, October 3, 1935.

37. Dyson, *Red Harvest*, p. 100; Remele, "North Dakota Farm Strike," pp. 17–18; Shover, *Cornbelt Rebellion*, p. 92; Schell, *History of South Dakota*, p. 287; *Lewistown Democrat-News*, August 30, 1933; Bates, "My Life," p. 4.

38. On this episode see Saloutos and Hicks, *Agricultural Discontent*, pp. 447–49; Schlesinger, *Coming of the New Deal*, pp. 42–44.

39. *Williston Herald*, November 10, 1932; Saloutos and Hicks, *Agricultural Discontent*, pp. 446–47 (quoted). An initiated measure abolishing mortgages on growing and unharvested crops had, however, been approved in June 1932. North Dakota Legislative Assembly, *Laws Passed at the Twenty-Third Session, 1933*, pp. 497, 506.

40. *Williston Herald*, October 11, 1934, January 9, 1936, and March 8, 1939; *Lewistown Democrat-News*, October 25, 1938; North Dakota Legislative Assembly, *Laws Passed at the Twenty-Third, 1933*, ch. 258, pp. 395–96 (March 3, 1933), and ch. 264, p. 417 (January 20, 1933); *Twenty-Fourth, 1935*, ch. 277, p. 412 (March 13, 1935), and ch. 280, pp. 415–16 (March 12, 1935); *Twenty-Fifth, 1937*, ch. 159, pp. 296–97 (March 1, 1937), ch. 161, pp. 299–304 (February 15, 1937), and ch. 240, p. 449 (March 16, 1937); *Twenty-Sixth, 1939*, ch. 165, pp. 255–59 (March 15, 1939), ch. 227, pp. 380–82 (March 14, 1939), ch. 236, p. 400 (January 28, 1939), and ch. 238, pp. 402–3 (February 18, 1939); *Twenty-Seventh, 1941*, ch. 190, pp. 281–83 (March 17, 1941), ch. 152, pp. 503–5 (March 15, 1941), and ch. 273, pp. 503–5 (March 15, 1941); *Twenty-Eighth, 1943*, ch. 168, pp. 235–36 (March 17, 1943), and ch. 250, pp. 348–49 (March 6, 1943); South Dakota Legislative Assembly, *Laws Passed at the Twenty-Third, 1933*, ch. 137, pp. 133–34 (February 17, 1933), ch. 138, pp. 134–35 (February 9, 1933), and ch. 198, pp. 234–36 (March 13, 1933); *Twenty-Fourth, 1935*, ch. 150, pp. 222–23 (February 26, 1935), ch. 178, pp. 287–90 (March 2, 1935), ch. 194, pp. 310–11 (February 11, 1935), and ch. 195, pp. 311–13 (March 14, 1935); *Twenty-Fifth, 1937*, ch. 163, p. 209 (February 11, 1937), ch. 207, pp. 274–77 (February 18, 1937), ch. 248, pp. 346–48 (March 6, 1937), and ch. 241, pp. 380–81 (March 6, 1937); *Twenty-Sixth, 1939*, ch. 145, pp. 178–81 (January 24, 1939), ch. 279, pp. 346–47 (March 7, 1939), and ch. 280, pp. 347–49 (March 3, 1939); *Twenty-Seventh, 1941*, ch. 37, pp. 59–62 (March 10, 1941), ch. 150, p. 179 (February 21, 1941), ch. 163, pp. 191–94 (February 21, 1941), and ch. 359, pp. 359–60 (February 27, 1941); *Twenty-Eighth, 1943*, ch. 306, pp. 303–4 (February 10, 1943); Montana Legislative Assembly, *Laws . . . by the Twenty-Third, 1933*, ch. 41, pp. 61–62 (March 2, 1933); *Twenty-Third, in Extraordinary Session, 1934*, ch. 45, pp. 128–31 (January 16, 1934); *Twenty-Fourth, 1935*, ch. 2, p. 2 (February 5, 1935), ch. 122, pp. 217–22 (March 13, 1935), and ch. 149, pp. 299–303 (March 13, 1935); *Twenty-Fifth, 1937*, ch. 70, pp. 120–21 (March 1, 1937), and ch. 73, pp. 136–39 (March 3, 1937); *Twenty-Sixth, 1939*, ch. 11, pp. 11–12 (February 7, 1939); *Twenty-Seventh, 1941*, ch. 16, pp. 22–25 (February 18,

1941), and ch. 19, p. 62 (February 25, 1941); *Twenty-Eighth, 1943*, ch. 159, pp. 292–93 (March 3, 1943).

A blanket moratorium protected property against tax deed action until October 1, 1939, in North Dakota. North Dakota Legislative Assembly, *Laws Passed at the Twenty-Sixth, 1939*, ch. 236, p. 400 (January 28, 1939). Thereafter the relief, as previously in Montana and South Dakota, was dependent upon whether and when the counties might already have sold the tax certificates. On the significance of this difference, cf. Schickele and Engelking, *Land Values in North Dakota*, pp. 38, 41–42; Elwyn B. Robinson, *History of North Dakota* (Lincoln, Nebr.: University of Nebraska Press, 1966), p. 406; but cf. *Williston Daily Herald*, February 29, 1940.

41. 47, 48, 49 *U.S. Stat.* 1470–47 (March 3, 1933), 1289–91 (June 28, 1934), 942–45 (August 28, 1935); Arthur M. Schlesinger, Jr., *The Age of Roosevelt: The Politics of Upheaval* (Boston, Mass.: Houghton Mifflin Co., 1960), pp. 554–55, 559–61. On the effect of the legislation, see Ernest Feder and J. A. Munger, "Talking Over Farmer-Debtor Relief Legislation," SDAES, *Annual Report, 1953–54*, pp. 79–82, 99–100.

42. Kumlien, *Graphic Summary of the Relief Situation in South Dakota*, pp. 23, 56 (appendix table 2); *Williston Herald*, June 25, August 6, December 3, 31, 1931, January 21, 1932, March 29, April 13, 1939, and May 21, 1942; *Adams County Record*, November 21, 1935; *Lewistown Democrat-News*, December 11, 1930, May 6, July 23, August 15, 1931, and March 19, 1932; *Hill County Journal*, September 24, 1931; Montana Legislative Assembly, *Laws . . . by the Twenty-Fourth, 1935*, chs. 22, 29, pp. 28, 43 (February 13, 16, 1935). See also North Dakota Legislative Assembly, *Laws Passed at the Twenty-Fifth, 1937*, ch. 122, p. 235 (March 1, 1937).

43. *Lewistown Democrat-News*, October 10, 17, 28, 1931; *Hill County Journal*, October 8, 1931; *Williston Herald*, December 17, 1931; Dyson, *Red Harvest*, pp. 69–70.

44. *Lewistown Democrat-News*, August 16, September 24, November 5, 7, December 17, 1931, and February 10, March 13, September 23, 1932; *Williston Herald*, June 25 and July 9, 1931; *Hill County Journal*, April 16, July 23, and November 5, 1931; *Adams County Record*, November 12, 1931, and February 4, March 17, October 13, 1932.

45. *Williston Herald*, August 18, September 22 (quoted), 1932; *Lewistown Democrat-News*, November 3, 13, 27, 28, 1932, and February 22, 1933; *Adams County Record*, April 20, 1933.

46. *Lewistown Democrat-News*, February 22, March 10, 11, October 27, November 13, 1932, and March 11, 1933; *Hill County Journal*, October 9, 23, 1930, and January 27, 1932; *Williston Herald*, March 10, 1932; resolution of North Dakota FHA, April 8, 1933, in *Congressional Record*, 77:2 (73d Cong., 1st sess.): 1706; Montana Legislative Assembly, *Laws . . . by the Twenty-Third . . . in Extraordinary Session, 1933–1934*, Sen. Jt. Mem. no. 4, pp. 188–89 (January 1, 1934).

47. Wall, *Federal Seed-Loan Financing*, pp. 4–5, 14, 39–40; *Hill County Journal*, May 22, 1930; and *Lewistown Democrat-News*, April 11, 1930.

48. *Hill County Journal*, August 31, 1933; *Lewistown Democrat-News*, March 16, 1932; *Adams County Record*, July 7, 1932. See also Kifer and Stewart, *Farming Hazards in the Drought Area*, pp. 34–35. Wall reports slightly higher total sums but without detail on the number of seed-grain loans. Wall, *Federal Seed-Loan Financing*, pp. 6–10 figs. 1–5, and 45–46 table 14.

49. *Hill County Journal*, September 17, 1931, and August 31, 1933; *Lewistown Democrat-News*, March 16, 20, October 26, 1932; *Adams County Record*, March 24, 1932.

50. *Williston Herald*, July 16, 23, August 13, September 17, 24 (quoted), October 29, 1931, and February 18, March 3, 1932; *Hill County Journal*, March 12, July 9, August 6, 1931; *Lewistown Democrat-News*, April 24, 1932.

51. *Williston Herald*, April 28, May 5, 1932. For a time the North Dakota morato-

rium on seizures under crop mortgages again delayed the processing of loan applications in the spring of 1933, but the state legislature reacted to the threatened denial of federal seed loans by exempting debts to the U.S. government from the provisions of the measure (*Williston Herald*, March 2, 1933).

52. 77 *U.S. Stat.* 5–12, 709–24 (January 22, July 21, 1932); James Stuart Olson, *Herbert Hoover and the Reconstruction Finance Corporation, 1931–1933* (Ames, Iowa: The Iowa State University Press, 1977), pp. 39, 73, 86.

53. *Lewistown Democrat-News*, September 17, 1932, and January 13 (quoted), February 9, 1933; *Adams County Record*, August 18, September 1, 1932, and March 16, 1933; *Williston Herald*, November 24, 1932; Landis, *Rural Relief in South Dakota*, pp. 9, 13; Kumlien, *Graphic Summary of the Relief Situation in South Dakota*, pp. 17, 56 (appendix table 3).

54. Murray R. Benedict, *Farm Policies of the United States: A Study of Their Origins and Development* (New York: The Twentieth Century Fund, 1953), pp. 281–82; Lundy, *Farm-Mortgage Experience in South Dakota*, pp. 28–29 tables 6, 7; William G. Murray, *Agricultural Finance: Principles and Practice of Farm Credit*, (Ames, Iowa: The Iowa State University Press, 1941), pp. 207–8, 214–19; *Hill County Journal*, May 25, June 1, 1933; *Adams County Record*, October 26, 1933, January 4, 1934, and June 30, 1938; *Williston Herald*, October 5, 1933; *Lewistown Democrat-News*, August 31, 1936.

55. *Williston Daily Herald*, May 8, 1936; *Lewistown Democrat-News*, May 31, December 1, 1934; Baldur H. Kristjanson and Jacob A. Brown, *Credit Needs of Beginning Farmers in Selected Areas of North Dakota*, NDAES, *Bulletin* no. 386 (Fargo, N.D., 1953), p. 7 table 1; Kifer and Stewart, *Farming Hazards in the Drought Area*, p. 144 table 20. See also *Hill County Journal*, January 11, 1934; *Lewistown Democrat-News*, May 31, June 2, 1934, and September 15, 1936; *Williston Daily Herald*, January 19, 1937; Farm Credit Administration, *Seventh Annual Report, 1939* (Washington, D.C.: Government Printing Office, 1940), p. 142, appendix table 4.

56. *Lewistown Democrat-News*, November 18, 1933, January 17, 20, 1934, and August 31, 1936; *Adams County Record*, February 1, 1934; *Hill County Journal*, May 17, 1934; Olson, *Herbert Hoover and the Reconstruction Finance Corporation*, pp. 86–87; Farm Credit Administration, *Seventh Annual Report, 1939*, pp. 7, 9.

57. Douglas E. Bowers, Wayne D. Rasmussen, and Gladys L. Baker, *History of Agricultural Price Support and Adjustment Programs, 1933–84*, USDA, Economic Research Service, *Agricultural Information Bulletin* no. 485 (Washington, D.C., 1984), p. 13; Leonard J. Arrington, "Western Agriculture and the New Deal," *Agricultural History* 44 (October, 1970), pp. 340–41 table 2; *Lewistown Democrat-News*, January 6, May 19, and October 13, 1938; USDA, map of REA development, dated September 1, 1939, in MHS.

A resolution of the Montana legislative assembly as late as 1941 noted that application of the Federal Housing Act had been denied to rural localities in the state in nearly all cases (Montana Legislative Assembly, *Laws, 1941*, pp. 452–53 [H. Jt. Mem. no. 10]).

58. Bowers, Rasmussen, and Baker, *History of Agricultural Price Support*, p. 10; Edwin G. Nourse, Joseph S. Davis, and John D. Black, *Three Years of the Agricultural Adjustment Administration* (Washington, D.C.: The Brookings Institution, 1937), pp. 20–24, 125–28, 289, 328.

59. Perkins, *Crisis in Agriculture*, p. 130; James H. Marshall and Stanley Voelker, *Land-Use Adjustment in the Buffalo Creek Grazing District, Yellowstone County, Montana*, USDA, BAE, *Land Economics Report* no. 6 (Washington, D.C.: mimeographed, 1940), pp. 38–39; Kifer and Stewart, *Farming Hazards in the Drought Area*, pp. 3, 4; Arrington, "Western Agriculture and the New Deal," p. 339 table 1; *Williston Herald*, December 30, 1933, and January 5, 1935; *Adams County Record*, August 16, 1934, and March 27,

1941; *Hill County Journal*, November 30, 1933. Somewhat higher payments for the Dakotas are reported in Kristjanson and Heltemes, *Handbook of Facts about North Dakota Agriculture, Revised*, p. 56 table 29; W. F. Kumlien, *Basic Trends of Social Change in South Dakota: II, Rural Life Adjustments*, SDAES, *Bulletin* no. 357 (Brookings, S.D., 1941), p. 16 table 6; *Adams County Record*, March 27, 1941.

On "sod busting," see *Williston Herald*, November 30, 1933.

60. *Lewistown Democrat-News*, April 1, July 12, 1930, October 16, 1931, and July 15, 28, 1933; *Hill County Journal*, March 12, 1931; *Adams County Record*, February 26, 1931, and July 26, 1934; *Williston Herald*, April 4, 8, 1930, and February 5, March 5, August 20, 1931; William D. Rowley, *M. L. Wilson and the Campaign for the Domestic Allotment* (Lincoln, Nebr.: University of Nebraska Press, 1970), pp. 56–57, 107, 110; Richard S. Kirkendall, *Social Scientists and Farm Policy in the Age of Roosevelt* (Columbia, Mo.: University of Missouri Press, 1966), pp. 28–29.

61. *Lewistown Democrat-News*, August 29, 1933; *Adams County Record*, August 3, 1933; *Hill County Journal*, November 16, 1933; Lawrence Svobida, *Farming the Dust Bowl: A First-Hand Account from Kansas* (Lawrence, Kans.: University Press of Kansas, 1986; orig. pub. as *Empire of Dust*, Caxton Printers, 1940), p. 78.

62. *Lewistown Democrat-News*, August 11, 17, September 17, November 4, 12, 1933, and January 24, March 20, 1934; *Adams County Record*, September 28, 1933; *Hill County Journal*, August 31, December 7, 1933; *Williston Herald*, November 16, December 30, 1933. Cf. Theodore Saloutos, *The American Farmer and the New Deal* (Ames, Iowa: The Iowa State University Press, 1982), pp. 76, 195; Perkins, *Crisis in Agriculture*, pp. 127–28.

63. *Adams County Record*, November 2, 1933; February 22, April 26, May 3, 1934, and April 25, October 31, 1935.

64. *Adams County Record*, September 27, November 22, 1934, and March 28, April 11, August 22, 1935; *Lewistown Democrat-News*, April 3, August 15, 1935; *Hill County Journal*, September 14, 1933.

65. Nourse, Davis, and Black, *Three Years of the Agricultural Adjustment Administration*, pp. 20–27, 29–31n (quoting the Topeka speech and the speech of October 25, 1935), 358, 364–70, 373–74, 382–85; *Lewistown Democrat-News*, April 11, June 9, 1936; *Adams County Record*, June 18, 1936, and January 21, 1937. Cf. Randall A. Kramer and Sandra S. Batie, "Cross Compliance Concepts in Agricultural Programs: The New Deal to the Present," *Agricultural History* 59 (April 1985): 309–11.

66. Franklin Delano Roosevelt, statement upon signing the Soil Conservation and Domestic Allotment Act, March 1, 1936, Edgar B. Nixon, comp. and ed., *Franklin D. Roosevelt and Conservation, 1911–1945*, 2 vols. (New York: Arno Press, 1972), 1: 491; *Adams County Record*, July 16, 1936, and April 8, 1937; *Lewistown Democrat-News*, July 8, 1936; *Williston Daily Herald*, July 8, 1936.

67. Dana H. Myrick, *All-Risk Crop Insurance: Principles, Problems, Potentials*, MAES, *Bulletin* no. 640 (Bozeman, Mont., 1970), p. 8; Benedict, *Farm Policies*, pp. 381–83; *Williston Daily Herald*, May 10, 1939; *Adams County Record*, October 11, 1928, May 26, 1938, and February 29, July 25, 1940; *Hill County Journal*, July 5, 1934; *Lewistown Democrat-News*, June 18, 1935, May 14, October 12 (quoted), 1938, June 9, 1939, and May 24, 1940. There is considerable disagreement on the number of policy-holders for North Dakota in 1939, the variants ranging upward to a total of twenty-nine thousand. Cf. *Adams County Record*, October 12, 19, 1939, and January 4, 1940.

Complaint that the premium rate for wheat, based on returns over the period 1930 to 1935, failed to give a representative record had led to an adjustment for 1940, based on the longer term 1919 through 1938, and lower rates for more newly settled districts (Marion Clawson, M. H. Saunderson, and Neil W. Johnson, *Farm Adjustments in Montana, Study of Area IV: Its Past, Present, and Future*, MAES, *Bulletin* no. 377 [Bozeman, Mont., 1940], p. 50).

For discussion of the use of crop insurance as an adjustment of risk aversion in dry-land management, see pp. 208–10 in this volume.

68. The controversy relating to employment of farmers under the WPA was occasioned by efforts to reduce work-relief rolls with the termination of the FERA and to bring the assistance for needy farmers under the programs of the newly established Resettlement Administration. See Asch and Mangus, *Farmers on Relief*, pp. xix-xx, 93; Peck and Forster, *Six Rural Problem Areas*, pp. 32, 36; Olaf F. Larson, *"Ten Years of Rural Rehabilitation in the United States," Summary of a Report* (Bombay: Indian Society of Agricultural Economics, 1950), pp. 118–19; National Resources Planning Board, *The Northern Great Plains* (Washington, D.C., 1940), p. 2 fig. 1; *Williston Daily Herald*, September 5, October 3, 10, 1935, June 22, September 9, November 28, December 10, 1936, July 19, 20, 23, 1937, August 6, September 17, 27, 1937, September 7, 1938, and February 3, 1943; but cf. *Lewistown Democrat-News*, November 28, 1933, November 7, 1935, July 9, August 9, 1936, and May 10, 1941; oral history interviews with Selmer Helland and Art Olson, MHS. President Roosevelt conceded the necessity for work relief for farmers, even while urging revision of farm policy, upon returning from a survey of the drought situation in the summer of 1936. Speech, September 6, 1936, in Nixon, ed., *Franklin D. Roosevelt and Conservation*, 1:569; *Williston Daily Herald*, December 10, 1936, but cf. *Adams County Record*, September 24, 1936, and February 25, 1937.

Data on expenditures for work relief vary considerably in their coverage. The per capita figures given are calculated on the basis of a table in Leonard Arrington, "The New Deal in the West: A Preliminary Statistical Inquiry," *Pacific Historical Review* 38 (August 1969): 315 table 3. They appear to be somewhat high as compared with data given in Federal Works Agency, Works Project Administration, *Report on Progress of the WPA Program, June 30, 1939*, p. 166 table 9; the final report on CWA and FERA expenditures for North Dakota as published in *Adams County Record*, November 21, 1935; and a comprehensive statement of expenditure of federal funds for Montana from March 4, 1933 through June 30, 1938, published under a Helena dateline in *Lewistown Democrat-News*, October 13, 1938. The fact that Arrington's data were based on reports solicited in preparation for the re-election campaign of President Roosevelt in 1940 may have contributed to some inflation. The FERA sums here included would have covered grants as well as work relief.

Public works expenditures, not included in the Arrington listing, amounted in Montana through June 1938 to $65,626,258 under the PWA, $25,041,598 under the Bureau of Public Roads, and $2,753,157 under the Bureau of Reclamation. *Lewistown Democrat-News*, October 13, 1938. Expenditures of $6,328,675 were added for Montana during the next fiscal year under the special PWA Appropriation Act of 1938. Those for North Dakota and South Dakota under this last legislation were much less, at $2,997,875 and $1,806,722, respectively (*Report of the President . . . to Congress Showing the Status of Funds and Operations under the Emergency Relief Appropriations . . . to 1941, Inclusive* [Washington, D.C.: U.S. Government Printing Office, January, 1941— T63.114], p. 551). Labor on the Fort Peck Dam, under the U. S. Army Corps of Engineers, also provided employment for some 10,000 workers in Montana through the mid-thirties (*Williston Daily Herald*, June 23, 1936).

69. Benedict, *Farm Policies*, p. 335; *Lewistown Democrat-News*, September 4, 1934, October 9, 1940, and May 9, 1941; *Adams County Record*, September 24, 1936, April 25, 1940, and April 3, 1941; Arrington, "Western Agriculture and the New Deal," p. 339

table 1. For analysis of the policy considerations that dictated operation of this program, see C. Roger Lambert, "Want and Plenty: The Federal Surplus Relief Corporation and the AAA," *Agricultural History* 46 (July, 1972): 390–99.

70. *Hill County Journal*, May 25, 1933, May 3, 1934, and August 31, 1935; *Lewistown Democrat-News*, March 12, August 2, 23, 1934, and December 20, 1936; *Williston Herald*, March 2, 1933, March 13, 24, 1936, February 1, 1937, February 21, 1939, and February 13, 1941.

71. Wall, *Federal Seed-Loan Financing*, p. 13 fig. 8, amended to cover 1935; Renne, *Montana Farm Bankruptcies*, p. 49; *Williston Daily Herald*, April 13, 1936. In an effort to bring such borrowing under the rehabilitation program of the Resettlement Administration, President Roosevelt had vetoed the appropriation bill for seed loans in 1936 and shifted the funding from the FCA to the Rural Resettlement Administration. The attendant delays occasioned strong local protest. See *Williston Daily Herald*, March 13, April 2, 13, 1936.

72. Wall, *Federal Seed-Loan Financing*, pp. 45–46 table 14; Philip J. Thair, *Stabilizing Farm Income against Crop Yield Fluctuations*, NDAES, *Bulletin* no. 362 (Fargo, N.D., 1950), p. 10 table 3; *Williston Daily Herald*, March 23, 1939; Stewart, *Natural and Economic Factors Affecting Rural Rehabilitation Problems in Southwestern North Dakota . . . Hettinger County*, p. 13; computations from Farm Credit Administration, *Seventh Annual Report, 1939*, pp. 206, 207, appendix table 64.

The exceptionally low repayment record in North Dakota was occasioned in large part by Governor Langer's declaration of a moratorium on repayment of the obligations in 1937 (*Williston Daily Herald*, March 23, 1939).

73. Wall, *Federal Seed-Loan Financing*, pp. 45–46 table 14; Farm Credit Administration, *Seventh Annual Report, 1939*, pp. 88, 208, appendix table 65.

74. Theodore Saloutos, "The New Deal and Farm Policy in the Great Plains," *Agricultural History* 43 (July 1969): 374; C. Roger Lambert, "The Drought Cattle Purchase, 1934–1935: Problems and Complaints," *Agricultural History* 45 (April 1971): 85; Kifer and Stewart, *Farming Hazards in the Drought Area*, p. 27 fig. 9; Stewart, *Natural and Economic Factors Affecting Rural Rehabilitation . . . in Southwestern North Dakota*, p. 9; *Williston Herald*, July 26, September 6, October 18, 1934; *Lewistown Democrat-News*, November 5, December 9, 1934, and July 1, 9, 1936; computations from *Sixteenth Census of the United States: 1940, Agriculture*, vol. 1, pt. 2, pp. 397–401, 486–91, and pt. 6, pp. 31–35—county table 4.

75. *Williston Daily Herald*, March 17, 1936, August 29, 1938, April 14, 1939, and February 6, 1940; Farm Security Administration, *Report of the Administrator, 1941*, pp. 28–29 table 1; Joe J. King, "The Farm Security Administration and Its Attack on Rural Poverty," *Rural Sociology* 7 (June 1942): 158–60; *Land Policy Circular*, January 1937, p. 9; Marshall and Voelker, *Land-Use Adjustment in the Buffalo Creek Grazing District*, p. 41; Olaf F. Larson, "Lessons from Rural Rehabilitation Experience," *Land Policy Review* 9 (Fall 1946): 13–14; Larson, ed., "Ten Years of Rural Rehabilitation in the United States," pp. 7–11, 29–31, 78, 81–89, 92–94, 120. See also *Adams County Record*, May 19, 1938, March 7, 1940, and January 2, April 17, 1941; *Williston Daily Herald*, December 21, 1936, January 11, 26, March 25, 1937, and October 24, 25, 1939; *Lewistown Democrat-News*, February 10, 1936.

76. Montana Legislative Assembly, *Laws . . . by the Twenty-Third . . . in Extraordinary Session, 1934*, ch. 56, p. 168 (January 20, 1934); *Twenty-Fourth, 1935*, ch. 109, pp. 190–95 (March 11, 1935); North Dakota Legislative Assembly, *Laws Passed at the Twenty-Third, 1933*, ch. 261, pp. 397–412 (March 7, 1933); *Twenty-Fourth, 1935*, ch. 276, pp. 401–11 (March 11, 1935); Jt. Res. B., p. 470 (February 8, 1935); *Twenty-Fifth, 1937*, ch. 6, pp. 20–23 (March 6, 1937), ch. 214, pp. 391–93 (March 16, 1937), ch. 249, pp. 464–77 (March 12, 1937), and ch. 84, pp. 139–40 (January 16, 1937); *Twenty-Sixth, 1939*, ch.

234, p. 391 (March 14, 1939); *Twenty-Seventh, 1941*, ch. 283, pp. 519–22 (March 14, 1941); South Dakota Legislative Assembly, *Laws Passed at the Twenty-Third, 1933*, ch. 184, pp. 201–211 (March 3, 1933), and ch. 185, pp. 211–12 (March 6, 1933); *Twenty-Third . . . in Special Session, 1933*, ch. 12, pp. 24–25 (August 5, 1933); *Twenty-Fourth, 1935*, ch. 52, pp. 66–68 (March 15, 1935), and ch. 205, pp. 326–52 (March 14, 1935). To reduce property tax levies and finance aid to the common schools, South Dakota also enacted a controversial gross income tax in 1933, which was supplanted two years later by net income and retail sales taxes.

77. Computations are based on Works Progress Administration, *Report on Progress of the WPA Program, June 30, 1938*, pp. 114 table 73, and 116 table 75; supplemented by data covering the period from January 1938 through June 30, 1939, derived from Federal Works Agency, Work Projects Administration, *Report on Progress of the WPA Program, June 30, 1939*, pp. 166 table 9 and 170, table 13. The basic data cover expenditures of the CWA, WPA, NYA, Civilian Conservation Corps, rural rehabilitation loans and grants, and aid for blind, child, and aged dependents. The Montana State Planning Board reported the figure for that state, covering the period July 1, 1932, to June 30, 1939, as "in excess of $132,000,000" ("Preliminary Report on Development of Economic Opportunities in Montana for Migratory and Stranded Families," p. 1, attachment to Montana State Planning Board, *Annual Report . . . as of December 31, 1940* [Helena, Mont., 1940]).

Other useful indicators of the relationship between federal and local expenditures were published in *Hill County Journal*, June 7, 1936, reporting that the final FERA statement showed expenditure in Montana of over $25.5 million, of which the federal government contributed almost $22.7 million, and in *Williston Daily Herald*, July 5, 1939, reporting that of $115.9 million expended for relief in North Dakota through fiscal 1939, $103.3 million represented the federal contribution.

Until enactment of the Emergency Relief Appropriation Acts of 1935 through 1939 there was no centralized accounting, and then it was limited to operations covered by that specific legislation. See *Report of the President . . . to Congress, Showing the Status of Funds and Operations under the Relief Appropriation Acts . . . to 1941*, p. 118.

78. *Spokane Review* quoted in *Williston Daily Herald*, October 24, 1935.

79. *Williston Daily Herald*, October 24, 1935, and February 23, 1939.

CHAPTER FOUR: LAND-USE PLANNING FOR REGIONAL ADJUSTMENT: THE RESETTLEMENT PROGRAM

1. See pp. 13–14 and 19 in this volume; Elwood Mead to R. F. Walter, Chief Engineer, May 15, 1930; H. A. Parker to Elwood Mead, May 21, and June 4, 1930; H. H. Johnson to Elwood Mead, June 9, 1930; A. W. Walker to Elwood Mead, June 29, 1930—all Federal Records Center, Denver, Col., RG 115, Office of Chief Engineer, General Correspondence File, 42-A4, FRC 220868, 220870, file pp. 02547–48, 08581, 08590–93, 08598–99, 08611.

2. *Fergus County Argus* (Lewistown, Mont.), January 2, 1920, supp.; David F. Houston, "Annual Report of the Secretary," USDA, *Yearbook, 1919*, pp. 28–29 (quoted); Henry C. Wallace, "Annual Report of the Secretary," USDA, *Yearbook, 1923*, p. 21; Rexford Guy Tugwell, "The Problem of Agriculture," *Political Science Quarterly* 39 (December 1924): 577–85; James H. Shideler, *Farm Crisis: 1919–1923* (Berkeley, Calif.: University of California Press, 1957), pp. 88, 112–16.

3. Wallace, "Annual Report, 1923," p. 25; O. E. Baker, "Government Research in Aid of Farmers in the Northern Great Plains of the United States," W. L. G. Joerg, ed., *Pioneer Settlement, Cooperative Studies by Twenty-Six Authors* (New York, N.Y.: Amer-

ican Geographical Society, 1932), pp. 64–70 (quotation on p. 68); see pp. 58–59 in this volume.

4. Albert Z. Guttenberg, "The Land Utilization Movement of the 1920s," *Agricultural History* 50 (July 1976): 479: L. C. Gray and others, "The Utilization of Our Lands for Crops, Pasture, and Forests," USDA, *Yearbook, 1923*, pp. 417, 430, 437–38, 502–3.

5. Gray and others, "Utilization of Our Lands," p. 505 (quoted).

6. U.S. General Land Office, *Annual Report of the Commissioner, 1923*, pp. 3–8, *1924*, p. 11, *1925*, p. 34, *1926*, p. 33, *1927*, p. 13, and *1928*, pp. 13–14; James H. Marshall, "Montana Grazing Districts," *Land Policy Circular*, March 1938, p. 10; and pp. 44–45 in this volume.

7. Roy M. Robbins, *Our Landed Heritage: The Public Domain, 1776–1936* (Lincoln, Nebr.: University of Nebraska Press, 1962), pp. 413–23; William A. Hartman, "State Policies in Regulating Land Settlement Activities," *Journal of Farm Economics* 13 (April 1931): 259–69.

8. *Lewistown Democrat-News* (Mont.), November 29, 1934; Montana Legislative Assembly, *Laws . . . Passed by the Twenty-Third, 1933*, ch. 66, pp. 123–28 (March 7, 1933); *Twenty-Fourth, 1935*, chaps. 194, 195, pp. 425–37 (March 18, 1935); North Dakota Legislative Assembly, *Laws Passed at the Twenty-Fourth, 1935*, ch. 106, pp. 122–28 (March 13, 1935); South Dakota Legislative Assembly, *Laws Passed at the Twenty-Fourth, 1935*, ch. 71, pp. 97–100 (March 14, 1935); R. J. Penn and C. W. Loomer, *County Land Management in Northwestern South Dakota*, SDAES, *Bulletin* no. 326 (Brookings, S.D., 1938), pp. 26–27.

9. Wallace, "Annual Report of the Secretary, 1923," p. 19; O. E. Baker, "Land Utilization in the United States: Geographical Aspects of the Problem," *Geographical Review* 13 (January 1923): 20 (quoted), 23, 26; O. E. Baker, "The Potential Supply of Wheat," *Economic Geography* 1 (March 1925): 50–51.

10. O. E. Baker, *Do We Need More Farm Land? Address, Agricultural Extension Conference, University of Minnesota . . . December 13 and 14, 1928* (St. Paul, Minn., 1928), pp. 1, 29–30 (all quoted); O. E. Baker, "Changes in Production and Consumption of Our Farm Products and the Trend in Population," American Academy of Political and Social Science, *Annals* 142 (March 1929): 146 (quoted); L. C. Gray and O. E. Baker, *Land Utilization and the Farm Problem*, USDA, *Miscellaneous Publication* no. 97 (Washington, D.C., 1930), p. 53 (quoted) et passim.

11. See p. 29 in this volume; L. C. Gray, "Research Relating to Policies for Submarginal Areas," *Journal of Farm Economics* 16 (April 1934): 300 (quoted), 301–2 (quoted), 303 (quoted).

12. L. C. Gray, "National Land Policies in Retrospect and Prospect," *Journal of Farm Economics* 13 (April, 1931): 234 (quoted). For a more extended discussion of Gray's views, see Richard L. Kirkendall, "L. C. Gray and the Supply of Agricultural Land," *Agricultural History* 37 (October 1963): 206–14.

13. E. G. Nourse, "The Apparent Trend of Recent Economic Changes in Agriculture," American Academy of Political and Social Science, *Annals* 149 (May 1930): 48–50 (quotation, p. 50); R. G. Tugwell, "Reflections on Farm Relief," *Political Science Quarterly* 43 (December 1928): 481, 490; R. G. Tugwell, "Farm Relief and a Permanent Agriculture," American Academy of Political and Social Science, *Annals* 142 (March 1929): 282 (quoted).

14. Kirkendall, "L. C. Gray and the Supply of Agricultural Land," pp. 210–13.

15. M. L. Wilson, "A Land Use Program for the Federal Government," *Journal of Farm Economics* 15 (April 1933): 218–20.

16. Wilson, "Land Use Program," pp. 218, 220; Gray, "National Land Policies," pp. 232–33; M. L. Wilson, "Mechanization, Management and the Competitive Position of Agriculture," *Agricultural Engineering* 13 (January 1932): 3–5; William D.

Rowley, M. L. *Wilson and the Campaign for the Domestic Allotment* (Lincoln, Nebr.: University of Nebraska Press, 1970), p. 89; *Lewistown Democrat-News*, March 24, 1931 (quoted).

17. Rowley, M. L. *Wilson*, p. 94. Wilson had also organized and chaired a post-conference committee of fourteen, restricted to agricultural economists, designed to promote specific legislative proposals. It achieved little. See Rowley, M. L. *Wilson*, pp. 99–102.

18. Wilson, "Land Use Program," pp. 222–27 (quotations on pp. 224–25).

19. 48 *U.S. Stat.* 205 (June 16, 1933); Richard S. Kirkendall, *Social Scientists and Farm Politics in the Age of Roosevelt* (Columbia, Mo.: University of Missouri Press, 1966), p. 71.

20. Kirkendall, *Social Scientists*, pp. 74–75; Edgar B. Nixon, comp. and ed., *Franklin D. Roosevelt and Conservation, 1911–1945*, 2 vols. (New York, N.Y.: Arno Press, 1972), 1: 159, 196–97, 327.

21. Kirkendall, *Social Scientists*, pp. 75, 82 (quoting Gray), 83; U.S. National Resources Board, "Statement of the Advisory Commission," *Progress Report, 1938*, p. 2n; U.S. National Resources Board, *A Report on National Planning and Public Works in Relation to Natural Resources and Including Land Use and Water Resources with Findings and Recommendations, December 1, 1934* (Washington, D.C.: U.S. Government Printing Office, 1934), pp. v–vi; Arthur M. Schlesinger, Jr., *The Age of Roosevelt: The Coming of the New Deal* (Boston, Mass.: Houghton Mifflin Co., 1959), pp. 350–51; Nixon, comp. and ed., *Franklin D. Roosevelt and Conservation*, 1: 317, 318n.

22. U.S. National Resources Board, *Report . . . December 1, 1934*, pp. 2–6 (p. 3 quoted).

23. U.S. National Resources Board, *Report . . . December 1, 1934*, pp. 159–60, 175, 176 fig. 44, 180.

24. U.S. National Resources Board, *Report . . . December 1, 1934*, pp. 156–57.

25. Cf. U.S. National Resources Board, *Report . . . December 1, 1934*, p. 177 fig. 45; U.S. National Resources Board, Land Planning Committee, *Maladjustments in Land Use in the United States, Supplementary Report*, pt. 6 (Washington, D.C.: U.S. Government Printing Office, 1935), facing p. 1 fig. 1).

26. Cf. Francis D. Cronin and Howard W. Beers, *Areas of Intense Drought Distress, 1930–1936*, U.S. WPA, Division of Social Research, *Research Bulletin*, ser. 5, no. 1 (Washington, D.C., 1937), pp. 26, 30, 43–44, 52 table 8; P. G. Peck and M. C. Forster, *Six Rural Problem Areas: Relief—Resources—Rehabilitation*, U.S. FERA, Division of Research Statistics, *Research Monograph* no. 1 (Washington, D.C., 1935), p. 137 fig. 3; W. F. Kumlien, *A Graphic Summary of the Relief Situation in South Dakota, 1930–1935*, SDAES, *Bulletin* no. 310 (Brookings, S.D., 1937), cover, pp. 20 fig. 21, 21 fig. 22, and 48 fig. 47; W. F. Kumlien, Robert L. McNamara, and Zetta E. Bankert, *Rural Population Mobility in South Dakota, 1928–1935*, SDAES, *Bulletin* no. 315 (Brookings, S.D., 1938), pp. 14–15; Berta Asch and A. R. Mangus, *Farmers on Relief and Rehabilitation*, U.S. WPA, Division of Social Research, *Research Monograph* no. 8 (Washington, D.C.: U.S. Government Printing Office, 1937), p. 20 fig. 4.

Because of broken terrain, dry-land farming in Toole and Blaine counties was limited to relatively small but prosperous districts. The mapping of proposed retirement zones made no such distinctions, however.

27. U.S. National Resources Board, *Report . . . December 1, 1934*, pp. 103–4 (quoted), 159; Mary W. M. Hargreaves, *Dry Farming in the Northern Great Plains: 1900–1925*, Harvard Economic Studies, vol. 101 (Cambridge, Mass.: Harvard University Press, 1957), p. 18 fig. 7. Cf. E. A. Starch, "Type of Farming Modifications Needed in the Great Plains," *Journal of Farm Economics* 21 (February 1939): 117–18.

28. Sherman E. Johnson, *From the St. Croix to the Potomac—Reflections of a Bureaucrat* (Bozeman, Mont.: Big Sky Books, 1974), pp. 115, 117, 120 (quoted), 123, 127, 130; H. H. Wooten, *The Land Utilization Program, 1934 to 1964: Origin, Development, and Present Status*, USDA, Economic Research Service, *Agricultural Economic Report* no. 85 (Washington, D.C., 1965), pp. 74–75, appendix table 12.

Areas added to the central Montana project after enactment of the Bankhead-Jones Act were recommended for purchase by county planning bodies and the boards of directors of local grazing associations. They were described as usually tracts too small for profitable livestock operations apart from access to supplementary range, sites needed for stock-water development, or units owned by sheepmen "whose operations complicated the grazing economy of the area" (James H. Marshall and Stanley Voelker, *Land-Use Adjustment in the Buffalo Creek Grazing District, Yellowstone County, Montana*, USDA, BAE, *Land Economics Report* no. 6 [Washington, D.C.: mimeographed, 1940], p. 42). Idealized objectives for selection of such areas were summarized in C. F. Clayton, "Program of the Federal Government for the Purchase and Use of Submarginal Land," *Journal of Farm Economics* 17 (February 1935): 57–58.

29. Nixon, comp. and ed., *Franklin D. Roosevelt and Conservation*, 2: 283n, 323–24n, 569–70n.

30. Wooten, *Land Utilization Program*, pp. 6, 10, 12, 18 table 4, 36; L. C. Gray, "Federal Purchase and Administration of Submarginal Land in the Great Plains," *Journal of Farm Economics* 21 (February, 1939): 126.

31. Wooten, *Land Utilization Program*, pp. 20, 24; John J. Haggerty, *Public Finance Aspects of the Milk River Land Acquisition Project (LA-MT-2) Phillips County, Montana*, USDA, Resettlement Administration, *Land Use Planning Publication* no. 18-a (Washington, D.C.: mimeographed, 1937), p. 5; *Fifteenth Census of the United States: 1930, Agriculture*, vol. 2, pt. 3, pp. 136–40—county table 5; M. H. Saunderson and others, *An Approach to Area Land Use Planning*, USDA, Resettlement Administration, *Land Use Planning Publication* no. 16 (Washington, D.C., USDA, Resettlement Administration, Division of Land Utilization in cooperation with MAES, 1937), pp. 55, 58–59, appendix fig. 4; M. R. Benedict, "Production and Control in Agriculture and Industry, *Journal of Farm Economics* 18 (August 1936): 465 (quoted). See also Marshall and Voelker, *Land Use Adjustment in the Buffalo Creek Grazing District*, p. 16; Roy E. Huffman and James L. Paschal, "Integrating the Use of Irrigated and Grazing Land in the Northern Great Plains," *Journal of Land and Public Utility Economics* 18 (February 1942): 23, 27.

32. Wooten, *Land Utilization Program*, pp. 24, 47, 74–75 table 12; Loyd [sic] Glover, *The Future of Federal Land Use Purchase Projects in South Dakota*, SDAES, *Bulletin* no. 464 (Brookings, S.D., 1957), p. 5; *Lewistown Democrat-News*, March 13, 1938, and January 2, 1939, Gray, "Federal Purchase and Administration of Submarginal Lands," p. 128; M. B. Johnson, "Land Use Adjustment in North Dakota," *Soil Conservation* 5 (March 1940): 221. Reports on the size of the range areas differ, dependent upon whether data cover only federally purchased tracts (the Land Utilization projects), authorized purchase areas, or a combination of the holdings of cooperative grazing districts, Taylor grazing reserves, and federally purchased tracts.

Additional government purchases of land in Montana were projected to include 135,000 acres for the Fort Peck, Fort Belknap, and Blackfeet Indian reservations, close to 50,000 acres for migratory waterfowl refuges near Malta, Medicine Lake, and Dillon, and nearly 60,000 acres to enlarge Yellowstone Park (Will W. Alexander, "Resettlement in Montana," *Montana Farmer* 24 (October 1, 1936): 10; *Lewistown Democrat-News*, January 16, 1935). Such collateral purchases were also made in other states to establish national parks, monuments, game and recreational areas, forest reserves, and Indian reservations. See Mont H. Saunderson, *Western Land and Water Use* (Norman, Okla.: University of Oklahoma Press, 1950), p. 99; *Lewistown Democrat-News*, April 29, 1935.

33. Data for this and the following paragraph drawn from Haggerty, *Public Finance Aspects*, pp. 5–6, 43; H. L. Lantz, "Readjustment of Population to Land Resources in Northern Montana," *Soil Conservation* 5 (February 1940): 212; *Lewistown Democrat-News*, January 2, 1939; Wooten, *Land Utilization Program*, p. 47; Glover, *Future of Federal Land Use Purchase Projects*, pp. 5, 13, 15, 17; Johnson, "Land Use Adjustment in North Dakota," p. 221; Gray, "Federal Purchase and Administration of Submarginal Land," p. 128. By the end of 1939 only 7.5 percent of the desired land in eastern Fergus County had been optioned and 17 percent of that in the older Musselshell-Petroleum County section, either optioned or bought. *Lewistown Democrat-News*, November 16, 1939.

34. Marshall and Voelker, *Land Use Adjustment in the Buffalo Creek Grazing District*, pp. 22–23 (quoted), 41, 43; Donald R. Rush, "Credit as a Factor in Land Policy," *Land Policy Circular*, March 1938, pp. 11–13; Donald R. Rush, "Coordination of Agricultural Credit and Land Use Policies," *Land Policy Circular*, April 1938, pp. 13–15; Donald R. Rush, "The Use of Agricultural Credit in a Land-Use Program," *Land Policy Review* 1 (May-June 1938): 12–16; M. B. Johnson, "Land Use Readjustment in the Northern Great Plains," *Journal of Land and Public Utility Economics* 13 (May 1937): 161; Tivis E. Wilkins and George B. McIntire, *An Analysis of the Land Acquisition Program under Title III of the Bankhead-Jones Farm Tenant Act*, USDA, SCS, *Miscellaneous Publication* no. 26 (Washington, D.C., 1942), pp. 13–14; R. J. Penn, W. F. Musbach, and W. C. Clark, *Possibilities of Rural Zoning in South Dakota: A Study of Corson County*, SDAES, *Bulletin* no. 345 (Brookings, S.D., 1940), p. 14; *Lewistown Democrat-News*, February 10, 1936. See also U.S. Great Plains Committee, *The Future of the Great Plains, Report* (Washington, D.C.: U.S. Government Printing Office, 1936), p. 98; Gray, "Federal Purchase and Administration of Submarginal Land," p. 130.

35. M. B. Johnson, "Land Use Adjustment in North Dakota," p. 226 (quoted); *Lewistown Democrat-News*, July 12, 1936; [U.S.] National Resources Board, *State Planning, a Review of Activities and Progress* (Washington, D.C.: U.S. Government Printing Office, 1935), pp. 8–12, 55, 71, 85; Sherman E. Johnson, *From the St. Croix to the Potomac*, p. 119 (quoted); *Williston Daily Herald* (N.D.), August 1, 1938; Montana Legislative Assembly, *Laws . . . Enacted by the Twenty-Sixth, 1939*, Sen. Jt. Res. no. 5, pp. 719–20 (March 1, 1939).

36. Sherman E. Johnson, "Land Use Readjustments in the Northern Great Plains," *Journal of Land and Public Utility Economics* 13 (May 1937): 156–57 (quoted).

Montana analysts questioned whether even stockmen promoted demand for the program. As of January 1941 only nine of forty grazing associations in that state were wholly outside the federal land acquisition areas, and only one of the nine was outside the boundaries of a Taylor grazing district. This geographic coincidence, they noted, suggested a close relationship between the development of "so-called 'voluntary' collective tenure devices" and the program of the federal agencies (Glenn H. Craig and Charles W. Loomer, *Collective Tenure on Grazing Land in Montana*, MAES, *Bulletin* no. 406 (Bozeman, Mont., 1942), p. 7).

37. *Lewistown Democrat-News*, January 25 (quoted), July 26, 1934, and July 12, 1936 (quoted).

38. *Lewistown Democrat-News*, July 26 (quoted), and 28 (quoted), 1934; *Hill County Journal* (Havre, Mont.), August 2, 1934; Elwood Mead to Tom Berry, July 27, 1934, DRC, RG 115, Office of the Chief Engineer, General Correspondence File, 42, Lands General, FRC 220868, file p. 02027.

39. R. G. Tugwell, "Down to Earth," *Current History* 44 (July 1936): 33.

40. *Lewistown Democrat-News*, July 15, 1936 (quoted); *Hill County Journal*, June 14, 1934, quoting *Montana Standard: Land Policy Circular*, O1ctober, 1936, pp. 8–9, 11–12; Kirkendall, *Social Scientists*, pp. 118–23. See also advice by S. D. Director of Extension Service, *Adams County Record* (Hettinger, N.D.), March 18, 1937.

41. *Adams County Record*, January 26, 1939, and January 25, 1940; *Willison Daily Herald*, January 24 and August 15, 1940; Roland R. Renne, "Probable Effects of Federal Land Purchase on Local Government," *National Municipal Review* 25 (July 1936): 403; North Dakota Legislative Assembly, *Laws Passed at the Twenty-Sixth, 1939*, House Concurrent Res. no. 179, pp. 562–63 (March 4, 1939); Montana Legislative Assembly, *Laws . . . Enacted by the Twenty-Seventh, 1941*, H. Jt. Mem. no. 6, pp. 447–48 (February 28, 1941); *Laws . . . Enacted by the Twenty-Eighth, 1943*, Sen. Jt. Mem. no. 14, pp. 613–14 (March 3, 1943). See also U.S. Great Plains Committee, *The Future of the Great Plains*, pp. 78–79; *Lewistown Democrat-News*, December 19, 1948.

42. Renne, "Probable Effects," pp. 403, 405; Haggerty, *Public Finance Aspects*, pp. 10–11, 45; Hugo C. Schwartz, "Governmental Tax Immunity: I. The Problem," *Land Policy Review* 2 (January-February, 1939): 33–34; 50 *U.S. Stat.* 525–26 (July 22, 1937); Wooten, *Land Utilization Program*, pp. 48, 51; *Lewistown Democrat-News*, March 25, 1935, April 3, 10, 18, and July 11, 1937; Marshall and Voelker, *Land Use Adjustment in the Buffalo Creek Grazing District*, pp. 22–23.

Leases for oil and gas exploration, developing in the early fifties, greatly increased the revenues of the federal Land Utilization projects. Such income amounted to $541,000, compared to $329,000 for grazing utilization, in total revenues of approximately $900,000 paid into the federal treasury from the Land Utilization projects in Montana and the Dakotas in 1959. Twenty counties shared a total payment of $224,368 that year (*Dickinson Press*, April 9, 1960).

43. Haggerty, *Public Finance Aspects*, pp. 10–11; Marshall and Voelker, *Land Use Adjustment in the Buffalo Creek Grazing District*, pp. 22–23; *Lewistown Democrat-News*, March 25, 1935, and April 3, 10, 18, July 11, 1937.

44. Roland R. Renne, *Montana Land Ownership: An Analysis of the Ownership Pattern and Its Significance in Land Use Planning*, MAES, *Bulletin* no. 322 (Bozeman, Mont., 1936), p. 10 n. 2; *Land Policy Circular*, March 1938, p. 14; Lantz, "Readjustment of Population to Land Resources," p. 212; computations from data in Glover, *Future of Federal Land Use Purchase Projects*, p. 32, appendix table 1; U.S. National Resources Board, Land Planning Committee, *Maladjustments in Land Use . . . , Supplementary Report*, pt. 6, map facing p. 14; *Sixteenth Census of the United States: 1940, Agriculture*, vol. 1, pt. 2, pp. 382, 384, 469, 472, and pt. 6, pp. 16, 19, 20—county table 1. Initially, prices for land purchase in the Milk River project area were quoted at $0.50 to $2.00 an acre for grazing land and $3.00 to $7.00 an acre for "cultivated land" and meadows; the average, however, was cited as $2.50 (*Hill County Journal*, January 20, 1935). An average of $2.75 an acre was the anticipated cost estimated at the Lincoln, Nebraska, regional office (*Williston Herald*, February 2, 1935).

45. C. B. Baldwin, "Statement," May 11, 1943, U.S. Congress, 78th Cong., 1st sess., *Hearings before the Select Committee of the House Committee on Agriculture, to Investigate the Activities of the Farm Security Administration*, 4 pts. (Washington, D.C.: U.S. Government Printing Office, 1943), pt. 1, pp. 14–15; Marshall and Voelker, *Land Use Adjustment in the Buffalo Creek Grazing District*, pp. 42–43; C. F. Clayton, "The Land Utilization Program Begins Its Second Year," *Land Policy Review* 1 (no. 3): 12; Murray R. Benedict, *Farm Policies of the United States: A Study of Their Origins and Development* (New York: The Twentieth-Century Fund, 1953), p. 326; Marion Clawson, "Resettlement Experience on Nine Selected Resettlement Projects," *Agricultural History* 52 (January 1978): 50; *Lewistown Democrat-News*, March 25, 1935, and July 28, 1936, quoting Tugwell; Rexford G. Tugwell, "The Resettlement Idea," *Agricultural*

History 33 (October 1959): 161–64; *Adams County Record*, June 20, 1935; Penn and Loomer, *County Land Management*, p. 16; and pp. 84, 154, 155 in this volume. Although a majority of the county commissioners of Phillips County, Montana, favored acceptance of the government's offer for tax-sale lands, the issue occasioned extended debate (*Hill County Journal*, March 10, 1935). Cf., however, *Williston Daily Herald*, July 8, 1936, or *Lewistown Democrat-News*, July 7, 1936, both carrying an Associated Press dispatch quoting the opposition of Mayor John Hipple, of Pierre, S.D.; editorial in *Hill County Journal*, September 6, 1936.

46. See p. 101; T. E. Hayes, "Sub-Marginal Land," *Adams County Record*, April 26, 1934 (quoted). A growing recognition of the social implications of removal, evidenced in the later planning literature, also testifies to the dissatisfaction. See, in particular, L. C. Gray, "Our Land Policy Today," *Land Policy Review* 1 (May-June 1938): 7–8; Carl C. Taylor, "The Human Aspects of Land-Use Planning," *Land Policy Review* 1 (September-October 1938): 10; Robin M. Williams, "Planning for People, Not for Plans," *Land Policy Review* 4 (January 1941): 30–34.

Two bitter accounts document the removal story in other sections of the West during this period and provide much insight on the distress occasioned. Ann Marie Low, in *Dust Bowl Diary* (Lincoln, Nebr.: University of Nebraska Press, 1984) tells of her family's struggle to remain and their eventual removal from a farm taken for a waterfowl refuge in eastern North Dakota. Paul Bonnifield in *The Dust Bowl: Men, Dirt, and Depression* (Albuquerque, N.M.: University of New Mexico Press, 1979), discusses the program as applied in the southern Plains.

47. Wooten, *Land Utilization Program*, pp. 20–21; Lantz, "Readjustment of Population to Land Resources," p. 212; Wendell L. Lund, "Bought Out by the Government," *Land Policy Review* 2 (May-June, 1939): 22; *Hill County Journal*, October 18, 1934, quoting Harry Hopkins.

48. Lantz, "Readjustment of Population to Land Resources," p. 213; Gladys R. Costello, "Irrigation History and Resettlement on Milk River Project, Montana," *Reclamation Era* 30 (1940): 136–42, 170–71; Hargreaves, *Dry Farming . . . 1900–1925*, pp. 457, 459.

49. U.S. Congress, 78th Cong., 1st sess., *Hearings . . . to Investigate . . . the Farm Security Administration*, pt. 3, pp. 1078–79, and pt. 4, p. 1803. In a letter to the author, November 21, 1977, Elmer Starch noted that the governmental administrators had been "burned" in resettlement on the valley lands of the Malta project because the soils were poor. They had had to "relieve those who tried it there."

50. Ernest Eugene Melvin, "Problems Related to the Development of the Greenfields Irrigation District" (Ph.D. diss., Northwestern University, 1952), pp. 14, 93, 96–98; Hargreaves, *Dry Farming . . . 1900–1925*, pp. 456–57; Fred J. Martin, "Resettlement and Readjustment in Montana," *Montana Farmer* 27 (April 15, 1940): 30; U.S. Congress, 78th Cong., 1st sess., *Hearings . . . to Investigate . . . the Farm Security Administration*, pt. 3, pp. 1077, 1121; Clawson, "Resettlement Experience," pp. 3, 12.

51. Clawson, "Resettlement Experience," pp. 17, 56–70 passim.

52. Clawson, "Resettlement Experience," pp. 52–53; Lantz, "Readjustment of Population to Land Resources," p. 213; Melvin, "Problems Related to . . . Greenfields Irrigation District," p. 98; Ralph E. Ward and M. M. Kelso, *Irrigation Farmers Reach Out into the Dry Land*, MAES, *Bulletin* no. 464 (Bozeman, Mont., 1949), pp. 7, 29, 31–32; P. L. Paschal and P. L. Slagsvold, "Irrigation Development and Area Adjustment in the Great Plains," *Journal of Farm Economics* 25 (May 1943): 434–35, 441. See also Glover, *Future of Federal Land Use Purchase Projects*, pp. 21–23; South Dakota Legislative Assembly, *Laws Passed at the Twenty-Fourth, 1935*, ch. 71, pp. 97–100 (March 14, 1935).

The U.S. Forest Service also experienced difficulty in adjusting the conflicting demands of large and small operators for grazing leases (*Williston Daily Herald*, February 18, 21, 1936).

53. U.S. Congress, 78th Cong., 1st sess., *Hearings . . . to Investigate . . . the Farm Security Administration*, pt. 3, pp. 1078, 1125; Clyde E. Stewart and D. C. Myrick, *Control and Use of Resources in the Development of Irrigated Farms: Buffalo Rapids and Kinsey, Montana*, MAES, Bulletin no. 476 (Bozeman, Mont., 1951), pp. 45–46, 48, 50. Several of the Kinsey farms were abandoned because of unsuitable land.

54. For data cited in this and the following paragraph, see 54 *U.S. Stat.* 439 (quoted); Paschal and Slagsvold, "Irrigation Development," pp. 441–43; John C. Page, "Statement . . . before the Special Committee of the House of Representatives . . . , September 17, 1940," p. 26, DRC, RG 115, Office of the Chief Engineer, General Correspondence File, 42-A7, Lands Migration, FRC 220870, file p. 09275; Stewart and Myrick, *Control and Use of Resources . . . Buffalo Rapids and Kinsey*, pp. 9, 45–46, 48–50; Clyde E. Stewart and D. C. Myrick, *Irrigated Farm Development in Buffalo Rapids and Kinsey, Montana*, MAES, Circular no. 195 (Bozeman, Mont., 1951), p. 15 table 2. Little integrated farming-grazing development actually resulted on the Buffalo Rapids Project (Roy E. Huffman, *Irrigation Development and Public Water Policy* [New York, N.Y.: Ronald Press Co., 1953], p. 146).

55. For detail on these projects, see Hargreaves, *Dry Farming . . . 1900–1925*, pp. 131, 458, 460; Huffman, *Irrigation Development and Public Water Policy*, pp. 143–44; Roy E. Huffman and James L. Paschal, "Integrating the Use of Irrigated and Grazing Land in the Northern Great Plains," *Journal of Land and Public Utility Economics* 18 (February 1942): 17–27 passim; Stanley W. Voelker, *Settlers' Progress on Two North Dakota Irrigation Projects: A Study of Farm Development and Resource Accumulation on the Buford-Trenton and Lewis and Clark Projects*, NDAES, Bulletin no. 369 (Fargo, N.D., 1951), pp. 5–9; *Williston Daily Herald*, July 10, 1940, July 28, December 13, 1943, November 30, 1944, February 7, April 11, October 13, 14, 1947, and November 20, 1951; Page, "Statement . . . September 17, 1940," file p. 09275. In most cases estimates of irrigable acreage varied greatly according to the source and date of the prospectus. Those here stated are given by Voelker and are lower than those earlier anticipated.

56. U.S. Congress, 78th Cong., 1st sess., *Hearings . . . to Investigate . . . the Farm Security Administration*, pt. 3, pp. 1130, 1136.

57. Glover, *Future of Federal Land Use Purchase Projects*, p. 14; Herbert S. Schell, *History of South Dakota* (Lincoln, Nebr.: University of Nebraska Press, 1961), pp. 294–95; U.S. Congress, 78th Cong., 1st sess., *Hearings . . . to Investigate . . . the Farm Security Administration*, pt. 3, pp. 1126, 1131, 1133, 1136; Edward G. Grest, "Management of Lands Held under Title III of the Bankhead-Jones Farm Tenant Act," *Soil Conservation* 5 (February 1940): 220; John C. Page, Commissioner to Michael W. Straus, undated memorandum in fall 1942, on postwar plans, appendix A, p. 2, DRC, RG 115, Office of the Chief Engineer, General Correspondence File, 42—Lands, FRC 220867. The Rapid Valley project would have required readjustment of 165 farm families, of which 60 were to have been rehabilitated in place and 60, by resettlement in the project area; 45 others would have been moved elsewhere ("Schedule of Farm Family Adjustments," in Page, "Statement," September 17, 1940, file p. 09275).

Early in 1934 the State Rural Relief Corporation, financed by $650,000 from the federal government, had bought 10,000 acres under the Belle Fourche project for subdivision into ten-acre units, each with a new house constructed by the unemployed. In August farmers were reported on the move to this experiment, described as "rural socialism," by "auto, truck, and train" (*Williston Herald*, August 16, 1934). No subsequent reference to it has been found. For a somewhat similar "subsistence homestead" project in North Dakota, see *Williston Herald*, June 14, 21, 1934.

58. Clawson, "Resettlement Experience," pp. 37–43, 62–63. The contract for

houses built on Kinsey Flats called for two to three bedrooms, a combination kitchen and dining area, living room, unfinished space for modern bathroom, and a full basement (*Lewistown Democrat-News*, April 4, 1939).

The difference in expectations on living standards was indicated when a county planning committee for Judith Basin County in 1940 questioned whether the planning "should be on the basis of opportunity for a plane of living which is considered adequate by local people, or whether this plane of living should be determined by technicians, national thinking, or other means?" Montana State Agricultural Advisory Committee, "Minutes . . . September 20 and 21, 1940," DNA, RG 83, Montana file.

59. U.S. Farm Security Administration, *Annual Report . . . for 1942–43*, p. 22; *Postwar Developments in Farm Security, The Annual Report . . . for 1945–46*, p. 17; U.S. Congress, 78th Cong., 1st sess., *Hearings to Investigate . . . the Farm Security Administration*, pt. 3, pp. 1077–79, 1092–93, 1101. The cost of the houses to be built on the Milk River project had been estimated at five thousand dollars per unit, as originally announced (*Hill County Journal*, September 29, 1935).

60. 50 *U.S. Stat.* 525–26 (July 22, 1937); *Lewistown Democrat-News*, July 23, 1935, quoting Wheeler; Wilkins and McIntire, *Analysis of the Land Acquisition Program*, p. 15; FDR Press Conference, July 7, 1936, Nixon, comp. and ed., *Franklin D. Roosevelt and Conservation*, 1: 539 (quoted).

CHAPTER FIVE: ALTERNATIVE PLANNING

1. U.S. Great Plains Drought Area Committee, *Report* (Washington, D.C., 1936), pp. 2 (quoted), 17; U.S. National Resources Committee, Water Resources Committee, *Drainage Policy and Projects, Report of Special Subcommittee* (Washington, D.C., 1936); U.S. Mississippi Valley Committee of the Public Works Administration, *Report . . . Submitted October 1, 1934.* (Washington, D.C.: U.S. Government Printing Office, 1934), pp. 5–6, 169, 179, 234; H.S. Person, with cooperation of E. Johnston Coil and Robert T. Beall, *Little Waters, Study of Headwaters Streams and Other Little Waters, Their Use and Relations to the Land* (Washington, D.C., 1936); Kenneth E. Trombley, *The Life and Times of a Happy Liberal: A Biography of Morris Llewellyn Cooke* (New York, N.Y.: Harper and Brothers, 1954), ch. 8 and pp. 138–39. Gray served as technical adviser to the Great Plains Drought Area Committee (*Williston Daily Herald* [N.D.], August 11, 1936).

2. Frederick A. Delano to Harold L. Ickes, January 13, 1937, Edgar B. Nixon, comp. and ed., *Franklin D. Roosevelt and Conservation*, 2 vols. (New York, N.Y.: Arno Press, 1972), 1: 612–14; Gilbert F. White, "*The Future of the Great Plains* Re-Visited," *Great Plains Quarterly* 6 (Spring 1986); 84 (quoted).

3. U.S. Great Plains Drought Area Committee, *Report*, pp. 10–12, 14, 15 (quoted); U.S. Great Plains Committee, *The Future of the Great Plains, Report* (Washington, D.C.: U.S. Government Printing Office, 1936), pp. 10–11 (quoted), 76–77, 85–86.

4. Sherman Johnson, "Irrigation Policies and Programs in the Northern Great Plains Region," *Journal of Farm Economics* 18 (August 1936): 544, 552; Roy E. Huffman, *Irrigation Development and Public Water Policy* (New York, N.Y.: Ronald Press Co., 1953), pp. 85–86; *Land Policy Circular*, July 1935, p. 28; *Williston Daily Herald*, November 1, 1939.

5. Johnson, "Irrigation Policies," pp. 545–48; H. H. Johnson to Commissioner [Elwood Mead], August 3, 1935, DRC, RG 115, Office of the Chief Engineer, General Correspondence File, 42 Lands General, FRC 220867, file pp. 01526–27; *Hill County Journal* (Havre, Mont.), July 24, 1930; Marvin P. Riley, W. F. Kumlien, and Duane Tucker, *Fifty Years Experience on the Belle Fourche Irrigation Project*, SDAES, *Bulletin* no.

450 (Brookings, S.D., 1955), pp. 18–20, 32, 51; Huffman, *Irrigation Development*, pp. 68–69; P. L. Slagsvold, *Readjusting Montana's Agriculture: VI, Montana's Irrigation Resources*, MAES, *Bulletin* no. 315 (Bozeman, Mont., 1936), pp. 13–16; P. L. Slagsvold, *An Analysis of the Present Status of Agriculture on the Sun River Irrigation Project*, MAES, *Bulletin* no. 321 (Bozeman, Mont., 1936), pp. 5, 23, 43–46; P. L. Slagsvold, *An Analysis of Agriculture on the Valier Irrigation Project*, MAES, *Bulletin* no. 330 (Bozeman, Mont., 1936), p. 5 (quoted); P. L. Slagsvold, *Agriculture on the Huntley Project*, MAES, *Bulletin* no. 342 (Bozeman, Mont., 1937), pp. 6–7, 11; MAES, *Forty-First Annual Report, 1934*, p. 22.

6. U.S. National Resources Board, *A Report on National Planning and Public Works in Relation to Natural Resources and Including Land Use and Water Resources with Findings and Recommendations, December 1, 1934* (Washington, D.C.: U.S. Government Printing Office, 1934), p. 194 table 24; Montana Legislative Assembly, *Laws . . . Passed by the Extraordinary Session of the Twenty-Third, 1934*, ch. 39, pp. 116–17 (January 16, 1934); *Twenty-Fourth, 1935*, ch. 176, pp. 363–67 (March 14, 1935); Mary W. M. Hargreaves, "Land-Use Planning in Response to Drought: The Experience of the Thirties," *Agricultural History* 50 (October 1976): 570–71; *Lewistown Democrat-News* (Mont.), September 25, 1935, and March 12, 19, August 25, October 5, 1936; *Williston Daily Herald*, October 13, 1936, and January 2, April 29, 1937.

Of 112,424 acres described as "irrigable" on the Milk River project in 1934, only 43,927 acres were irrigated (*Hill County Journal*, January 13, 1935).

The Federal Bureau of Reclamation during the summer of 1935 listed only 629 sites, 50,528 acres, on fifteen projects in nine states, as available for the "rural rehabilitation program" (Mead to Francis R. Kenny, July 2 [with attachment], 27, 1935, DRC, RG 115, Office of the Chief Engineer, General Correspondence File, 42 Lands General, FRC 220867, file pp. 01569, 01570, 01545).

7. *Lewistown Democrat-News*, October 13, 1935, and October 7, 1938—both under Washington, D.C., datelines; Johnson, *Irrigation Policies*, pp. 44–49; U.S. National Resources Board, *Report on National Planning and Public Works*, pp. 194–95; U.S. National Resources Committee, Water Resources Committee, *Drainage Policy and Projects*, passim; U.S. Great Plains Committee, *Future of the Great Plains*, pp. 76–77, 85–86; U.S. National Resources Planning Board, *The Northern Great Plains* (Washington, D.C., 1940), pp. 4, 6 (quoted)–20.

8. *Sixteenth Census of the United States: 1940, Irrigation of Agricultural Lands* (Washington, D.C.: U.S. Government Printing Office, 1942), pp. 340, 454, 524—county table 1; *Lewistown Democrat-News*, October 3, 15, 19, 1933; *Williston Daily Herald*, March 25, 1936, September 12, 1938, February 24, March 25, September 24, 1941, and January 14, May 8, 1942; L. F. Gieseker, *Soils of Valley County: Soil Reconnoissance of Montana, Preliminary Report*, MAES, *Bulletin* no. 198 (Bozeman, Mont., 1926), p. 56; Richard Lowitt, *The New Deal and the West* (Bloomington, Ind.: Indiana University Press, 1984), p. 87; Elwyn B. Robinson, *History of North Dakota* (Lincoln, Nebr.: University of Nebraska Press, 1966), pp. 462–66.

Organization of the campaign for diversion of the Missouri River to the Devil's Lake district had begun at least as early as 1927 (*Williston Herald*, December 22, 1927). The North Dakota State Planning Board estimated in 1936 that "complete and permanent development of all stream basins" in the state would cost approximately $65 million, including $30,507,940 for the Missouri River diversion project (*Williston Daily Herald*, September 2, 1936). The Army Corps of Engineers, who disapproved of that project, then estimated that it would cost $54 million and would afford nowhere near the benefit (*Ibid.*, May 3, 1937). On the subsequent development of the Missouri River Basin program, see pp. 212–18 in this volume.

9. *Hill County Journal*, April 4, June 6, 1937; *Lewistown Democrat-News*, November

27, 1934, November 10, 20, 1939, and January 11, 1940; *Williston Daily Herald*, November 10, 1939, and February 24, 1941; L. F. Gieseker, *Soils of Hill County: Soil Reconnoissance of Montana, Preliminary Report*, MAES, *Bulletin* no. 246 (Bozeman, Mont., 1931), p. 45. For a history of the projected Chain-of-Lakes development, see *Hill County Journal*, September 7, 1933.

10. *Williston Daily Herald*, July 28, 1938, July 10, 1940, and September 30, 1941; *Adams County Record* (N.D.), February 8, 1934, September 15, 1940, and April 3, 1941; *Lewistown Democrat-News*, December 22, 1939; Riley, Kumlien, and Tucker, *Fifty Years Experienced on the Belle Fourche Irrigation Project*, pp. 18–21; S. O. Harper to Val Huska, March 6, 1941; "Settlement Opportunities on Federal Reclamation Projects, December, 1941"; "Bureau of Reclamation, Normal and Potential Construction Program . . . , Fiscal Years 1942, 1943, and 1944 . . . "—MSS. in DRC, RG 115, Office of the Chief Engineer, General Correspondence File, 42 Lands General, FRC 220867, file pp. 00491, 00772, 00594. See pp. 214, 220 in this volume.

11. *Hill County Journal*, November 24, 1932, February 2, 1933; *Lewistown Democrat-News*, May 19, 1931, January 4, June 13, November 24, December 18, 1932, February 28, April 27, June 26, December 13, 1933, April 16, September 16, 1934, November 8, 1936, and October 7, 1938; article by Alfred Atkinson, carried by the Associated Press, in *Lewistown Democrat-News*, September 21, 1936; Johnson, "Irrigation Policies," p. 554; James H. Marshall and Stanley Voelker, *Land-Use Adjustment in the Buffalo Creek Grazing District, Yellowstone County, Montana*, USDA, BAE, *Land Economics Report* no. 6 (Washington, D.C.: mimeographed, 1940), p. 19; Elmer A. Starch to author, November 21, 1977; Huffman, *Irrigation Development*, pp. 141–46; Roy E. Huffman, "Montana's Contributions to New Deal Farm Policy," *Agricultural History* 33 (October 1959): 167; Roy E. Huffman and James L. Paschal, "Integrating the Use of Irrigated and Grazing Land in the Northern Great Plains," *Journal of Land and Public Utility Economics* 18 (February 1942): 17–27; Commissioner [Elwood Mead] to Chief Engineer, May 15, 1930; H. A. Parker to Commissioner, May 21, 1930; Mead to R. F. Walter, June 3, 1930; H. H. Johnson to Mead, June 9, 1930; A. W. Walker to Commissioner, June 29, 1930—all DRC, RG 115, Office of the Chief Engineer, General Correspondence File, 42-A4, FRC 220870, file pp. 08598–99, 08611, 08596–97, 08590–93, 08581.

12. Alexander Joss, "Reclamation Experience on Federal Reclamation Projects," *Journal of Farm Economics* 27 (February 1945): 154, 156, 157; U.S. Secretary of the Interior, *Annual Report . . . for the Fiscal Year Ended June 30, 1940*, pp. 7–8; *Lewistown Democrat-News*, January 7 and February 15, 1937. By the end of the decade, public employment and direct relief funding had provided for construction of 180 irrigation reservoirs, 6,000 stock-water reservoirs, 1,100 flood irrigation systems, and 200 pumping installations in Montana (Montana State Planning Board, "Preliminary Report on Development of Economic Opportunities for Migratory and Stranded Families" [Helena, Mont., October 1939], p. 2, attachment to *Annual Report . . . as of December 31, 1940*).

13. 50 *U.S. Stat.* 841–42, 869–70 (August 26, 28, 1937); 52 *U.S. Stat.* 748 (June 16, 1938); 53 *U.S. Stat.* 976 (June 30, 1939), 54 *U.S. Stat.* 565 (June 25, 1940); *Land Policy Circular*, November 1937, pp. 7, 10.

14. 53 *U.S. Stat.* 719, 1418–19 (May 10, August 11, 1939); 54 *U.S. Stat.* 1119–28 (October 14, 1940).

15. White, "*Future of the Great Plains* Revisited," p. 92; MAES, *Fortieth Annual Report, 1933*, p. 25; *Land Policy Circular*, December 1936, p. 4; *Lewistown Democrat-News*, March 24, July 11, December 20, 1936, and February 20, 1939; *Adams County Record*, February 10, 17, 1938, and July 18, August 15, 1940; *Williston Daily Herald*, March 4, June 28 (quoted), 1937, and September 19, 1940. The Montana Extension

Service reported the addition of only twenty-nine pumping projects, for an unspecified acreage, in 1940 (*Lewistown Democrat-News*, March 3, 1941). For mapping of the aquifers in the Plains, see GPC, *Proceedings, 1960*, p. 74 fig. 3.

16. Cf. Olaf F. Larson, "Lessons from Rural Rehabilitation Experience," *Land Policy Review* 9 (Fall 1946): 13–18; Olaf F. Larson, ed., "*Ten Years of Rural Rehabilitation in the United States*," *Summary of a Report* (Bombay, India: Indian Society of Agricultural Economics, 1950); Clarence A. Wiley, "Settlement and Unsettlement in the Resettlement Administration Program, *Law and Contemporary Problems* 4 (October 1937); 464 (quoted)–69; Marion Clawson, "Resettlement Experience on Nine Selected Resettlement Projects," *Agricultural History* 52 (January 1978): 56–70; John D. Black, "Agricultural Credit Policy in the United States, 1945," *Journal of Farm Economics* 27 (August 1945): 592–93, 602–8; *Adams County Record*, April 6, 1939, December 24, 1941; *Williston Daily Herald*, May 3, 1939.

17. John C. Page, "Statement of Commissioner . . . Bureau of Reclamation, Prepared for Presentation before the Special Committee of the House of Representatives to Investigate Interstate Migration of Destitute Citizens . . . September 17, 1940," p. 25, DRC, RG 115, Office of the Chief Engineer, General Correspondence File, 42-A7, Lands Migration, FRC 220870, file p. 09274 (quoted); North Dakota Legislative Assembly, *Laws Passed at the Twenty-Sixth, 1939*, Sen. Concurrent Res. no. 42, pp. 535–36 (February 2, 1939); Joe J. King, "The Farm Security Administration and Its Attack on Rural Poverty," *Rural Sociology* 7 (June 1942): 158 (quoted); Larson, "Lessons from Rural Rehabilitation Experience," p. 15; Larson, ed., "*Ten Years of Rural Rehabilitation*," pp. 10–11, 14; *Adams County Record*, January 25, March 7, 1940, January 2, 1941, and January 1, 1942; *Williston Daily Herald*, October 22, 1942. Cf., however, Title II of the Bankhead-Jones Act, which authorized loans to "eligible individuals for the purchase of livestock, farm equipment, supplies, and for other farm needs (including minor improvements and minor repairs to real property)" (50 *U.S. Stat.* 524–25). The reported lack of availability of such loans in North Dakota appears to have been occasioned by administrative policy.

18. Larson, ed., "Ten years," pp. 122, 133–34. See also *Williston Daily Herald*, March 23, 1939.

19. *Lewistown Democrat-News*, September 10, 1937, and July 9, 10, 15, August 27, 1938; *Williston Daily Herald*, July 13, 15, 1938.

20. MAES, *Fortieth Annual Report, 1932–33*, pp. 20, 21 (both quoted); *Lewistown Democrat-News*, January 1, 1940; Wilson quoted in *Williston Daily Herald*, July 23, 1937; Marion Clawson, M. H. Saunderson, and Neil W. Johnson, *Farm Adjustments in Montana, Study of Area IV: Its Past, Present and Future*, MAES, *Bulletin* no. 377 (Bozeman, Mont., 1940), p. 51; M. H. Saunderson and others, *An Approach to Area Land Use Planning*, USDA, RA, *Land Use Planning Publication* no. 16 (Washington, D.C.: USDA, RA, Division of Land Utilization, Land Use Planning Section in cooperation with MAES, 1937), pp. 51–52; Neil W. Johnson, *Farm Adjustments in Montana, Study of Area VII: Its Past, Present and Future*, MAES, *Bulletin* no. 367 (Bozeman, Mont., 1939), pp. 34–35, 39–44; Neil W. Johnson, *Farm Adjustments in Montana, Graphic Supplement* (Washington, D.C.: USDA, BAE in cooperation with MAES and with assistance of WPA, 1940), p. 7. See also E. A. Starch, "Type of Farming Modifications Needed in the Great Plains," *Journal of Farm Economics* 21 (February 1939): 117. On the reported increase in barley acreage, see Table 1.1, p. 21.

21. *Hill County Journal*, October 25, 1934, and October 24, 1937; *Lewistown Democrat-News*, April 18, 23, 1935; *Williston Daily Herald*, September 8, 1936, May 19, June 24, 1937, and July 13, 1938.

22. For detail of this and the following paragraph, see *Williston Daily Herald*, September 8, 1936, and June 24, 1937; *Hill County Journal*, October 24, 1937; *Lewistown*

Democrat-News, October 8, 1937; *Soil and Water Conservation in the Northern Great Plains*, (USDA, SCS, Region Nine, 1937), pamphlet filed DRC, RG 115, Office of the Chief Engineer, General Correspondence File, FRC 220867, doc. no. 042-01219; and Herbert S. Schell, *History of South Dakota* (Lincoln, Nebr.: University of Nebraska Press, 1961), p. 347.

President Roosevelt's program to establish a band of trees from north to south through the prairie states had extended 35 miles in North Dakota and 28 miles in South Dakota by the summer of 1935, but was restricted to a zone east of the Missouri River, not encompassed in this study. See *Williston Daily Herald*, June 24, 1935; North Dakota Agricultural Advisory Council, "Some Brief Notes on the Meeting . . . December 9 and 10, 1941," p. 6, DNA, RG 83, North Dakota file.

23. See pp. 48–51; Mary W. M. Hargreaves, *Dry Farming in the Northern Great Plains, 1900–1925*, Harvard Economic Studies, vol. 101 (Cambridge, Mass.: Harvard University Press, 1957), p. 320n; O. R. Mathews and John S. Cole, "Special Dry-Farming Problems," USDA, *Soils and Men, Yearbook, 1938*, p. 692; *Campbell's 1902 Soil Culture Manual: Explains How the Rain Waters Are Stored and Conserved in the Soil* (Holdrege, Nebr.: H. W. Campbell, 1902), pp. 10–11; Thomas Shaw, *Dry Land Farming* (St. Paul, Minn.: The Pioneer Co., 1911), pp. 364–65; *Lewistown Democrat-News*, January 20, 1931, January 12, 1933, March 22, 1936, and January 18, 1938; *Williston Herald*, April 9, 1931, and April 18, 1935; *Adams County Record*, April 25, 1935. See also *Hill County Journal*, May 14, 1931.

24. *Adams County Record*, May 23, 1935; *Lewistown Democrat-News*, December 16, 1937; *Williston Daily Herald*, August 25, 1939, and September 16, 1940; M. P. Hansmeier, *Soil Drifting on Cropland in the Plains Area of Montana*, MExtS, *Bulletin* no. 176 (Bozeman, Mont., 1939), p. 25.

25. *Adams County Record*, September 15, November 24, 1938, reprinting article by Dean E. Erlandson from *Dakota Farmer*; *Williston Daily Herald*, June 16, 1938; H. G. Davis, "Farm Equipment for 1937, Both New and Improved," mimeographed press release by Farm Equipment Institute, Farm Research Department; Mathews and Cole, "Special Dry-Farming Problems," pp. 688–89; Earle H. Bell, *Culture of a Contemporary Rural Community: Sublette, Kansas*, USDA, BAE, *Rural Life Studies* no. 2 (Washington, D.C., 1942), p. 44; Paul Bonnifield, *The Dust Bowl: Men, Dirt, and Depression* (Albuquerque, N.M.: University of New Mexico Press, 1979), pp. 155, 162.

26. Roosevelt to Thomas D. Campbell, August 24, 1936, Nixon, ed., *Franklin D. Roosevelt and Conservation*, 1:556; George S. Wehrwein and others, "The Remedies: Policies for Private Lands," USDA, *Soils and Men: Yearbook, 1938*, pp. 248 (quoted)–51; *Lewistown Democrat-News*, December 18, 1938. On the fundamental planning goals, see also Harry A. Steele and John Muehlbeier, "Land and Water Development Programs in the Northern Great Plains," *Journal of Farm Economics* 32 (August 1950) 436.

27. Montana Legislative Assembly, *Laws . . . Passed by the Twenty-Sixth, 1939*, ch. 72, pp. 124–53 (February 28, 1939); North Dakota Legislative Assembly, *Laws Passed at the Twenty-Fifth . . . in Special Session, 1937*, ch. 9, pp. 29–45 (March 16, 1937); South Dakota Legislative Assembly, *Laws Passed at the Twenty-Fifth Session, 1937*, ch. 19, pp. 20–35 (March 5, 1937); Wehrwein and others, "Remedies," pp. 251–52; Marshall and Voelker, *Land-Use Adjustment in the Buffalo Creek Grazing District*, p. 28; North Dakota State Agricultural Advisory Council, "Minutes, September 6, 1939," p. 6; "Some Brief Notes . . . December 9 and 10, 1941," p. 9; South Dakota Land Use Program Committee, "Minutes, January 30–31, 1940," p. 14—MSS. in DNA, RG 83, respective state files; *Williston Daily Herald*, September 21, 1938, and January 13, 1941; *Lewistown Democrat-News*, September 23, 1940, and January 25, 1941. On the controversy between the state and federal governments in framing the program for soil conserva-

tion districts, see J. C. Headley, "Soil Conservation and Cooperative Extension," *Agricultural History* 59 (April 1985): 290–306.

28. South Dakota Legislative Assembly, *Laws Passed at the Twenty-Fifth Session, 1937*, ch. 19, pp. 27 (quoted), 28–29, 30 (quoted); George S. Wehrwein, "Enactment and Administration of Rural Zoning Ordinances," *Journal of Farm Economics* 18 (August 1936): 522; F. F. Elliott, comments on George S. Wehrwein, "Public Control of Land Use in the United States," *Journal of Farm Economics* 21 (February 1939): 86–87; U.S. National Resources Committee conference on rural zoning, December 13–14, 1937, reported in *Land Policy Circular*, January 1938, p. 14; R. J. Penn, W. F. Musbach, and W. C. Clark, *Possibilities of Rural Zoning in South Dakota: A Study in Corson County*, SDAES, *Bulletin* no. 345 (Brookings, S.D., 1940), pp. 6–7; Glenn H. Craig and Charles W. Loomer, *Collective Tenure on Grazing Land in Montana*, MAES, *Bulletin* no. 406 (Bozeman, Mont., 1942), p. 24. Cf., however, Charles M. Hardin, *The Politics of Agriculture: Soil Conservation and the Struggle for Power in Rural America* (Glencoe, Ill.: The Free Press, 1952), pp. 73–75, which notes that only two soil conservation districts outside Colorado, one being the Cedar Creek District in North Dakota, had to that date used the authority to issue regulations. The Colorado experience in attempting enforcement led to legislative restriction on the regulatory power and transfer of administration of "blow-land" ordinances to county commissioners rather than the SCS district supervisors.

The organization and program of soil conservation districts in southwestern North Dakota, as in Montana, operated very much like grazing districts. See Craig and Loomer, *Collective Tenure on Grazing Land*, pp. 9, 20–21; Glenn K. Rule, *Soil Conservation Districts in Action on the Land*, USDA, *Miscellaneous Publication* no. 448 (Washington, D.C.: U.S. Government Printing Office, 1941), pp. 2–11; Edwin E. Ferguson, "Nation-Wide Erosion Control: Soil Conservation Districts and the Power of Land-Use Regulation," *Iowa Law Review* 34 (January 1949): 181; *Adams County Record*, February 29, 1940.

29. Douglas E. Bowers, Wayne D. Rasmussen, and Gladys L. Baker, *History of Agricultural Price-Support and Adjustment Programs, 1933–84*, USDA, ERS, *Agricultural Information Bulletin* no. 485 (Washington, D.C., 1984), p. 11; *Lewistown Democrat-News*, May 30, June 9, December 20, 1936, and February 24, September 9, 1937; *Adams County Record*, April 16, May 7, June 18, July 16, 1936, and March 14, 1940; *Williston Daily Herald*, May 15, September 15, 1936, July 8, 1937, and February 21, 1939; Marshall and Voelker, *Land-Use Adjustment in the Buffalo Creek Grazing District*, p. 39.

30. Bowers, Rasmussen, and Baker, *History of Agricultural Price-Support . . . Programs*, pp. 12–15; *Lewistown Democrat-News*, February 24, April 17, May 11, July 29, 1938; *Adams County Record*, January 6, 1938.

31. *Lewistown Democrat-News*, November 12, 1937, and May 7, 1938; Marshall and Voelker, *Land-Use Adjustment in the Buffalo Creek Grazing District*, pp. 39–40 (quoted); USDA, *Agricultural Statistics, 1941*, pp. 636–37 table 757; *1942*, pp. 750–51 table 815; L. C. Gray, "Federal Purchase and Administration of Submarginal Land in the Great Plains," *Journal of Farm Economics* 21 (February 1939): 125 (quoted). A distinction is made here between "restoration" land under the AAA program and grasslands under the land-purchase program. Cf. MAES, *Forty-Sixth and Forty-Seventh Annual Reports, 1939–1941*, p. 55.

32. For data in this and the following paragraph, see USDA, *Agricultural Statistics, 1942*, pp. 750–51 table 815, and pp. 752–55 table 816; *Lewistown Democrat-News*, November 9, 1937, March 4, 6, 1940, and March 14, 1941; *Adams County Record*, July 25, 1940; *Williston Daily Herald*, July 14, 1937, and July 31, 1939. On the speed of the increase in strip farming in Fergus County, see the reports of H. R. Stucky, county agent, *Lewistown Democrat-News*, December 20, 1936, and December 12, 1938. On the

beginning of the practice in western North Dakota, see *Williston Daily Herald*, April 12, 1935. The first contour pasture furrowing in Williams County, North Dakota, by a cooperator in the Little Muddy Soil conservation district, was reported in *Williston Daily Herald*, August 25, 1939.

33. 49 *U.S. Stat.* 1148–49 (February 29, 1936); Montana Legislative Assembly, *Laws . . . Passed by the Twenty-Fifth, 1937*, ch. 134, pp. 404–14 (March 16, 1937); North Dakota Legislative Assembly, *Laws Passed at the Twenty-Fifth . . . in Special Session, 1937*, ch. 2, pp. 3–9 (March 12, 1937); South Dakota Legislative Assembly, *Laws Passed at the Twenty-Fifth, 1937*, ch. 18, pp. 14–19 (March 6, 1937); H. R. Tolley, "County Planning Gets Its Second Wind," *Extension Service Review* 7 (September 1936): 129–30; *Hill County Journal*, November 8, 1936; *Lewistown Democrat-News*, May 30, 1936, and July 7, 11, November 7, 1937; *Williston Daily Herald*, June 15, 1937, and October 1, 1940; *Adams County Record*, April 9, 1936.

34. *Lewistown Democrat-News*, March 13, 19, 24 (quoted), 27, 31, 1931.

35. U.S. National Resources Board, *State Planning: A Review of Activities and Programs . . . June, 1935* (Washington, D.C.: U.S. Government Printing Office, 1935), pp. 55, 71, 85.

36. *Lewistown Democrat-News*, March 22, 27, April 19, 21 (quoted), 28, 30, May 1, 1934; *Hill County Journal*, May 17, 1934 (quoted). See also *Lewistown Democrat-News*, October 27, 1934.

37. Montana Legislative Assembly, *Laws . . . of the Twenty-Fourth, 1935*, ch. 176, pp. 363–67 (March 14, 1935); North Dakota Legislative Assembly, *Laws Passed at Twenty-Fourth, 1935*, ch. 217, pp. 303–5 (March 6, 1935); South Dakota Legislative Assembly, *Laws Passed at Twenty-Fourth, 1935*, ch. 191, pp. 307–8 (March 1, 1935); U.S. National Resources Committee, Advisory Committee, *Progress Report 1938: Statement* (Washington, D.C.: U.S. Government Printing Office, 1939), p. 4.

38. Montana State Planning Consultant, *Progress Report . . . to the National Resources Board, June 16, 1935* (mimeographed), pp. 4a, 5, 8 (quoted); *Lewistown Democrat-News*, January 2, 1937.

39. Montana State Planning Consultant, *Progress Report . . . to the National Resources Committee, June 16, 1936* (mimeographed), pp. 13–15, 19.

40. U.S. National Resources Board, *State Planning: A Review . . . June 1935*, pp. 71 (quoted), 85.

41. *Adams County Record*, September 13, 1934, August 29, 1935, and June 11, 1936; *Williston Herald*, January 19, 1935 (quoted); Governor Elmer Holt's address to the Northern Montana Fair Association, *Lewistown Democrat-News*, January 28, 1936 (quoted). Cf. also *Williston Daily Herald*, February 10, 1936.

42. *Lewistown Democrat-News*, April 14, 1938, and December 22, 1940.

43. *Lewistown Democrat-News*, April 5, May 24, 1936.

44. *Lewistown Democrat-News*, December 16, 24, 1936, January 8, November 20, December 10, 1937, February 11, 12, 1938, October 26, November 10, December 8, 1939, and January 13, November 2, December 22, 1940.

45. *Lewistown Democrat-News*, February 19, 1938; *Williston Daily Herald*, March 10, October 13, 1936, September 21, 1937, November 22, 1938, November 2, 3, 7, 1939, and September 11, November 14, 1940; *Adams County Record*, March 5, 12, 1936, December 21, 1937, January 27, 1938, December 12, 1940, and January 2, 1941; H. W. Herbison, "Planning in Schools," *Land Policy Review* 5 (March 1942): 33–34; H. W. Herbison to [North Dakota] State Agricultural Advisory Council, November 15, 1939, attachment, p. 5; North Dakota Agricultural Advisory Council, "Minutes of the Meeting . . . October 22–23, 1940," p. 2—DNA, RG 83, North Dakota file; Ralph E. Johnston to Ross D. Davies, March 9, 1940, pp. 1–5, DNA, RG 83, South Dakota file.

46. See, e.g., *Williston Daily Herald*, June 9, September 8, 1937; *Lewistown Demo-*

crat-News, August 13, 1936; also Bushrod W. Allin, "County Planning Project—A Cooperative Approach to Agricultural Planning," *Journal of Farm Economics* 22 (February 1940): 292–94; and Richard S. Kirkendall, *Social Scientists and Farm Politics in the Age of Roosevelt* (Columbia, Mo.: University of Missouri Press, 1966), pp. 152–57.

47. *Adams County Record*, April 28, 1938; U.S. National Resources Committee, Advisory Committee, *Progress Report 1938: Statement*, pp. 9, 12–13; U.S. Great Plains Committee, *Future of the Great Plains*, p. 88 (quoted); Great Plains Council, *Proceedings . . . at Fort Collins, Colorado, August 2–5, 1953, Eighteenth Annual Session: Achievements and Activities*, pp. 36–37.

48. Detail of this and the next two paragraphs has been drawn from Allin, "County Planning Project," pp. 295–96; Anon., *Planning for a Permanent Agriculture, Including a Summary of the Programs Administered by the Department of Agriculture that Influence the Use of the Land: Prepared for Community and County Land Use Program Planning Committees*, USDA, *Miscellaneous Publication* no. 351 (Washington, D.C., 1939), pp. 4–8 (quotation, p. 5); Ellery A. Foster and Harold A. Vogel, "Cooperative Land Use Planning—A New Development in Democracy," USDA, *Farmers in a Changing World, Yearbook of Agriculture: 1940*, pp. 1138–56.

49. John D. Lewis, "Democratic Planning in Agriculture," *American Political Science Review* 35 (April and June 1941): 244–49, 454–63; Neil C. Gross, "A Post Mortem on County Planning," *Journal of Farm Economics* 25 (August 1943): 625–55, 659. See also Bryce Ryan, "Democratic Telesis and County Agricultural Planning," *Journal of Farm Economics* 22 (November 1940): 698–99.

50. Foster and Vogel, "Cooperative Land Use Planning," pp. 1146–47; Lewis, "Democratic Planning," pp. 245–46; Anon., *Planning for a Permanent Agriculture*, p. 4; North Dakota State Agricultural Advisory Council, "Minutes of Meeting . . . September 6, 1939," p. 1; Herbison to North Dakota State Agricultural Advisory Council, November 15, 1939, attachment, p. 5. Cf. North Dakota Agricultural Advisory Council, "Minutes of the Meeting . . . December 5, 1939," pp. 1–2—all MSS. in DNA, RG 83, North Dakota file.

51. South Dakota State Land Use Program Committee, "Minutes . . . September 30, 1940," pp. 13–15; "Minutes . . . January 20–21, 1941," pp. 1–2; North Dakota State Agricultural Advisory Council, "Minutes . . . September 6, 1939," p. 6; "Minutes . . . December 5, 1939," p. 3; E. J. Haslerud to County Extension Agents, January 21, 1942 (quoted); Montana State Agricultural Advisory Committee, "Minutes . . . February 7–8, 1941," p. 1—all in DNA, RG 83, respective state files.

52. Anon., *Planning for a Permanent Agriculture*, p. 5 (quoted); and cf. North Dakota State Agricultural Advisory Council, "Minutes . . . September 6, 1939," attachment no. 1; "Minutes . . . December 5, 1939," pp. 1, 3; "Minutes . . . October 22–23, 1940," p. 3; "Some Brief Notes on the Meeting, June 16, 17, 1942," p. 2; Bushrod W. Allin to John Muehlbeier, February 9, 1940; Muehlbeier to Allin, February 21, 1940—all DNA, RG 83, North Dakota file.

In an unpublished paper presented to the Agricultural History Society, meeting at Philadelphia in April 1987, "Agricultural Planning and the Limits of New Deal Democracy," Christopher Clarke-Hazlett presented a strong repudiation of the view that the New Deal program provided for popular input in policy development and cited particularly the county planning program as a tool of the scientific elite.

53. South Dakota Land Use Program Committee, "Minutes . . . July 12, 1940," p. 2, DNA, RG 83, South Dakota file; Charles Gooze, "Progress in Rural Land Classification in the United States," *Land Policy Circular*, December 1935, Supplement, p. 23; F. B. Linfield, "The Montana Land Problem," in Montana State Planning Board, *Staff Report, Period Ending Dec. 31, 1936* (mimeographed), p. 35; Montana State Land Use

Advisory Committee, "Minutes . . . May 16, 17, 1939," attached report by Ray B. Haight, E. H. Aicher, and H. G. Bolster on proposals for unified county development; "Minutes . . . August 25–26, 1939"; "Minutes . . . March 29, 30, 1940"—all in DNA, RG 83, Montana file; Ray B. Haight, "The Montana Land-Use Study," U.S. Great Plains Committee, *Future of the Great Plains*, appendix 5, pp. 146–57; *Lewistown Democrat-News*, July 17, 1938; Johnson, *Farm Adjustments in Montana: Graphic Supplement*.

54. North Dakota State Agricultural Advisory Council, "Minutes, April 5, 6, 1939," pp. 2–5; Herbison to North Dakota State Agricultural Advisory Council, November 15, 1939, attachment pp. 1–2 (quoted), DNA, RG 83, North Dakota file; Lloyd E. Jones and Olav Rogeness, *Farmers Face Farm Management Facts in Ward County (North Dakota)*, USDA, BAE, *Farm Management Report* no. 15 (Washington, D.C., 1941), pp. 56–58. Land-use capability classification, primarily for tax assessment purposes, had been initiated by the U.S. Soil Survey and NDAES, covering McKenzie and Billings counties and part of Montana, by December 1935 (Gooze, "Progress in Rural Land Classification," p. 19).

55. SDAES, *Annual Report . . . Ending 1935*, p. 8 (quoted); R. B. Westbrook, *Tax Delinquency and County Ownership of Land in South Dakota*, SDAES, *Bulletin* no. 322 (Brookings, S.D., 1938), p. 5; R. J. Penn and C. W. Loomer, *County Land Management in Northwestern South Dakota*, SDAES, *Bulletin* no. 326 (Brookings, S.D., 1938), pp. 31, 47; South Dakota Land Use Program Committee, "Minutes, July 12, 1940," pp. 4–6 (quoted); South Dakota Land Use Program Committee, Soil and Land Investigations Committee, "Minutes . . . August 31, 1940," pp. 1–8; South Dakota Land Use Program Committee, "Minutes . . . January 20–21, 1941," pp. 8–9—DNA, RG 83, South Dakota file.

56. Montana State Agricultural Advisory Committee, "Minutes . . . March 29, 30, 1940," p. 4; "Minutes . . . February 7, 8, 1941," p. 4; "Minutes . . . May 26–28, 1941," p. 5; "Minutes . . . June 27–28, 1941," pp. 5, 6; "Minutes . . . September 19, 20, 1941," p. 5; Herbison to North Dakota State Agricultural Advisory Council, November 15, 1939, attachment p. 5; Ralph E. Johnston to Ross D. Davies and A. L. Ford, March 9, 1940; South Dakota Land Use Planning Committee, "Minutes . . . July 12, 1940," pp. 3, 11; "Minutes . . . January 20, 21, 1941," pp. 2–8—all DNA, RG 83, respective state files. The Hand County, South Dakota, unified plan was listed as "under way" but had been sent to Washington for final approval.

Summaries of the plans for Teton and Ward counties were published in Foster and Vogel, "Cooperative Land-Use Planning," pp. 1146–51. See also Jones and Rogeness, *Farmers Face Farm-Management Facts in Ward County*.

57. Herbison to North Dakota State Agricultural Advisory Council, November 15, 1939, attachment p. 25 (quoted); Allin, "County Planning Project," pp. 303–4 (quoting Black's critique).

58. North Dakota State Agricultural Advisory Council, "Minutes . . . April 5 and 6, 1939," p. 6 (quoted); "Minutes . . . September 6, 1939," p. 4; "Minutes . . . November 8, 1939," pp. 2–3; "Minutes . . . June 13, 1940," p. 3; Montana State Land Use Advisory Committee, "Minutes . . . May 16, 17, 1939," attachment on Fallon County meeting, April 12, 1939; "Minutes . . . August 25–26, 1939," pp. 4, 5; "Minutes . . . November 29, 30, 1940," pp. 10–11; "Minutes . . . February 7, 8, 1941," including Ray B. Haight to D. A. Fitzgerald; "Minutes . . . June 27–28, 1941," p. 3; South Dakota Land Use Program Committee, "Minutes . . . January 30–31, 1940," pp. 2 (quoted), 9–12; "Conference on Land Use Planning Problems, September 10, 1940," p. 4—all DNA, RG 83, respective state files.

59. Montana State Agricultural Planning Committee, "Minutes . . . February 5, 6, 7, 1942," pp. 8–11; North Dakota Agricultural Council, Executive Committee,

"Minutes . . . November 8, 1939," p. 3 and attachments 1 and 2; "Minutes . . . May 15, 1940," p. 2; BAE-N.D. College Committee, "Minutes . . . July 7, 1939," p. 2; South Dakota Land Use Program Committee, "Minutes . . . September 30, 1940," p. 11 (quoted)—all DNA, RG 83, respective state files.

A subcommittee of the South Dakota state committee, at what appears to have been that body's last meeting, in March 1941 also complained that farms were "being made larger at the expense of the smaller operator, thereby crowding people off the farms" (South Dakota Land Use Program Committee, "Minutes . . . March 27–28, 1941, Report of Subcommittee A," p. 16 and attachment pp. 4–6, DNA, RG 83, South Dakota file).

60. South Dakota Land Use Program Committee, "Tour, October 17, 18, and 19, 1940," p. 4 (quoted), NDA, RG 83, South Dakota file; see p. 74.

61. Roy I. Kimmel, "Post-War Planning—The Department Acts," *Land Policy Review* 4 (no. 9, September, 1941): 17–23; Montana State Agricultural Advisory Committee, "Minutes . . . February 7 & 8, 1941," "Minutes . . . March 21 and 22, 1941," with attachments; John Muehlbeier to North Dakota Agricultural Advisory Council, January 24, 1941; North Dakota Agricultural Advisory Council, "Minutes . . . March 27 and 28, 1941"; South Dakota Land Use Program Committee, "Minutes . . . January 20–21, 1941," pp. 7–8; "Minutes . . . March 27–28, 1941," pp. 3–7 and attachment—all DNA, RG 83, respective state files; Sherman Johnson, *From the St. Croix to the Potomac—Reflections of a Bureaucrat* (Bozeman, Mont.: Big Sky Books, 1974), p. 153; Kirkendall, *Social Scientists and Farm Politics*, p. 204.

62. Montana Land Grant College-BAE Committee, "Minutes . . . September 15, 1941" (quoted), DNA, RG 83, Montana File.

63. North Dakota Agricultural Advisory Council, "Some Brief Notes on the Meeting . . . December 9 and 10, 1941," p. 9, with T. L. Gaston to Walter M. Rudolph, June 7, 1941 (quoted), DNA, RG 83, North Dakota file.

64. For extended discussion of the "Destroying of the Planning Committees," on which this and the following paragraph are largely based, see Kirkendall, *Social Scientists and Farm Politics*, ch. 11. With particular reference to the role of the American Farm Bureau Federation in this action, see Christiana McFadyen Campbell, *The Farm Bureau and the New Deal: A Study of the Making of National Farm Policy* (Urbana, Ill.: University of Illinois Press, 1962), pp. 174–78; and Grant McConnell, *The Decline of Agrarian Democracy*, First Atheneum Edition (University of California Press, 1953; Atheneum, N.Y., 1969), pp. 118–22.

65. North Dakota Agricultural Advisory Council, "Some Brief Notes on the Meeting . . . June 16 and 17, 1942," p. 5; E. J. Haslerud remarks, Great Plains Agricultural Council, *Proceedings, 1956*, pp. 39–40.

66. Hargreaves, "Land-Use Planning," pp. 573–74; Black's comments attached to Allin, "County Planning Project," p. 304; South Dakota Land Use Program Committee, "Minutes . . . January 30–31, 1940" (quoted). See also Herbison to North Dakota State Agricultural Advisory Council, November 15, 1939, attachment pp. 19–20, 25 et passim; Montana State Agricultural Advisory Committee, "Minutes . . . June 14 and 15, 1940," report on Roosevelt County Program—DNA, RG 83, respective state files.

67. See, e.g., Johnson, *Farm Adjustments in Montana, Study of Area VII*, pp. 20–22, 52; Clawson, Saunderson, and Johnson, *Farm Adjustments in Montana, Study of Area IV*, pp. 39–50 passim; SDAES, *Annual Report . . . 1941–42*, p. 5; Raymond B. Hile, *Farm Size as a Guide to Planning in the Tri-County Soil Conservation District*, USDA, BAE, *Farm Management Report* no. 16 (Washington, D.C., 1940), p. 14; R. S. Kifer and H. L. Stewart, *Farming Hazards in the Drought Area*, WPA, Division of Social Research, *Research Monograph* no. XVI (Washington, D.C., U.S. Government Printing Office, 1938)

pp. 106–9 passim; Jones and Rogeness, *Farmers Face Farm-Management Facts in Ward County,* pp. 17–23, 25–26, 49–51 table 25. The last of these studies reported, however, with recommendation that the finding be tested elsewhere, that on 480 acres general farming, with addition of feed grains and livestock to cash grain production, more than doubled the income from cash grains alone (Ibid., p. 30). Cf. also Baldur H. Kristjanson and C. J. Heltemes, *Handbook of Facts about North Dakota Agriculture,* NDAES, *Bulletin* no. 357 (Fargo, N.D., 1950), p. 13.

68. Starch, "Type of Farming Modifications Needed," pp. 115–19 (quotation, p. 117). While conceding that wheat would continue "an important enterprise in the region," E. C. Johnson of the FCA challenged Starch's advocacy of wheats before the symposium and emphasized the need for increased production of livestock, with reduction of the acreage in cultivation (Ibid., pp. 121–22). Cf., also, Johnson, *Farm Adjustments in Montana, Study of Area VII,* p. 35; Bell, *Culture of a Contemporary Rural Community,* p. 41.

69. Computations from data in Table 1.1. See also *Adams County Record,* August 4, 16, 1938; and C. M. Studness, "Development of the Great Plains Grain Farming Economy: Frontier to 1953" (Ph.D. diss., Columbia University, 1963), pp. 168–71, 180–81.

Extensive use of cotton cake and linseed oil meal for stock feed was initiated in western South Dakota in 1936. The increase in number of sheep was a temporary phenomenon, dictated by the need for more rapid growth in income than was possible with cattle (Aaron G. Nelson and Gerald E. Korzan, *Profits and Losses in Ranching, Western South Dakota, 1931–1940,* SDAES, *Bulletin* no. 352 [Brookings, S.D., 1941], pp. 10–11).

70. E. A. Starch, *Readjusting Montana's Agriculture: VII. Montana's Dry-Land Agriculture,* MAES, *Bulletin* no. 318 (Bozeman, Mont., 1936), p. 14; Clawson, Saunderson, and Johnson, *Farm Adjustments in Montana . . . Area IV,* p. 44; Johnson, *Farm Adjustments in Montana . . . Area VII,* pp. 41, 44; Hile, *Farm Size as a Guide,* pp. 25, 26; Kifer and Stewart, *Farming Hazards in the Drought Area,* p. 108; H. L. Stewart, *Natural and Economic Factors Affecting Rural Rehabilitation Problems in Southwestern North Dakota (as Typified by Hettinger County),* Resettlement Administration, *Research Bulletin* K-4 (Washington, D.C., mimeographed, 1936), p. 17; H. L. Stewart, *Natural and Economic Factors Affecting Rural Rehabilitation Problems in Northwestern North Dakota and Northeastern Montana (as Typified by Divide County, North Dakota),* Resettlement Administration, *Research Bulletin* no. K-7 (Washington, D.C., mimeographed, 1936), p. 18.

71. Russell Lord, *The Wallaces of Iowa* (Boston, Mass.: Houghton Mifflin Co., 1947), pp. 352–53. On Tugwell's comments, see above, pp. 108–9.

72. U.S. National Resources Committee, *Progress Report 1939: Statement of the Advisory Committee* (Washington, D.C.: U.S. Government Printing Office, 1939), pp. 99, 101, 117; U.S. National Resources Planning Board, *Progress Report 1940–1941* (Washington, D.C.: U.S. Government Printing Office, 1941), pp. 56–58, 60–61; Montana State Planning Board, *Annual Report . . . as of December 31, 1940,* pp. 1–8 and attachment; A. M. Eberle, "Achievements and Activities of the Great Plains Council, 1935–1953," Great Plains Agricultural Council, *Proceedings . . . August 2–5, 1953,* pp. 36–37; Henry J. Tomasek, "The Great Plains Agricultural Council" (Ph.D. diss., University of Chicago, 1959), pp. 30–35.

73. On the early organizational history of the Great Plains Agricultural Advisory Council, see Tomasek, "Great Plains Agricultural Council," pp. 13–35. Leaders of extension work at the agricultural colleges in the states of the northern and southern Great Plains, as two separate bodies, had met at the call of the head of the federal extension service as early as April 1934. Continuing formal meetings, they had re-

ported on their sessions to the Great Plains Drought Area Committee, which subsequently recommended continuance of such regional planning activities. The separate northern and southern councils began meeting jointly in 1939, and in 1946 largely consolidated their programs.

CHAPTER SIX: PROSPERITY RETURNS

1. Average annual precipitation computed from data in Lloyd D. Teigen and Florence Singer, *Weather in U.S. Agriculture: Monthly Temperature and Precipitation by State and Farm Production Region, 1950–86*, USDA, ERS, *Statistical Bulletin* no. 765 (Washington, D.C., 1988), pp. 69, 145, 177, 205, tables of precipitation by states, weighted by areas of harvested cropland. For production data, see Jim Langley and Suchada Langley, *State-Level Wheat Statistics, 1949–88*, USDA, ERS, *Statistical Bulletin* no. 779 (Washington, D.C., 1989), p. 42 table 38.

2. See Table 8.1 in this volume; Langley and Langley, *State-Level Wheat Statistics*, p. 42 table 38.

3. See below, p. 269; USDA, *The Second RCA Appraisal: Soil, Water, and Related Resources on Nonfederal Land in the United States, Analysis of Condition and Trends* (Washington, D.C.: U.S. Government Printing Office, 1989), p. 151 fig. 82.

4. W. A. Laycock, "History of Grassland Plowing and Grass Planting on the Great Plains," John E. Mitchell, ed., *Impacts of the Conservation Reserve Program in the Great Plains, Symposium Proceedings, September 16–18, 1987, Denver, Colorado*, USDA, Forest Service, *General Technical Report* RM-158 (Fort Collins, Colo.: Forest Service, Rocky Mountain Forest and Range Experiment Station, 1987), p. 7; and below, p. 269. The program drew a much greater percentage of support in the Dakotas, but because of the limited scale of the coverage as extended to those states and the difficulty of distinguishing the quality of the specific localities involved, the impact on dry-land development there is less measurable.

5. *Annual and Seasonal Precipitation at Six Representative Locations in Montana*, MAES, *Bulletin* no. 447 (Bozeman, Mont., 1947), pp. 6–7, 12–13, 18–19, 24–25; Gabriel Lundy, *Graphic Views of Changes in South Dakota Agriculture*, SDAES, *Circular* no. 78 (Brookings, S.D., 1949), p. 10 fig. 7; *Adams County Record* (N.D.), July 14, 1954, reprinting *Mott Pioneer Press*; *Williston Daily Herald* (N.D.), June 12, July 1, December 26, 1941, January 5, 1944, and January 3, 1946; *Lewistown Democrat-News* (Mont.), April 11, 1948.

6. Computations from data in Table 1.1; *Lewistown Democrat-News*, August 12, 13, 1941, August 19, 1942, September 1, 17, 1943, May 12, August 23, 1944, and March 9, 1945; *Williston Daily Herald*, August 23, 1940, September 19, 1941, July 8, 1942, September 1, 1943, December 15, 1944, and August 24, 1945; *Adams County Record*, January 6, 1944.

7. *Lewistown Democrat-News*, September 27, 1942; May 1, September 11, November 17, December 1, 1943, and January 17, 1946; *Williston Daily Herald*, November 27, December 17 (quoted), 1942; August 5, 10, 1943, and July 26, 1944; *Adams County Record*, July 9, November 5, 1942, and May 9, 1946 (quoted); Montana Legislative Assembly, *Laws, Resolutions, and Memorials . . . Passed by the Twenty-Eighth, 1943*, S. Mem. no. 1, p. 597; S. Jt. Mem. no. 2, pp. 599–600 (February 3, 1943); Laura Gessaman reminiscences, OH 187, MHS.

Farm machinery quotas were raised by one-third in the summer of 1944 and terminated that autumn, and production controls were lifted in August 1945. Because of the increased acreage under cropping and the shortage of farm labor, demand continued to exceed supply (*Lewistown Democrat-News*, July 25, 1944; Bela Gold, *Wartime Economic Planning in Agriculture: A Study in the Allocation of Resources* [New York, N.Y.: Columbia University Press, 1949], pp. 216, 222. Gold, pp. 208–39, especially p. 228, strongly questions the need for larger allocations).

8. USDA, *Agricultural Statistics, 1943*, through ibid., *1952*, table 3; above, Table 1.1; Baldur H. Kristjanson and C. J. Heltemes, *Handbook of Facts about North Dakota Agriculture*, NDAES, *Bulletin* no. 357 (Fargo, N.D., 1950), pp. 38 table 12 and fig. 19, 39 figs. 20, 21; ibid., revised ed., NDAES, *Bulletin* no. 382 (Fargo, N.D., 1952), p. 48 figs. 26, 27; Lundy, *Graphic Views*, pp. 16–21; *Lewistown Democrat-News*, December 29, 1941, and June 14, 1942; *Williston Daily Herald*, December 8, 1942, and February 23, 1943; and *Adams County Record*, July 27, 1944, and January 4, 1945.

9. *Lewistown Democrat-News*, September 11, 1939, and January 8, 1941; *Williston Daily Herald*, May 17, 1940; *Historical Statistics of the United States, 1789–1945, A Supplement to the Statistical Abstract of the United States* (Washington, D.C.: U.S. Government Printing Office, 1949), pp. 99, ser. 88–104, 106, ser. E 181–94; USDA, *Agricultural Statistics, 1943*, through ibid., *1952*, table 3; Kristjanson and Heltemes, *Handbook of Facts*, NDAES, *Bulletin* no. 357, pp. 65, 71, 72. Regionally prices received in the northern Plains were a few cents lower than the national level and the parity ratios, about ten points lower.

By January 1944 the price of hard wheat was so closely aligned to parity that the Office of Price Administration instituted controls for the first time. There were still bonuses for high protein ratings ranging from one to seven cents at 13 to 16.5 percent or more (*Williston Daily Herald*, January 5, 1944; *Adams County Record*, August 27, 1942, and September 7, 1944).

10. Computed from price and production data in USDA, *Agricultural Statistics, 1943*, through ibid., *1952*, table 3; Gary Lucier, Agnes Chesley, and Mary Ahearn, *Farm Income Data: A Historical Perspective*, USDA, ERS, *Statistical Bulletin* no. 740 (Washington, D.C., 1986), pp. 180 table 47, 220 table 55, 255 table 62.

11. USDA, *Agricultural Statistics, 1946*, p. 606 table 660; *Lewistown Democrat-News*, March 26, 1945, citing federal land bank of Spokane; April 5, 1942; *Williston Daily Herald*, September 3, 1943 (quoted); September 13, 1946; USDA, *Postwar Developments in Farm Security: The Annual Report of the Farm Security Administration for 1945–46*, appendix table 1; Rainer Schickele and Reuben Engelking, *Land Values and the Land Market in North Dakota*, NDAES, *Bulletin* no. 353 (Fargo, N.D., 1949), p. 41. Cf., however, *Williston Daily Herald*, January 2, 1945, which notes that legislation had permitted "compromise" of unpaid debts by negotiation, the Frazier-Lemke Act being one measure under which such clearance had been obtained. Even so, some $362 million of the total of $469 million lent in feed and seed loans nationally since 1918 had been repaid.

12. *Lewistown Democrat-News*, July 4, 1942; computations based on Fred R. Taylor, C. J. Heltemes, and R. F. Engelking, *North Dakota Agricultural Statistics*, NDAES, *Bulletin* no. 408 (Fargo, N.D., 1957), p. 91 table 84.

13. *Williston Daily Herald*, June 18, 1935; *Adams County Record*, August 8, 1935; *Lewistown Democrat-News*, April 2, 14 (reprinting *Montana Standard*), 1935, and May 9, 1937.

14. *Lewistown Democrat-News*, March 8 (quoted), December 28 (quoted), 1940, May 31, August 14 (quoted), 1942, and June 10, 1946 (quoted).

15. *Adams County Record*, October 23, 1941, November 9, 1944 (quoted); *Lewistown Democrat-News*, July 27, 1946, July 9, 1947 (quoted); South Dakota Legislative Assembly, *Laws Passed at the Twenty-Eighth, 1943*, ch. 270, pp. 273–74 (March 5, 1943).

16. North Dakota Legislative Assembly, *Laws Passed at the Twenty-Third Session, 1933,* chap. 89, pp. 122–23 (March 4, 1933); Morris H. Taylor and Raymond J. Penn, *Management of Public Land in North Dakota,* NDAES, *Bulletin* no. 312 (Fargo, N.D., 1942), pp. 3–4, 9–10, 17–18; Schickele and Engelking, *Land Values . . . in North Dakota,* pp. 32, and 33 table 8; Gabriel Lundy and Ray F. Pengra, *Land Market Trends in South Dakota, 1941–1950,* SDAES, *Bulletin* no. 413 (Brookings, S.D., 1951), pp. 10, 15–17 fig. 5, 18, 22, 23 fig. 8; Layton S. Thompson, *Changing Aspects of the Farm Real Estate Situation in Montana, 1940 to 1946,* MAES, *Bulletin* no. 440 (Bozeman, Mont., 1947), pp. 17, 18 fig. 3, 19; Elwyn B. Robinson, *History of North Dakota* (Lincoln, Nebr.: University of Nebraska Press, 1966), p. 426; *Lewistown Democrat-News,* August 5, 1939, and November 29, 1941; *Williston Daily Herald,* July 11, 1941, June 1, December 31, 1942, and January 2, 1946.

In 1940 the Bank of North Dakota had 5,593 farms for sale; by the summer of 1947, it had only 350 remaining as it sought to wind down the liquidation process (*Williston Daily Herald,* July 11, 1947).

17. Taylor and Penn, *Management of Public Land in North Dakota,* pp. 10, 12, 13, 19, 27; Lundy and Pengra, *Land Market Trends in South Dakota,* pp. 12, and 22–23 fig. 8; Thompson, *Changing Aspects of the Farm Real Estate Situation in Montana,* pp. 18 fig. 3, and 19 tables 8, 9; *Lewistown Democrat-News,* January 12, July 2, 1944.

18. *Lewistown Democrat-News,* January 19, 1940 (quoted); *Williston Daily Herald,* June 1, 1938, and July 11, 1941 (quoted); Montana State Agricultural Advisory Committee, "Minutes, December 4, 5, 6, 1941," p. 2, DNA, RG 83, Mont.; Taylor and Penn, *Management of Public Land in North Dakota,* p. 13; North Dakota Legislative Assembly, *Laws Passed at the Twenty-Seventh, 1941,* ch. 134, p. 201 (quoted). On local sales for "investment," see, for examples, *Williston Daily Herald,* March 18, 1943; *Lewistown Democrat-News,* November 9, 1943.

19. Montana Legislative Assembly, *Laws . . . Enacted by the Twenty-Third, 1933,* ch. 67, sec. 5, p. 130 (March 7, 1933, quoted); *Laws . . . Enacted by the Twenty-Sixth, 1939,* ch. 193, p. 487 (March 17, 1939); South Dakota Legislative Assembly, *Laws Passed at the Twenty-Fifth, 1937,* ch. 64, pp. 83–84 (March 3, 1937); *Laws Passed at the Twenty-Sixth, 1939,* ch. 25, pp. 32–33, 36 (March 7, 1939); North Dakota Legislative Assembly, *Laws Passed at the Twenty-Sixth, 1939,* ch. 237, pp. 400–1 (March 15, 1939); *Laws Passed at the Twenty-Seventh, 1941,* ch. 134 (March 17, 1941); *Laws Passed at the Twenty-Eighth, 1943,* ch. 120, pp. 161–62 (March 9, 1943); Taylor and Penn, *Management of Public Land in North Dakota,* pp. 13–14. The North Dakota statute permitted ten-year grazing leases also to incorporated soil conservation districts and to resident individuals; the South Dakota statute extended them generally. Even the South Dakota law, however, authorized reclassification of grazing lands every five years, which could then remove them from the long-term leasing provision. A North Dakota study found that although former owners and tenants and neighboring farmers and ranchers were generally given preference in land transactions, their bids were rejected if they were unwilling to meet those of others. Few applications were denied because the bidder was a nonresident or even a nonfarmer (Taylor and Penn, *Management of Public Land in North Dakota,* pp. 3–4).

Local planning committees did occasionally attempt to influence sales arrangements. The Williams County, North Dakota, extension agent reported in December 1941 that the local body had won the agreement of state and county authorities that none of the property held by the Bank of North Dakota or the county would be sold without the consent of the land use committee, except land sold to an occupying tenant or previous owners (*Williston Daily Herald,* December 2, 1941).

20. Montana State Agricultural Advisory Committee, "Minutes . . . September

19, 1941," p. 15 (quoted); "Minutes . . . December 3–6, 1941," attachment no. 1; "Minutes . . . February 7, 1941"—DNA, RG 83, Mont. file.

21. North Dakota, Agricultural Advisory Council, "Minutes . . . March 27 and 28, 1941," p. 3; "Some Brief Notes on the Meeting . . . December 9 and 10, 1941," pp. 6, 8 (quoted)—DNA, RG 83, N.D. file.

22. Montana State Agricultural Advisory Committee, "Minutes . . . December 4, 5, 6, 1941," p. 2; "Minutes . . . May 22 and 23, 1942," attachment no. 2 (quoted); "Minutes . . . June 19 and 20, 1942," pp. 2–5.

23. R. J. Penn, W. F. Musbach, and W. C. Clark, *Possibilities of Rural Zoning in South Dakota: A Study in Corson County,* SDAES, *Bulletin* no. 345 (Brookings, S.D., 1940), pp. 5–6, 11–13; South Dakota Land Use Program Committee, Executive Committee, "Minutes . . . December 14, 1940," p. 3, DNA, RG 83, S.D. file.

24. Penn, Musbach, and Clark, *Possibilities of Rural Zoning,* pp. 16–21.

25. Ibid., pp. 22–23; South Dakota Land Use Program Committee, Executive Committee, "Minutes . . . March 11, 1941," attachment: "Preliminary Handbook of Plans for Developing a Unified South Dakota Agricultural Program," pp. 4–5; South Dakota Legislative Assembly, *Laws Passed at the Twenty-Seventh, 1941,* ch. 216, pp. 250 (quoted in next paragraph)–52 (March 12, 1941).

26. *Lewistown Democrat-News,* September 9, 1939, February 6, 18 (quoted), 1941; *Williston Daily Herald,* October 1, 1940, and April 10, 1941 (quoted); MAES, *Forty-Eighth and Forty-Ninth Annual Reports, 1942,* p. 9; A. M. Meyers, Jr., "Production Goals and Good Land Use," *Land Policy Review* 5 (June 1942): 3–4; debates of state land use advisory committees through spring 1941, passim, especially South Dakota Land Use Program Committee, "Minutes of the Meeting . . . March 27–28, 1941, pp. 4–6, 20–24.

27. *Lewistown Democrat-News,* September 1, 7, 1939, and April 22, 1943; R. Burnell Held and Marion Clawson, *Soil Conservation in Perspective* (Baltimore, Md.: Resources for the Future, Inc., by The Johns Hopkins Press, 1965), pp. 176, 183; USDA, *Agricultural Statistics, 1946,* p. 664 table 714; J. C. Dykes, "Soil Conservation in Wartime and After Victory," *Land Policy Review* 6 (Spring 1943): 36 (quoted).

28. *Lewistown Democrat-News,* September 1, 7, 1939, September 5, 1940, January 8, May 23, June 2, October 15, November 4, 1941, and February 4, April 1, November 30, December 7, 1942; *Williston Daily Herald,* July 27, 1939, September 18, 1941, and January 12, 1942; *Adams County Record,* June 19, October 2 (quoted), 1941; Douglas E. Bowers, Wayne D. Rasmussen, and Gladys L. Baker, *History of Agricultural Price-Support and Adjustment Programs, 1933–84,* USDA, ERS, *Agriculture Information Bulletin* no. 485 (Washington, D.C., 1984), p. 16.

29. Bowers, Rasmussen, and Baker, *History of Agricultural Price Support,* pp. 16–17; *Williston Daily Herald,* June 1, 1945, and August 21, 1946; *Lewistown Democrat-News,* November 6, 1941.

30. *Williston Daily Herald,* January 27, February 23, August 26, October 5, 1943; *Lewistown Democrat-News,* January 12, Bozeman dateline (quoted), May 27, 1942, and February 22, 24, April 1, July 26, October 22, 1943.

31. Great Plains Agricultural Council (hereafter cited as GPC), *Proceedings, 1953,* pp. 64a fig. 11, and 68 table 12; USDA, *Agricultural Statistics, 1946,* pp. 16 and 17 tables 10 and 11; *Williston Daily Herald,* February 26, 1943, and February 16, 1944; *Lewistown Democrat-News,* April 1, November 30, December 13, 1942, January 11, July 13, 20, 1943, March 10, December 9, 1944, March 4, 1945, and February 22, 1946; Gold, *Wartime Economic Planning,* pp. 123, 132, 137, 149, 156–57, 479–80. Cf. Walter W. Wilcox, *The Farmer in the Second World War* (Ames, Iowa: The Iowa State College Press, 1947), p. 58.

Gold is highly critical that in this period of "national crisis . . . the farmer and his government tutors were still following the lessons of acreage restriction learned since 1933" (p. 88). He notes that wheat was a crop that fell significantly short of production goals two years out of the four (1942–1945) and that a sizeable expansion could have been gained by shifting from the small plots of the East and Midwest to larger-scale operations in the Great Plains (p. 141). He particularly indicts the excessive emphasis upon livestock production (chap. 5), which occasioned the cropping orientation of the quotas.

32. *Lewistown Democrat-News*, November 7, 13, 23, 1943, October 4, December 9, 1944, January 22, March 4, 18, 1945, and February 22, 1946; *Williston Daily Herald*, November 19, 1940, July 2, 1941, and March 13, 1942; *Adams County Record*, February 11, 1943; MAES, *Varieties, 1944*, MAES, *Bulletin* no. 177, p. 9; *Varieties, 1945*, p. 10, Cf. Gold, *Wartime Economic Planning*, pp. 103, 110, 119.

33. Allen J. Matusow, *Farm Policies and Politics in the Truman Years*, Harvard Historical Studies, vol. 80 (Cambridge, Mass.: Harvard University Press, 1967), pp. 4–5; *Williston Daily Herald*, July 2, 1941 (quoted), and January 11, 1943 (quoted); *Lewistown Democrat-News*, January 5, 1943 (quoted), and August 5, 1944, Bozeman dateline (quoted); Thompson, *Changing Aspects of the Farm Real Estate Situation in Montana*, p. 5.

34. Karl Swanson, quoted in *Williston Daily Herald*, March 24, 1943; Donald C. Horton and Harald C. Larsen, "Federally Sponsored Farm Mortgage Credit Agencies: Wartime Operations and Postwar Prospects," *Agricultural Finance Review*, 8 (November, 1945): 31 (quoted); *Adams County Record*, March 2, 1942; *Lewistown Democrat-News*, March 6, 1946; Lundy and Pengra, *Land Market Trends in South Dakota*, pp. 8–9.

35. Gold, *Wartime Economic Planning*, pp. 229 table 35, 270–75; Black, in U.S. Senate, 78th Cong., 1st sess., *Hearings before the Subcommittee of the Committee on Appropriations . . . on H.R. 2481, A Bill Making Appropriations for the Department of Agriculture for the Fiscal Year Ending June 30, 1944*, p. 367. See also Baldur H. Kristjanson and Jacob A. Brown, *Credit Needs of Beginning Farmers in Selected Areas of North Dakota*, NDAES, *Bulletin* no. 386 (Fargo, N.D., 1953), pp. 6–7 table 1.

36. Kristjanson and Heltemes, *Handbook of Facts*, NDAES, *Bulletin* no. 382, p. 82 table 40; SDAES, *Annual Report, 1945–46*, p. 37; Lundy and Pengra, *Land Market Trends in South Dakota*, pp. 10, 13, 14 table 4, 15 table 5, 16; C. R. Hoglund and M. B. Johnson, *Ranching in Northwestern South Dakota*, SDAES, *Bulletin* no. 385 (Brookings, S.D., 1947), pp. 3, 12, 22, 27, but cf. p. 30; *Sixteenth Census of the United States: 1940*, vol. 1, pt. 2, pp. 470 county table 1, 498 county table 6; *1950 United States Census of Agriculture*, vol. 1, pt. 11, pp. 248 county table 1, 481 county table 5. For generalized analysis of the changing land use, as discussed here and in the next several paragraphs, see Tables 1.1 and 6.1 of this volume.

37. Analysis based on data in Schickele and Engelking, *Land Values . . . in North Dakota*, pp. 16–17, 22–23, figs. 3–5, 24; Kristjanson and Heltemes, *Handbook of Facts*, NDAES, *Bulletin* no. 357, p. 63 figs. 40, 41.

38. Taylor, Heltemes, and Engelking, *North Dakota Agricultural Statistics*, p. 92 table 85; Thompson, *Changing Aspects of the Farm Real Estate Situation in Montana*, p. 17; Lundy and Pengra, *Land Market Trends in South Dakota*, pp. 16, 17 fig. 5, 18; Schickele and Engelking, *Land Values . . . in North Dakota*, p. 33.

39. Thompson, *Changing Aspects of the Farm Real Estate Situation in Montana*, pp. 4, 8–10 table 2, 12 tables 4 and 5, 13, 20; Gordon E. Rodewald, Donald K. Larson, and D. C. Myrick, *Dryland Grain Farms in Montana: How They Started, Growth, and Control of Resources*, MAES, *Bulletin* no. 579, technical (Bozeman, Mont., 1963), p. 15; MAES, *Some Accomplishments of Ten Years of Agricultural Research in Montana, Fiftieth-*

Fifty-Ninth Annual Reports . . . July 1, 1942–June 30, 1952, p. 6. A graphic measurement of increased farm size by county for the entire region is presented in Bradley H. Baltensperger, "Farm Consolidation in the Northern and Central States of the Great Plains," *Great Plains Quarterly* 7 (Fall 1987): 261.

40. E. A. Starch, "Type of Farming Modifications Needed in the Great Plains," *Journal of Farm Economics* 21 (February 1939): 115. See also Elmer Starch, "The Future of the Great Plains Reappraised," *Journal of Farm Economics* 31 (November 1949) 919, 921.

41. For general state averages, weighted by geographic and harvested cropland areas, see Teigen and Singer, *Weather in U.S. Agriculture*, pp. 144–45, 176–77, 204–5. Localized detail is available in the "Crop Prospects" reports of the Great Plains Agricultural Council, *Proceedings, 1947*, table 3; *1948*, pp. 38 table 2, 39 table 3, *1949*, p. 5 table 2; *1953*, pp. 13, 59a fig. 7, 59b fig. 8; *1954*, pp. 11–12; *1956*, p. 5; *Adams County Record*, October 5, 1949, January 4, March 29, 1950, June 25, 1952, and July 14, 1954; *Dickinson Press* (N.D.) December 4, 1952, and June 27, July 12, 1956; *Williston Daily Herald*, October 26, 1949, August 21, September 3, December 3, 1952, and August 11, 31, 1956; *Lewistown Democrat-News*, July 25, September 11, 13, October 4, 1949.

42. Table 1.1; and computations from USDA, *Agricultural Statistics, 1958*, p. 3 table 3. Cf. also Langley and Langley, *State-Level Wheat Statistics*, pp. 42 table 38, 56 table 52, 67 table 63.

43. *Dickinson Press*, August 17, 1956. Between 1942 and 1948 the acreage seeded to wheat was increased from 100 to 150 percent in northwestern South Dakota and from 50 to 100 percent across northern Montana. GPC, *Proceedings, 1949*, p. 30 fig. 3.

44. Matusow, *Farm Policies . . . in the Truman Years*, pp. 14–37; Gold, *Wartime Economic Planning in Agriculture*, pp. 455–84; *Adams County Record*, April 25, May 2, August 29, 1946; *Williston Daily Herald*, April 23, 25, May 9, July 10, 1946; *Lewistown Democrat-News*, May 9, 11, 1946, and June 3, 1947.

45. *Williston Daily Herald*, March 18, 1947; *Lewistown Democrat-News*, March 19, September 19, 1947; Lucier, Chesley, and Ahearn, *Farm Income Data*, p. 29 table 8.

46. *Lewistown Democrat-News*, June 3, August 21, September 4, December 18, 1947, and September 2, 1948; *Adams County Record*, October 23, 1947, and May 28, 1948; *Williston Daily Herald*, December 22, 1947.

47. *Lewistown Democrat-News*, February 7, 11, December 19, 1948, and September 28, 1949; *Williston Daily Herald*, May 19, 1948; *Adams County Record*, May 28, 1948; Maurice C. Taylor, *Income Trends in Montana*, MAES, *Bulletin* no. 590 (Bozeman, Mont., 1964), cover.

48. Bowers, Rasmussen, and Baker, *History of Agricultural Price Support*, pp. 17–21; *Adams County Record*, August 4, 1949, and January 24, 1951; *Williston Daily Herald*, July 22, August 3, November 1, 1949, August 1, 1950, and February 2, 14, March 14, July 12, November 28, 1951; *Lewistown Democrat-News*, July 21, 1949; Lucier, Chesley, and Ahearn, *Farm Income Data*, pp. 29 table 8, 180 table 47, 220 table 55, 255 table 62.

49. *Williston Daily Herald*, July 21, November 10, 28, December 7, 1951, September 11, December 21, 1953, December 29, 1954, and December 20, 1955; *Dickinson Press*, July 18, 31, August 1, 6, 12, December 18, 1953, August 27, October 12, December 21, 1954, and May 16, 1956; *Adams County Record*, January 5, 1955; Langley and Langley, *State-Level Wheat Statistics*, pp. 42 table 38, 44 table 40, 45 table 41.

50. GPC, *Proceedings, 1953*, p. 64a fig. 11; *Dickinson Press*, April 14, 1954; *Adams County Record*, September 22, 1954; *Williston Daily Herald*, July 25, 1953, and September 15, 1954; Langley and Langley, *State-Level Wheat Statistics*, pp. 42 table 38, 56 table 52, 67 table 63; Taylor, *Income Trends in Montana*, p. 8 table 3.

On the program of reinstituted acreage controls developed after the Korean War, see below, pp. 228–37.

51. Computations from *United States Census of Agriculture: 1945*, vol. 1, pt. 11, pp. 18–28, 130–43, and pt. 27, pp. 18–29—county table 1 (pt. 1 of 2); *United States Census of Agriculture: 1954*, vol. 1, pt. 11, pp. 42–46, 226–31, and pt. 27, pp. 42–46—county table 1; Kristjanson and Heltemes, *Handbook of Facts*, NDAES, *Bulletin* no. 382, pp. 14–15 figs. 1–4, 82 table 40; *Farm Real Estate: Historical Series Data, 1950–85*, USDA, ERS, *Statistical Bulletin* no. 738 (Washington, D.C., 1985), pp. 23 table 21, 24 table 22, 41 table 41. The percentages relating to farm real estate values prior to 1950 are based on a series of index numbers (1912–1914 = 100); those for the later period, on state averages of dollar values. See also Rodewald, Larson, and Myrick, *Dryland Grain Farms in Montana*, pp. 3, 20–21, 47 table G; Baldur H. Kristjanson, *What about Our Large Farms in North Dakota?* NDAES, *Bulletin* no. 360 (Fargo, N.D., 1950), pp. 5–6, 12, 13, 23–25; C. R. Hoglund, *What Size Farm or Ranch for South Dakota?* SDAES, *Bulletin* no. 387 (Brookings, S.D., 1947), pp. 3, 5 table 1, 7–9, 11, 13 table 5, 16–17.

52. *Williston Daily Herald*, January 8, 1948, January 18, 1950, October 11, November 8, 1951, and January 3, 1952; *Dickinson Press*, January 13, April 1, 1959.

53. John O. Lyngstad to author, July 14, 1949; Schickele and Engelking, *Land Values . . . in North Dakota*, p. 33; John P. Johansen, *Population Trends in Relation to Resources Development in South Dakota*, SDAES, *Bulletin* no. 440 (Brookings, S.D., 1954), p. 26; Donald E. Anderson, Laurel D. Loftsgard, and Lloyd E. Erickson, *Characteristics and Changes of Land Ownership in North Dakota, 1945 to 1958*, NDAES, *Bulletin* no. 438 (Fargo, N.D., 1962), pp. 8–9, 11–12, 17; *Williston Daily Herald*, April 26, 1956; *Dickinson Press*, April 27, 1956, reporting findings by USDA, ERS.

54. *Williston Daily Herald*, July 7, 1958 (quoted). The railway corporations which had so heavily promoted dry-land settlement prior to the thirties remained committed to irrigation as the prudent basis for future development. See, e.g., *Williston Daily Herald*, July 31, August 31, 1944, and May 18, 1954; *Dickinson Press*, July 13, 1954; *Adams County Record*, April 23, 1942; *Lewistown Democrat-News*, December 12, 1944. On the advertising of the Greater North Dakota Association, see *Williston Daily Herald*, August 30, 1952; *Dickinson Press*, December 25, 1952, February 12, 1955, and August 10, 1957.

55. Taylor, Heltemes, and Engelking, *North Dakota Agricultural Statistics*, p. 92 table 85; Lundy and Pengra, *Land Market Trends in South Dakota*, p. 15 table 6.

56. Glenn H. Craig and Charles W. Loomer, *Collective Tenure on Grazing Land in Montana*, MAES, *Bulletin* no. 406 (Bozeman, Mont., 1942), p. 39; Harry A. Steele and John Muehlbeier, "Land and Water Development Programs in the Northern Great Plains," *Journal of Farm Economics* 32 (August 1950): 439; *Williston Daily Herald*, February 12, 1948, an Associated Press dispatch quoting extensively the Truman veto; *Dickinson Press*, February 21, 1953, under Glasgow, Mont., dateline (quoted); Mont H. Saunderson, *Western Land and Water Use* (Norman, Okla.: University of Oklahoma Press, 1950), p. 99; H. H. Wooten, *The Land Utilization Program, 1934 to 1964: Origin, Development, and Present Status*, USDA, ERS, *Agricultural Economic Report* no. 85, pp. 33–35; Paul W. Gates, *History of Public Land Law Development* (Washington, D.C., 1968), pp. 615–34.

57. GPC, *Proceedings, 1947*, p. 3, *1948*, p. 35, *1955*, p. 27. Data on South Dakota were lacking from most of these reports.

58. Analysis in this and the following paragraph drawn from Tables 1.1 and 6.1 in this volume.

59. *United States Census of Agriculture: 1950*, vol. 1, pt. 27, p. xiii (quoted).

60. See Table 1.1; Kristjanson and Heltemes, *Handbook of Facts*, NDAES, *Bulletin* no. 382, pp. 30 figs. 10, 11, 56 table 29, 57 table 30, 64 fig. 34, 66 fig. 37; GPC, *Proceed-*

ings, 1949, frontispiece; Taylor, Heltemes, and Engelking, *North Dakota Agricultural Statistics*, pp. 70, 77.
    61. See Table 6.2.

CHAPTER SEVEN: POSTWAR DROUGHT ADJUSTMENTS

    1. *Williston Daily Herald (N.D.)*, October 14, November 25, 1949, and December 16, 1954, citing marketing economist of NDExtS; *Adams County Record (N.D.)*, November 23, 1949; *Dickinson Press (N.D.)*, January 17, 1956; Baldur H. Kristjanson and C. J. Heltemes, *Handbook of Facts about North Dakota Agriculture*, NDAES, Bulletin no. 357 (Fargo, N.D., 1950), pp. 11, 13; ibid., revised edn., NDAES, *Bulletin* no. 382 (Fargo, N.D., 1952), pp. 56–57 tables 29, 30; M. B. Johnson, *Range Cattle Production in Western North Dakota*, NDAES, *Bulletin* no. 347 (Fargo, N.D., 1947), p. 32; Lloyd E. Jones, "Stabilizing Farming by Shifting Wheat Land to Grass in the Northern Great Plains, with Particular Emphasis on Annual Effects," *Journal of Farm Economics* 32 (August 1950): 378, 380–90; Leroy W. Schaffner, Laurel D. Loftsgard, and Duane C. Vockrodt, *Production and Income Variability for Farm Enterprises on Irrigation and Dryland*, NDAES, *Bulletin* no. 445, technical (Fargo, N.D., 1963), p. 12 table 4.
    2. NDAES, *Annual Report, 1952–53, Bulletin* no. 387, p. 13; Johnson, *Range Cattle Production*, p. 22; Schaffner, Loftsgard, and Vockrodt, *Production and Income Variability*, p. 24; *Dickinson Press*, January 17, 1956, quoting H. W. Herbison of NDExtS.
    3. Johnson, *Range Cattle Production*, p. 45; James R. Gray and Chester B. Baker, *Organization, Costs and Returns on Cattle Ranches in the Northern Plains, 1930–52*, MAES, *Bulletin* no. 495 (Bozeman, Mont., 1953), p. 23; Ronald G. Fraase and Gordon W. Erlandson, *Geographic Changes in the Production of Cattle and Calves in the North Central Region*, NDAES, *Bulletin* no. 483 (Fargo, N.D., 1969), pp. 7, 9; Rex D. Helfinstine, *Economic Potentials of Irrigated and Dryland Farming in Central South Dakota*, SDAES, *Bulletin* no. 444 (Brookings, S.D., 1955), pp. 15–16; Herbert S. Schell, *History of South Dakota* (Lincoln, Nebr.: University of Nebraska Press, 1961), p. 350; Layton S. Thompson, "Grain Storage and Feed Reserve in Relation to Stabilizing Great Plains Agriculture," Great Plains Agricultural Council (hereafter cited as GPC), *Proceedings, 1953*, p. 132; *Dickinson Press*, January 20, 1956; *Lewistown Democrat-News* (Mont.), July 1, 7, 1949.
    4. *Dickinson Press*, March 11, 1953, March 10, 20, November 25, December 29, 1954, and February 5, 1955; *Adams County Record*, September 17, 1952; John T. Schlebecker, *Cattle Raising on the Plains, 1900–1961*, (Lincoln, Nebr.: University of Nebraska Press, 1963), pp. 212–13.
    5. *Williston Daily Herald*, April 30, 1957 (quoted), and January 15, February 20, 1958; *Dickinson Press*, August 27, 1959. Large-scale feedlot enterprise failed to expand in the northern Plains to the degree described by Charles L. Wood (*The Kansas Beef Industry* [Lawrence, Kansas: The Regents Press of Kansas, 1986], pp. 283–93), largely because of more limited expansion of corn and sorghum production under well irrigation.
    6. GPC, *Proceedings, 1953*, p. 71b fig. 14; *United States Census of Agriculture: 1954*, vol. 1, pt. 11, pp. 42–46, 226–31, pt. 27, pp. 42–46—county table 1; John P. Johansen, *Population Trends in Relation to Resources Development in South Dakota*, SDAES, *Bulletin* no. 440 (Brookings, S.D., 1954), pp. 23–25; C. R. Hoglund and M. B. Johnson, *Ranching in Northwestern South Dakota*, SDAES, *Bulletin* no. 385 (Brookings, S.D., 1947), p.4; Kristjanson and Heltemes, *Handbook of Facts*, NDAES, Bulletin no. 357, pp. 18 figs. 3 and 4, 21 figs. 9 and 10.
    7. S. C. Litzenberger, *Compana and Glacier Barley, Two New Varieties for Montana*,

MAES, *Bulletin* no. 422 (Bozeman, Mont., 1944), pp. 3–4, 6, 16; MAES, *Varieties of Farm Crops for Montana, 1943,* MAES, *Circular* no. 171 (Bozeman, Mont., 1943), pp. 9–11, 18; *Recommended Varieties of Farm Crops for Montana,* MAES, *Circular* no. 191 (Bozeman, Mont., 1949), pp. 5, 13; *Field Crop Varieties in Montana,* MAES, *Circular* no. 198 (Bozeman, Mont., 1952), pp. 7, 23; *Field Crop Varieties in Montana,* MAES, *Circular* no. 208 (Bozeman, Mont., 1956), p. 8; *Cereal and Oil Seed Crop Varieties for Montana,* MAES, *Circular* no. 223 (Bozeman, Mont., 1959), pp. 19–20; *Grain and Oil Seed Crop Varieties for Montana,* MAES, *Circular* no. 240 (Bozeman, Mont., 1963), pp. 6, 18–25. Hereafter these *Circulars* will be cited as *Varieties* for the respective years.

8. NDAES, *Annual Report, 1951,* pp. 9–10, *1955,* pp. 5, 15, *1957,* NDAES, *Bulletin* no. 405, p. 4; SDAES, *Annual Report, 1948–49,* p. 8, *1951–52,* pp. 35, 108, *1956–57,* p. 3; *Williston Daily Herald,* February 12, 1954, February 24, 1955, February 11, 1959, and February 3, 1960; *Dickinson Press,* April 1, 1958.

9. Samuel Garver, *Alfalfa in South Dakota, Twenty-One Years Research at Redfield Station,* SDAES, *Bulletin* no. 383 (Brookings, S.D., 1946), pp. 12, 15–16; W. W. Worzella and others, *Grasses and Legumes, Production and Management in South Dakota,* SDAES, *Bulletin* no. 427 (Brookings, S.D., 1953), pp. 9–10; MAES, *Varieties, 1943,* pp. 4, 11–12, *1944,* pp. 5, 21, *1945,* p. 6, and *1956,* p. 9; *Hill County Journal* (Havre, Mont.), February 26, 1931; NDAES, *Annual Report, 1955,* p. 16. On Hansen's early plant introductions, see Mary W. M. Hargreaves, *Dry Farming in the Northern Great Plains, 1900–1925,* Harvard Economic Studies vol. 101 (Cambridge, Mass.: Harvard University Press, 1957), pp. 202–4.

10. *Lewistown Democrat-News,* March 13, April 8, 1939; *Williston Daily Herald,* September 23, 1940; MAES, *Varieties, 1943,* pp. 5–6, *1945,* p. 3, *1952–53,* p. 29, *1956,* pp. 28–29, *1959,* pp. 10, 29–30, *1963,* p. 31; NDAES, *Annual Report, 1952–53,* p. 7, *1954,* p. 7, *1955,* p. 6; SDAES, *Annual Report, 1946–47,* p. 7, *1951–52,* p. 109.

11. A. N. Hume and Clifford Franzke, *Sorghums for Forage and Grain in South Dakota,* SDAES, *Bulletin* no. 285 (Brookings, S.D., 1934), pp. 2–3, 7, 15, 34, 56; A. N. Hume, Edgar Joy, and Clifford Franzke, *Twenty-One Years of Crop Yields from Cottonwood Experiment Farm,* SDAES, *Bulletin* no. 312 (Brookings, S.D., 1937), pp. 15, 43; SDAES, *Annual Report, 1938,* p. 10, *1948–49,* p. 14, *1949–50,* pp. 1–2, *1950–51,* p. 5; MAES, *Varieties, 1944,* pp. 25–26, *1959,* p. 32, *1963,* p. 33.

12. George A. Rogler, "Grass and Legume Introductions in the Northern Great Plains," GPC, *Proceedings, 1967,* pp. 91–94; MAES, *Forty-Sixth and Forty-Seventh Annual Reports, 1939–1941,* p. 55; *Forty-Eighth and Forty-Ninth Annual Reports, 1942,* p. 14; *Some Accomplishments of Ten Years of Agricultural Research in Montana, Fiftieth-Fifty-Ninth Annual Reports, 1942–1952,* pp. 43, 46; MAES, *Varieties, 1945,* p. 7, and *1956,* p. 10, *Forage Crop Varieties in Montana: Species and Varieties for Cultivated Land,* MAES, *Circular* no. 242 (Bozeman, Mont.; revised, 1970), p. 23; *Williston Daily Herald,* March 20, 1935; Ralph M. Williams and A. H. Post, *Dry-Land Pasture Experiments at the Judith Basin Branch Station,* MAES, *Bulletin* no. 388 (Bozeman, Mont., 1941), p. 24 et passim; Hume, Joy, and Franzke, *Twenty-One Years of Crop Yields from Cottonwood,* pp. 21, 25, 26, 28; C. J. Franzke and A. N. Hume, *Regrassing Areas in South Dakota,* SDAES, *Bulletin* no. 361 (Brookings, S.D., 1942), pp. 25–28; J. W. Turelle and Arnold Heerwagen, "Land Use Adjustments for Conservation in the Great Plains—Conversion to Grass," GPC, *Proceedings, 1963,* pp. 142–43.

13. *Adams County Record,* February 8, 1940, and September 7, 1944; MAES, *Varieties, 1943,* p. 7. On the earlier development of wheats for dry-land adaptation, see Hargreaves, *Dry Farming . . . 1900–1925,* pp. 301–5, and chap. 46–47 in this volume.

14. K. H. W. Klages, *Small Grain and Flax Varieties in South Dakota,* SDAES, *Bulletin* no. 291 (Brookings, S.D., 1934), p. 19; Hume, Joy, and Franzke, *Twenty-One Years of Crop Yields from Cottonwood,* p. 75; S. P. Swenson, *Spring Wheat Varieties in South*

Dakota, SDAES, *Bulletin* no. 342 (Brookings, S.D., 1940), p. 4; H. L. Walster and P. A. Nystuen, *North Dakota Wheat Yields*, NDAES, *Bulletin* no. 350 (Fargo, N.D., 1948), p. 9; MAES, *Varieties*, 1943, pp. 3, 7–8, 17, 1944, p. 4, 1945, p. 5, 1956, p. 7, 1963, pp. 12, 13; MAES, *Annual Report, 1946–47*, p. 12; *Lewistown Democrat-News*, February 28, November 18, 1939; *Williston Daily Herald*, September 6, 1935, October 30, 1939, and February 26, 1940.

15. *Sawtana Spring Wheat*, MAES, *Circular* no. 237 (Bozeman, Mont., 1962), pp. 2–3; MAES, *Varieties*, 1963, p. 11; Fred R. Taylor, C. J. Heltemes, and R. F. Engelking, *North Dakota Agricultural Statistics*, NDAES, *Bulletin* no. 408 (Fargo, N.D., 1957), p. 20 table 13; NDAES, *Annual Report, 1950, Bulletin* no. 365 (Fargo, N.D., 1951), p. 18; *Adams County Record*, April 12, 1945, November 23, 1949, February 1, 22, 1950, and September 26, 1951; *Williston Daily Herald*, September 6, 1944, and December 18, 1947.

16. *Lewistown Democrat-News*, October 10, December 1, 19, 1946; *Williston Daily Herald*, December 18, 1947.

17. Taylor, Heltemes, and Engelking, *North Dakota Agricultural Statistics*, pp. 20 table 13, and 21 table 14; *Adams County Record*, January 2, 1952, and September 2, 1953; *Dickinson Press*, December 8, 1954, and April 6, 1956; *Willison Daily Herald*, August 30, 1955; NDAES, *Bimonthly Bulletin* 18 (September-October 1955): 28 table 1; *Annual Report, 1954*, p. 4; SDAES, *Annual Report, 1952–53*, pp. 104–5; MAES, *Varieties, 1956*, p. 14.

18. MAES, *Varieties, 1956*, p. 14; *Dickinson Press*, November 24, 1953, January 13, 30, August 19, October 8, 1954, March 26, 1955, and June 21, 1957; *Williston Daily Herald*, April 28, 1955; NDAES, *Annual Report, 1954*, p. 4, 1955, p. 4. Since so little of the early production was made available to western counties, a lottery was held to apportion the allotment in Adams County (*Adams County Record*, December 15, 1954).

19. *Dickinson Press*, August 12, 28, October 8, 1953, February 26, 1954, and May 16, 1956; *Williston Daily Herald*, August 29, 1950; July 14, 1953, and October 23, 1954; NDAES, *Annual Report, 1952–53*, p. 5; Jim Langley and Suchada Langley, *State-Level Wheat Statistics, 1949–88*, USDA, ERS, *Statistical Bulletin* no. 779 (Washington, D.C., 1989), p. 58 table 54.

20. Hume, Joy, and Franzke, *Twenty-One Years of Crop Yields from Cottonwood*, p. 76; MAES, *Varieties, 1947*, p. 29, 1956, p. 18, 1959, p. 17, 1963, pp. 8, 14; NDAES, *Bimonthly Bulletin* 18 (September-October 1955): 27; NDAES, *Annual Report, 1954*, p. 3, 1957, p. 3; *Dickinson Press*, August 19, 1954, September 27, 1955, May 11, 1956, and December 23, 1959; *Williston Daily Herald*, November 3, 1955, May 10, 1956.

21. NDAES, *Annual Report, 1951*, p. 12, and 1957, p. 3; *Dickinson Press*, June 21, 1957; MAES, *Varieties, 1956*, pp. 16–17, 1963, pp. 10, 15, 16; F. H. McNeal, *Sheridan Spring Wheat*, MAES, *Circular* no. 246 (Bozeman, Mont., 1966), p. 3; Gordon E. Rodewald, Jr., Donald K. Larson, and D. C. Myrick, *Dryland Grain Farms in Montana: How They Started, Growth, and Control of Resources*, MAES, *Bulletin* no. 579, technical (Bozeman, Mont., 1963), pp. 8–9; V. A. Dirks, "Winter Grains for South Dakota," SDAES, *Annual Report, 1954–55*, pp. 41–45.

22. USDA, *Agricultural Statistics, 1956*, p. 535 table 724; R. Burnell Held and Marion Clawson, *Soil Conservation in Perspective* (Baltimore, Md.: Published for Resources for the Future, Inc., by Johns Hopkins Press, 1965), pp. 70–72.

23. *Adams County Record*, March 1, 1950, February 20, December 24, 1952, March 18, 1953, March 3, 1954, and March 23, 1955.

24. *Adams County Record*, May 20, 1948, *Williston Daily Herald*, January 24, March 2, 1944, January 28, 1946, February 16, 1950, and January 10, 1951.

25. O. R. Mathews, *Place of Summer Fallow in the Agriculture of the Western States*,

USDA, *Circular* no. 886 (Washington, D.C., 1951), pp. 4, 5–6 table 2, 8; L. W. Schaffner, *Economics of Grain Farming in Renville County, North Dakota*, NDAES, *Bulletin* no. 367 (Fargo, N.D., 1951), pp. 18, 21, 22; computations from data in *Sixteenth Census of the United States: 1940, Agriculture*, vol. 1, pt. 2, pp. 382–86, 468–73, pt. 6, pp. 16–20—county table 1; *United States Census of Agriculture: 1945*, vol. 1, pt. 11, pp. 18–28, 130–43, pt. 27, pp. 18–29—county table 1; *United States Census of Agriculture: 1954*, vol. 1, pt. 11, pp. 42–46, 226–31, pt. 27, pp. 42–46—county table 1. Census tallies specifically included idle land with fallow until 1949, which accounts for the larger amounts during the earlier years.

26. Lester R. Bronson, "The Value of ACP Cost-Sharing in Promoting Stubble Mulch Farming," GPC, *Proceedings, 1961*, pp. 115–16.

27. *Lewistown Democrat-News*, February 26, 1945; *Williston Daily Herald*, January 28, 1946.

28. O. R. Mathews and John S. Cole, "Special Dry-Farming Problems," USDA, *Yearbook, 1938*, pp. 685, 688; Mathews, *Place of Summer-Fallow in Agriculture of the Western States*, pp. 16–17; M. A. Bell, *The Effect of Tillage Method, Crop Sequence, and Date of Seeding upon the Yield and Quality of Cereals . . . under Dry-Land Conditions in North-Central Montana*, MAES, *Bulletin* no. 336 (Bozeman, Mont., 1937), p. 31; M. A. Bell, "Place of Fallow in Great Plains Land Use Adjustment," GPC, *Proceedings, 1949*, pp. 47–48; A. E. Seamans, *Recommended Practices for Soil Erosion Control*, MAES, *Circular* no. 190 (Bozeman, Mont., 1948), p. 2; Torlief S. Aasheim, *The Effect of Tillage Method on Soil and Moisture Conservation and on Yield and Quality of Spring Wheat in the Plains Area of Northern Montana*, MAES, *Bulletin* no. 468 (Bozeman, Mont., 1949), pp. 12–18, 21–22; SDAES, *Annual Report, 1953–54*, pp. 148–49; NDAES, *Annual Report, 1951*, p. 51; *Dickinson Press*, January 8, 1953.

29. Mathews and Cole, "Special Dry-Farming Problems," p. 692; Bell, *Effect of Tillage Method*, pp. 52–56; *Adams County Record*, March 10, 1932, and April 18, 1951; Aasheim, *Effect of Tillage Method . . . Northern Montana*, pp. 11, 14, 18, 25–28.

30. *Williston Daily Herald*, February 16, 1950 (quoted). On the divergences in recommended practice under the Soil Conservation program, even where "physical factors may be much the same," see Held and Clawson, *Soil Conservation*, pp. 288–89.

31. H. H. Finnell, "Classification of Land According to Sensitivity to Erosion," GPC, *Proceedings, 1949*, pp. 87–88 (quoted).

32. SDAES, *Annual Report, 1947–48*, p. 8, *1951–52*, pp. 94–96; NDAES, *Annual Report, 1951*, p. 55; GPC, *Proceedings, 1950*, p. 14; H. W. Omodt and others, *The Major Soils of North Dakota*, NDAES, *Bulletin* no. 472 (Fargo, N.D., 1968); D. D. Patterson and others, *Soil Survey Report: County General Soil Maps, North Dakota*, NDAES, *Bulletin* no. 473 (Fargo, N.D., 1968), p. 2 (quoted). Part of Bottineau County, North Dakota, had been surveyed in 1915, under the standards then prevalent.

33. *Lewistown Democrat-News*, June 27, 1941, and November 19, 26, December 1, 1942; *Williston Daily Herald*, December 17, 1952; *Adams County Record*, February 15, 1956; *Dickinson Press*, August 30, 1957; Rainer Schickele and Reuben Engelking, *Land Values and the Land Market in North Dakota*, NDAES, *Bulletin* no. 353 (Fargo, N.D., 1949), p. 18; computations based on data in USDA, *Agricultural Statistics, 1956*, p. 537 table 726; and *United States Census of Agriculture: 1959*, vol. 1, pt. 18, pp. 106–10; pt. 19, pp. 112–17, and pt. 38, pp. 118–22—county table 1. Cf. GPC, *Proceedings, 1950*, pp. 18–19, 22–26.

34. Bell, *Effect of Tillage Method*, p. 59; MAES, *Fiftieth-Fifty-Ninth Annual Reports, 1942–52*, pp. 14, 36; SDAES, *Annual Report, 1948–49*, pp. 3–5, *1951–52*, pp. 106–7, *1952–53*, pp. 36–41, *1954–55*, pp. 62, 65, 74–75; NDAES, *Annual Report, 1954*, pp. 12–14, *1955*, p. 11; Walter G. Heid, Jr., and Donald K. Larson, *Fertilizer Use in Montana, with Comparison, 1954–1967*, MAES, *Bulletin* no. 628 (Bozeman, Mont., 1969), pp. 3–

6, 13, 15-16, 18 table 8, 41-43 appendix tables 2, 3, 50-51 appendix table 7; GPC, *Proceedings, 1979*, p. 10 table 3; *Lewistown Democrat-News*, December 19, 1939; *Dickinson Press*, November 8, 1952, June 11, 1954, August 14, 1956, and April 2, 1957; *Williston Daily Herald*, May 19, 1953, November 27, December 2, 1954, and May 19, 1955; *Adams County Record*, August 4, 1949, February 24, 1954, and April 13, 1955. For a summary of dry-farming methods advocated for the region at that time, including the recent research on use of commercial fertilizers, see E. B. Norum, B. A. Krantz, and H. J. Haas, "The Northern Great Plains," USDA, *Yearbook of Agriculture, 1957: Soil*, pp. 496-500.

35. *Lewistown Democrat-News*, July 8, 1935, June 8, 1945, and January 1, 1946; *Dickinson Press*, March 26, 1953, August 23, 1957, and May 14, 1959; *Adams County Record*, December 6, 1945, January 22, 29, February 12, March 18, 1948, and July 22, 1953; SDAES, *Annual Report, 1947-48*, p. 16, *1949-50*, pp. 47-48, *1955-56*, pp. 8, 126-27; MAES, *Fiftieth-Fifty-Ninth Annual Reports, 1942-52*, p. 11; NDAES, *Annual Report, 1955*, p. 13; GPC, *Proceedings, 1953*, p. 23 table 4; Seamans, *Recommended Practices*, p. 4; Aasheim, *Effect of Tillage Methods*, p. 34; Dana H. Myrick, *All-Risk Crop Insurance: Principles, Problems, Potentials*, MAES, *Bulletin* no. 640 (Bozeman, Mont., 1970), p. 37; Gale E. Peterson, "The Discovery and Development of 2,4-D," *Agricultural History* 41 (July 1967): 251-53.

36. Walter W. Wilcox, *The Farmer in the Second World War* (Ames, Iowa: The Iowa State University Press, 1947), p. 294; *Adams County Record*, December 6, 1945, June 19, July 3, 1947, and March 8, August 30, 1950; *Williston Daily Herald*, April 3, 1947; April 12, 1950, and June 28, 1956; *Dickinson Press*, June 28, 1956, April 26, 1957, and August 12, 1958; *Lewistown Democrat-News*, July 30, 1947, and April 24, 1949; Claude Wakeland, "Grasshopper Problems and Development," GPC, *Proceedings, 1951*, pp. 30-32; C. K. Rowland and Mel Dubnick, "Decentralization of Agriculture," Don F. Hadwiger and Ross B. Talbot, eds., *Food Policy and Farm Programs, Proceedings of the Academy of Political Science* 34 (no. 3; New York, N.Y., 1982), pp. 216-22.

By the seventies use of aldrin and dieldrin had also been barred and that of heptachlor and chlordane suspended because of fear of environmental hazards. On the importance of fungicides and insecticides and the problems raised by their use in recent tillage methodology of the region, see below, p. 272.

37. *Lewistown Democrat-News*, February 15, 1947, and August 13, 1948; *Williston Daily Herald*, June 13, 1939, April 12, August 11, 1950, March 16, 1951, and March 11, 1954; *Dickinson Press*, September 8, 1955, and January 9, 1958.

38. Rodewald, Larson, and Myrick, *Dryland Grain Farms in Montana*, p. 46 table F; Kristjanson and Heltemes, *Handbook of Facts*, NDAES, *Bulletin* no. 382, p. 94 figs. 56, 57; Johansen, *Population Trends*, pp. 27, 28 figs. 14-16; *Lewistown Democrat-News*, July 29, 1933, and June 7, 1935; *Williston Daily Herald*, August 3, 1940; SDAES, *Annual Report, 1949-50*, p. 73; Thomas D. Isern, *Custom Combining on the Great Plains: A History* (Norman, Okla.: University of Oklahoma Press, 1981), pp. 48, 52-57; *1987 Census of Agriculture*, vol. 1, pt. 26, p. 212, pt. 34, p. 212, and pt. 41, p. 221—all county table 8.

39. SDAES, *Annual Report, 1946-47*, p. 39, *1949-50*, pp. 73, 128-29; GPC, *Proceedings, 1979*, p. 10; Ted Worrall interview, OH 246, Reel 2, MHS; below, p. 272.

40. Wylie D. Goodsell and Isabel Jenkins, *Costs and Returns on Commercial Farms, Long-Term Study, 1930-57*, USDA, ERS, *Statistical Bulletin* no. 297 (Washington, D.C., 1961), pp. 186-89 table 59; Warren R. Bailey and D. C. Myrick, "Alternative Farming Programs—Effects on Individual Farms," GPC, *Proceedings, 1961*, p. 63-10; Don Bostwick, *Management Strategies for Variable Wheat Yields in Montana*, MAES, *Bulletin* no. 585 (Bozeman, Mont., 1964), p. 16 table 9; *1987 Census of Agriculture*, vol. 1, pt. 26, p. 212 table 8.

Estimates of total capital requirements per farm differ greatly in covering land and livestock costs. A study sponsored by the GPC placed the figure in 1957 for the spring-wheat area of western North Dakota at $53,720, for the winter-wheat area of the western Plains at $93,000, and for the northern range area of the western Plains at $82,190. The machinery estimates in this study were not notably higher, at $8,550 for spring-wheat farms and $8,070 for winter-wheat operations (Howard W. Ottoson and others, *Land and People in the Northern Plains Transition Area* [Lincoln, Nebr.: University of Nebraska Press, 1966], p. 185 table 44).

41. Leroy W. Schaffner, Laurel D. Loftsgard, and Wayne W. Owens, *Economics of Leasing Farm Machinery and Buildings*, NDAES, *Bulletin* no. 450 (Fargo, N.D., 1964), pp. 6–7, 14–16, 18; Isern, *Custom Combining*, pp. 74–78, 83–85; Rodewald, Larson, and Myrick, *Dryland Grain Farms in Montana*, p. 4; acreage computations from data in Table 1.1.

42. Philip J. Thair, *Stabilizing Farm Income against Crop Yield Fluctuations*, NDAES, *Bulletin* no. 362 (Fargo, N.D., 1950), pp. 8–9; Philip J. Thair, *Meeting the Impact of Crop-Yield Risks in Great Plains Farming*, NDAES, *Bulletin* no. 392 (Fargo, N.D., 1954), pp. 9, 15; Schaffner, *Economics of Grain Farming in Renville County, North Dakota*, p. 16 table 14; Myrick, *All-Risk Crop Insurance*, pp. 11–12; Bostwick, *Management Strategies*, pp. 5, 9, 11–13; C. R. Hoglund, *What Size Farm or Ranch for South Dakota?* SDAES, *Bulletin* no. 387 (Brookings, S.D., 1947), p. 15.

43. See Table 7.1.

44. Data for this and the next paragraph derived from Thomas D. Campbell, "Interview with Largest Wheat Grower," *U.S. News & World Report* 46 (no. 22; June 1, 1959): 66–71; Hiram M. Drache, *Beyond the Furrow: Some Keys to Successful Farming in the Twentieth Century* (Danville, Ill.: The Interstate Printers & Publishers, 1976), ch. 4. Drache notes that for a period from the mid-twenties until the introduction of the AAA program, Campbell shifted from the alternating crop-fallow to a rotation of two years in crop to one in fallow (p. 111); but cf. Campbell, "Interview," p. 66.

45. Thair, *Stabilizing Farm Income*, pp. 2, 4–5 table 1, 12; Kristjanson and Brown, *Farmers Home Administration Approach*, pp. 10, 13, 15 fig. 5, 17 fig. 6; Schickele and Engelking, *Land Values and the Land Market in North Dakota*, p. 38; Baldur H. Kristjanson and Jacob A. Brown, *Credit Needs of Beginning Farmers in Selected Areas of North Dakota*, NDAES, *Bulletin* no. 386 (Fargo, N.D., 1953), pp. 5–6; *Williston Daily Herald*, January 10, 1947, and May 11, 1951; *Dickinson Press*, October 31, 1956, and August 12, 1958.

For discussion of the theoretical and research concerns relating to risk studies at that time, see Philip J. Thair, Glenn L. Johnson, and Rainer W. Schickele, eds., *Proceedings of Research Conference on Risk and Uncertainty in Agriculture, Bozeman, Montana, August 10–15, 1953*, GPC, *Publication* no. 11, NDAES, *Bulletin* no. 400 (Fargo, N.D., 1955); Howard W. Ottoson and Robert Finley, "Strategies to Meet the Hazards of Farming and Ranching in the Plains," GPC, *Proceedings, 1960*, pp. 140–64.

46. John D. Black, "Agricultural Credit Policy in the United States, 1945," *Journal of Farm Economics* 27 (August 1945): 609–11; Kristjanson and Brown, *Credit Needs*, pp. 5–6, 13 table 4, 15; Kristjanson and Brown, *Farmers Home Administration Approach*, p. 30; *Dickinson Press*, August 12, 1958.

A study sponsored by the GPC described agricultural credit in the dry-land transition area of Nebraska in the mid-fifties, showing considerably more long-term lending by federal agencies but the same dependence upon commercial banks and low activity by the PCAs in meeting needs for short- and intermediate-term credit (Ottoson and others, *Land and People in the Northern Plains Transition Area*, pp. 181–83, 291–95).

47. Schickele and Engelking, *Land Values and the Land Market in North Dakota*, pp. 38–39; Kristjanson and Brown, *Credit Needs*, pp. 14–16.

48. Dan Morgan, *Merchants of Grain* (New York, N.Y.: The Viking Press, 1979), p. 267; Glenn D. Pederson, "Economic Strategies for Coping with Financial Problems Faced by Farmers and Ranchers in the Great Plains," GPC, *Proceedings, 1985*, pp. 43–44.

49. Pederson, "Economic Strategies," pp. 47–48; 92 *U.S. Stat.* 427–32 (August 4, 1978); *New York Times*, March 7, July 22, December 19, 1985. See also Glenn Hertzler, Jr., "Farmers Home Administration Policies Which Deal with Control of Inflation in Great Plains Agriculture," GPC, *Proceedings, 1981*, pp. 52–57.

50. *Williston Daily Herald*, October 28, 1946; *Dickinson Press*, June 26, 1954, October 2, 1956; John E. Thompson, "Property Taxes and Farm Income," SDAES, *Annual Report, 1955–56*, p. 84; Thair, *Stabilizing Farm Income*, p. 26; Bostwick, *Management Strategies*, p. 18; North Dakota Legislative Assembly, *Laws Passed at the Twenty-Eighth, 1943*, ch. 250, p. 348–49 (March 6, 1943).

51. Bruce L. Gardner, *The Governing of Agriculture* (Lawrence, Kans.: Published for the International Center for Economic Policy Studies and the Institute for the Study of Market Agriculture by The Regents Press of Kansas, 1981), p. 7.

52. Thair, *Stabilizing Farm Income*, pp. 18, 23; Thair, *Meeting the Impact of Crop-Yield Risks*, p. 18; Bostwick, *Management Strategies*, pp. 11–12; Herman W. Delvo and L. D. Loftsgard, *All-Risk Crop Insurance in North Dakota*, NDAES, *Bulletin* no. 468 (Fargo, N.D., 1967), pp. 10–12, 33–34; Gordon Rodewald, Jr., *Crop Insurance in Montana*, MAES, *Circular* no. 235 (Bozeman, Mont., 1961), pp. 3–4. Barley insurance was also offered under the federal program, beginning in western North Dakota in 1958 (*Williston Daily Herald*, March 27, 1958).

53. *Dickinson Press*, September 23, 1958; *Williston Daily Herald*, March 20, September 26, 1957, and September 23, 1958.

54. Thair, *Stabilizing Farm Income*, pp. 18–19; Thair, *Meeting the Impact of Crop-Yield Risks*, pp. 24–25, 28, 30; Delvo and Loftsgard, *All-Risk Crop Insurance in North Dakota*, pp. 7, 12, 15–16, 26, 28, 30, 33; Rodewald, Larson, and Myrick, *Dryland Grain Farms in Montana*, p. 30; Rodewald, *Crop Insurance in Montana*, pp. 3, 4, 6 fig. 3; Myrick, *All-Risk Crop Insurance*, pp. 32–35; Earll H. Nikkel, "Federal All-Risk Crop Insurance in the Great Plains," GPC, *Proceedings, 1958*, pp. 102–6; Dana H. Myrick, "Report . . . on Crop Insurance," GPC, *Proceedings, 1968*, p. 167 table 3; *Dickinson Press*, December 14, 1954, September 27, 1957, September 10, 1958, and September 22, 1959; *Williston Daily Herald*, March 9, 1957, and September 23, December 22, 1959.

55. NDAES, *Annual Report, 1955*, p. 17; Gary Lucier, Agnes Chesley, and Mary Ahearn, *Farm Income Data: A Historical Perspective*, USDA, ERS, *Statistical Bulletin* no. 740 (Washington, D.C., 1986), pp. 180 table 47, 220 table 55, and 255 table 62; Rainer Schickele, "Farm Business Survival under Extreme Weather Risks," *Journal of Farm Economics* 31 (November 1949): 943; James D. Deal, "The Role of Crop Insurance in Coping with Risk and Uncertainty in Farming," GPC, *Proceedings, 1979*, pp. 102–6; *Adams County Record*, March 29, 1950, December 10, 1952, and February 2, 1955; *Williston Daily Herald*, December 3, 1952, June 27, 1955, July 7, 12, 16, August 31, 1956, August 18, 1959, and July 22, 1960; *Dickinson Press*, December 4, 1952, April 22, 1955, September 27, 1957, and June 24, September 10, 1958. On the Great Plains Conservation Program of 1955, see below, pp. 232–35.

56. Gardner, *Governing of Agriculture*, p. 29; 91 *U.S. Stat.* 1373 (November 15, 1977).

57. 94 *U.S. Stat.* 1312–16, 1321 (Septemer 26, 1980); 99 *U.S. Stat.* 1522 (December 23, 1985).

58. "Problems and Benefits Arising from Rain-Increasing Programs," GPC, *Proceedings, 1951*, pp. 88–113 (Schaefer quotation, p. 92).

59. *Williston Daily Herald*, June 25, July 21, 1948, March 21, May 28, June 1, 19, 22, September 1, 7, 1951, April 21, May 6, 14, 29, 1952, and November 3, 1953; *Adams County Record*, May 2, 30, June 6, 27, July 25, August 22, 1951, April 16, 23, 1952, and January 14, February 4, June 3, 1953; *Dickinson Press*, January 16, February 25, November 6, 1953.

60. Harry Wexter, quoted in GPC, *Proceedings, 1951*, p. 98; Dwight B. Klein, quoted in *Williston Daily Herald*, March 20, 1957; A. S. Dennis, "Augmentation of Rainfall from Summer Cumulus Clouds," John F. Stone and Wayne O. Willis, eds., "Plant Production and Management under Drought Conditions, Papers Presented at the Symposium, 4–6 October 1982, Held at Tulsa," *Agricultural Water Management* 7 (1983): 3 (quoted)–14; Lewis O. Grant, "Utilization and Assessment of Operational Weather Modification Programs for Augmenting Precipitation," ibid., pp. 23–35.

61. USDA, ACP Service, *Agricultural Conservation Program: Summary 1955* (Washington, D.C., 1956), p. 95 table 13; USDA, *Agricultural Statistics, 1956*, p. 539 table 728; *Adams County Record*, October 13, 1954. Relatively little such construction had been done for irrigation in the Dakotas, but over 19,000 dams had been built for erosion control and for livestock water during the period in North Dakota and 93,542 in South Dakota.

62. Detail for this and the next two paragraphs is drawn from Missouri Basin-Interagency Committee, *The Missouri: A Great River Basin of the United States, Its Resources and How We Are Using Them*, Public Health Service Publication no. 604 (Washington, D.C.: U.S. Government Printing Office, 1958), passim, esp. the end map; Robert G. Athearn, *High Country Empire: The High Plains and Rockies* (Lincoln, Nebr.: University of Nebraska Press, 1960), pp. 285–87; Herbert S. Schell, *History of South Dakota* (Lincoln, Nebr.: University of Nebraska Press, 1961), pp. 306, 354–55; Johansen, *Population Trends*, pp. 38–43; Edgar B. Nixon, comp. and ed., *Franklin D. Roosevelt and Conservation, 1911–1945*, 2 vols. (New York, N.Y.: Arno Press, 1972), vol. 2, pp. 608–18 passim; *Williston Daily Herald*, February 16, 23, 1944, October 19, 22, November 7, 1945, June 25, July 2, 1946, October 30, 1947, August 22, October 4, 21, 1949, August 3, 11, 17, 1950, June 2, 21, November 20, 1951, December 3, 4, 1953, and January 7, 1954; *Adams County Record*, August 21, October 30, 1947; *Dickinson Press*, February 4, 1954; *Lewistown Democrat-News*, April 15, 1945, January 10, 1946, August 30, October 21, 1947, and May 8, November 4, December 2, 29, 1949.

63. *Williston Daily Herald*, June 17, 1942, October 25, December 12, 1944, November 28, 1949, January 9, March 1, 24, July 12, 1950, April 4, 1951, February 27, 1952, and March 2, 1954; *Dickinson Press*, January 12, June 23, 1954. The best account of the history of the Garrison Diversion Project is provided in U.S. Congress, House, Committee on Interior and Insular Affairs, *Garrison Diversion Unit Reformulation Act of 1986*, 99th Cong., 2d sess., 1986, *H. Rept.* 99–525, pt. 1, pp. 8–15. The relevant statutes were 58 *U.S. Stat.* 887–907 (December 22, 1944), 79 *U.S. Stat.* 433–35 (August 5, 1965), and 100 *U.S. Stat.* 418–26 (May 12, 1986). See also Marc Reisner, *Cadillac Desert: The American West and Its Disappearing Water* (New York, N.Y.: Viking Penguin, Inc., 1986), pp. 189–202. For mapping of the project as contemplated in 1944, see Map 7.2.

64. On support for the project, see U.S. Congress, House, Committee on Interior and Insular Affairs, Subcommittee on Water and Power Resources: *Hearing on H. R. 1116, Recommendations of the Garrison Diversion Unit Commission*, 99th Cong., 1st sess., February 28, 1985, pp. 452–65; *Lewistown Democrat-News*, November 10, December 15, 1943, November 15, December 7, 1944, and June 26, 1947; *Williston Daily Herald*, August 29, December 12, 1944 (quoted), October 12, 1951, February 8, 27, August 5, 1952, May 10, November 14, 1955, and June 10, 1960; *Dickinson Press*, June 23, 1954;

Adams County Record, April 23, 1942; SDAES, *Annual Report, 1949–50,* pp. 34–39.

65. See Map. 7.3.

66. Except as specifically indicated, the development of local opposition as recounted in this and the next two paragraphs is based on *Williston Daily Herald,* June 11, August 13, 1947, May 20, 1948, January 21, March 28, May 19, December 1, 1949, February 28, March 31, April 3, November 16, 17, 1950, April 3, 15, August 16, November 10, 1952, March 13, 1953, February 17, 1954, January 20, February 1, 16 (quoted), April 14, May 13, June 15 (quoted), 1955, and January 8, August 23, October 30, 1957; *Dickinson Press,* May 30, 1953, February 16, 1955, and August 8, 1957; Schaffner, *Economics of Grain Farming in Renville County,* p. 5. The Farmers' Union strongly opposed construction of both the Garrison Dam and the diversion project. Favoring establishment of a Missouri Valley Authority, they questioned the proposals of both the Bureau of Reclamation and the Army Corps of Engineers (*Williston Daily Herald,* May 29, 1947, January 9, 1950). Conflicting views were evident among agricultural analysts, as evidenced in George F. Will, *Tree Ring Studies in North Dakota,* NDAES, *Bulletin,* no. 338 (Fargo, N.D., 1946), and Director H. L. Walster's appended statement, p. 24.

67. *Williston Daily Herald,* May 13, 1955 (quoted).

68. U.S. Cong., 99th Cong., 2d. sess., *H. Rept.* 99-525, pt. 1, pp. 18–19 table 1, p. 34 table 2. For an account of the controversy centered on South Dakota proposals in the early eighties to sell waters of the Oahe Reservoir for a coal slurry from the Powder River Basin to Louisiana, with the possibility that some water might be made available for supplementary irrigation in southwestern South Dakota, see John Ferrell, "Developing the Missouri: South Dakota and the Pick-Sloan Plan," *South Dakota History* 19 (Fall 1989): 334–41. The project failed because of the constitutional conflict with federal authority and the need to preserve downstream water flow.

69. Computations for this and the next paragraph based on *Sixteenth Census of the United States: 1940,* vol. 1, pt. 2, pp. 382–86, 468–73, pt. 6, pp. 16–20—county table 1; *United States Census of Agriculture: 1959,* vol. 1, pt. 18, pp. 106–15, pt. 19, pp. 112–23, pt. 38, pp. 118–27—county tables 1 and 1a; *1987 Census of Agriculture,* vol. 1, pt. 26, pp. 204–11, pt. 34, pp. 204–11, pt. 41, pp. 212–20—county table 7. See also *Dickinson Press,* March 7, 1958; Wallace McMartin and Ronald O. Bergan, *Irrigation Practices and Costs in North Dakota,* NDAES, *Bulletin* no. 474 (Fargo, N.D., 1968), pp. 8, 9, 14.

70. NDAES, *Annual Report, 1951,* pp. 66–67; L. W. Schaffner, *An Economic Analysis of Proposed Irrigation in Northern North Dakota,* NDAES, *Bulletin* no. 404 (Fargo, N.D., 1956), pp. 10–11, 19, 21; Schaffner, Loftsgard, and Vockrodt, *Production and Income Variability for Farm Enterprises on Irrigation and Dryland,* pp. 3–4, 31.

71. SDAES, *Annual Report, 1950–51,* p. 89, *1951–52,* p. 144; NDAES, *Annual Report, 1955,* p. 19; Schaffner, *Economic Analysis of Proposed Irrigation,* pp. 9–10; Schaffner, Loftsgard, and Vockrodt, *Production and Income Variability for Farm Enterprises on Irrigation and Dryland,* pp. 3–4, 7–8, 11, 12 table 4, 13 table 5, 14; Gray and Baker, *Organization, Costs and Returns on Cattle Ranches,* p. 80; Roy E. Huffman, *Irrigation Development and Public Water Policy* (New York, N.Y.: Ronald Press Co., 1953), p. 219; Howard M. Olson, "Irrigation Results and Promises," Carle C. Zimmerman and Seth Russell, eds., *Symposium on the Great Plains of North America* (Fargo, N.D.: The North Dakota Institute for Regional Studies, 1967), pp. 109–10; McMartin and Bergan, *Irrigation Practices and Costs,* pp. 7, 19, 34. Cf. Rex D. Helfinstine, *Economic Potentials of Irrigated and Dryland Farming in Central South Dakota,* SDAES, *Bulletin* no. 444 (Brookings, S.D., 1955), pp. 20–26.

72. Stanley W. Voelker, *Settlers' Progress on Two North Dakota Irrigation Projects, A Study of Farm Development and Resource Accumulation on the Buford-Trenton and Lewis and Clark Projects,* NDAES, *Bulletin* no. 369 (Fargo, N.D., 1951), pp. 7–8; Clyde E.

Stewart and D. C. Myrick, *Control and Use of Resources in the Development of Irrigated Farms: Buffalo Rapids and Kinsey, Montana*, MAES, *Bulletin* no. 476 (Bozeman, Mont., 1951), p. 4; Clyde E. Stewart and Roy E. Huffman, *Resource Needs and Income Potentials of Newly Irrigated Family-Operated Farms: Lower Marias Project, Montana*, MAES, *Bulletin* no. 521 (Bozeman, Mont., 1956), pp. 6, 9, 73–74. See also Johansen, *Population Trends*, p. 43; GPC, *Proceedings, 1955*, p. 35. By the early sixties increased attention to production of forage and row crops on irrigated land was evident on the Lower Yellowstone project, but there was still little irrigated pasture (L. W. Schaffner, Laurel D. Loftsgard, and Norman Dahl, *Integrating Irrigation with Dryland Farming*, NDAES, *Bulletin* no. 433 [Fargo, N.D., 1961], pp. 6, 8, 11). The Malta Division of the Milk River project was exceptional in that 71 percent of the irrigated land was devoted to raising hay, primarily alfalfa (Ralph E. Ward and M. M. Kelso, *Irrigation Farmers Reach Out into the Dry Land*, MAES, *Bulletin* no. 464 [Bozeman, Mont., 1949], pp. 5, 9, 28–29, 31–32).

73. SDAES, *Annual Report, 1951–52*, pp. 17–18; Ottoson and others, *Land and People in the Northern Plains Transition Area*, pp. 142 table 27, 143 table 28; Elmer Starch, "The Future of the Great Plains Reappraised," *Journal of Farm Economics* 31 (November 1949): 927; Elmer Starch, "The Great Plains–Missouri Valley Region," Merrill Jensen, ed., *Regionalism in America* (Madison, Wisc.: The University of Wisconsin Press, 1951; paperback edn., 1965), p. 364 (quoted). See also GPC, *Proceedings, 1948*, pp. 41, 43–44, *1949*, pp. 91–92, *1953*, p. 42; True D. Morse, "Future of Farming in the Great Plains," GPC, *Proceedings, 1954*, p. 13; M. B. Johnson, *Range Cattle Production in Western North Dakota*, pp. 14–15; Harry A. Steele and John Muehlbeier, "Land and Water Development Programs in the Northern Great Plains," *Journal of Farm Economics* 32 (August 1950): 444–45; Max Myers, "Men and Land," SDAES, *Annual Report, 1951–52*, pp. 17–18; Thair, *Meeting the Impact of Crop-Yield Risks*, p. 24; *Dickinson Press*, August 7, 1959.

74. Finnell remarks, GPC, *Proceedings, 1948*, p. 44.

75. Steele and Muehlbeier, "Land and Water Development Programs," p. 442; *Lewistown Democrat-News*, May 26, September 21, October 4, 1949; *Williston Daily Herald*, August 1, 1950.

76. GPC, *Proceedings, 1949*, p. 64; O. J. Scoville, "Costs of Transition from Wheat to Grass," ibid., p. 67; Wilkie Collins, Jr., "Bottlenecks in Diversion to Other Crops," ibid., pp. 81–86; R. O. Olson and Gabriel Lundy, "Overproduction Threatens Farm Income," SDAES, *Annual Report, 1949–50*, pp. 21–22; Starch, "Future of the Great Plains Reappraised," p. 924. See also GPC, *Proceedings, 1950*, p. 49.

77. Harold E. Myers, "Livestock and Feed Problems in the Great Plains," GPC, *Proceedings, 1954*, pp. 53, 55, 56; Starch, "Future of the Great Plains Reappraised," pp. 919–20, 923; Starch, reporting for Committee on Tenure and Economics, GPC, *Proceedings, 1950*, p. 50; Warren R. Bailey, "Adapting Farming to the Great Plains," GPC, *Proceedings, 1966*, p. 83; M. B. Johnson, *Range Cattle Production in Western North Dakota*, pp. 17–20, 25–26, 32; Gray and Baker, *Organization, Costs, and Returns on Cattle Ranches*, pp. 27–28; Thompson, "Grain Storage and Feed Reserve in Relation to Stabilizing Great Plains Agriculture," p. 132.

78. H. L. Stewart, "Impacts of Prospective Changes in Demand for Wheat on Great Plains Areas," GPC, *Proceedings, 1949*, pp. 34, 39 (quoted): also ibid., pp. 67–68, 86.

79. Starch, "Future of the Great Plains Reappraised," p. 922; Baldur H. Kristjanson, reporting on findings of the conference on credit, GPC, *Proceedings, 1954*, p. 85 (quoted); Huffman, *Irrigation Development and Public Water Policy*, pp. 250–66 (p. 251 quoted); Warren S. Thompson and P. K. Whelpton, *Population Trends in the United States* (New York, N.Y.: McGraw-Hill Book Co., 1933), pp. 5 table 3, 316 table 88, 319;

Joseph S. Davis, *The Population Upsurge in the United States, War-Peace Pamphlets* no. 12 (Stanford, Calif.: Food Research Institute, Stanford University, 1949), pp. 13 (quoted), 18–24, 39. Cf. Ray K. Linsley, "Report on the Hydrological Problems of the Arid and Semi-Arid Areas of the United States and Canada," UNESCO, *Reviews of Research on Arid Zone Hydrology* (1953), p. 128.

80. M. P. Hansmeier, "Long Time Adjustments, Recommendations to Great Plains Council Based on Discussion of Committee on Diverted Acres, August 10, 1949," GPC, *Proceedings, 1949*, p. 114. See also Roy E. Huffman, "Problems of the Plains," GPC, *Proceedings, 1958*, p. 21; Ottoson and Finley, "Strategies to Meet the Hazards of Farming," GPC, *Proceedings, 1960*, pp. 141–52 passim.

81. Starch, "Future of the Great Plains Reappraised," pp. 919–25 (quotation on p. 925).

82. GPC, *Proceedings, 1949*, p. 40.

CHAPTER EIGHT: THE "YO-YO" MARKETING PROBLEM

1. Lloyd D. Teigen and Florence Singer, *Weather in United States Agriculture: Monthly Temperature and Precipitation by State and Farm Production Region, 1950–86*, USDA, ERS, *Statistical Bulletin* no. 765 (Washington, D.C., 1988), pp. 145, 177, 205; *Lexington Herald-Leader* (Ky.), June 21, 1989.

2. For localized mapping as discussed in this and the following paagraph, see Great Plains Agricultural Council (hereafter cited as GPC), *Proceedings, 1956*, pp. 5–6, *1957*, pp. 8–9, 13–13a, *1958*, pp. 3, 7, *1959*, pp. 1, 4, 6–6a, *1960*, pp. 1, 5–5a, *1961*, pp. 2–4, 11, *1962*, appendix pp. 2–3, *1963*, p. 165 table 4, *1964*, pp. 5–6; *Williston Daily Herald* (N.D.), June 26, July 3, 7, 12, 16, August 11, 31, November 28, December 21, 1956, July 16, 1957, May 27, 28, June 17, 20, 27, July 2, August 1, 3, 18, September 25, 1959, and February 6, 16, April 13, 1960; *Dickinson Press* (N.D.), June 27, July 12, 1956, and August 1, 13, 1959; James T. Nichols, *Range Improvement Practices on Deteriorated Dense Clay Wheatgrass Range in Western South Dakota*, SDAES, *Bulletin* no. 552 (Brookings, S.D., 1969), p. 6.

3. C. W. Thornthwaite and J. R. Mather, "The Water Budget and Its Use in Irrigation," USDA, *Yearbook of Agriculture, 1955: Water*, p. 357 (quoted); Chester E. Evans and Edgar R. Lemon, "Conserving Soil Moisture," USDA, *Yearbook of Agriculture, 1957: Soil*, p. 342; and cf. USDA, *Climate and Man, Yearbook of Agriculture, 1941*, p. 711. For the research cited in this and the following paragraph, see O. R. Mathews and John S. Cole, "Special Dry-Farming Problems," USDA, *Yearbook of Agriculture, 1938: Soils and Men*, p. 684; Harry F. Blaney, "Climate as an Index of Irrigation Needs," USDA, *Yearbook of Agriculture, 1955: Water*, p. 345; Evans and Lemon, "Conserving Soil Moisture," p. 357; E. B. Norum, B. A. Krantz, and H. J. Haas, "The Northern Great Plains," USDA, *Yearbook of Agriculture, 1957: Soil*, p. 496; J. C. Hide, "Prevention of Evaporation from the Soil Surface under Dryland Farming," GPC, *Proceedings, 1960*, pp. 91–94 passim.

4. Warren R. Bailey and D. C. Myrick, "Alternative Farming Programs—Effects on Individual Farms," GPC, *Proceedings, 1961*, pp. 63-6 through 63-21 (quotations from pp. 63-8, 63-10, 63-17, 63-21). Operator return on wheat in central North Dakota in 1960 averaged $21.60 per acre; on barley, $8.45; on alfalfa hay, second to wheat in profitability, $12.60 (Fred R. Taylor and Ronald A. Anderson, *Wheat Statistics for North Dakota*, NDAES, *Agricultural Economics Report* no. 20 (Fargo, N.D., 1961), sec. I, p. 3).

5. Bailey and Myrick, "Alternative Farming Programs," p. 63-15 (quoted); War-

ren R. Bailey, "Adapting Farming to the Great Plains," GPC, *Proceedings, 1966*, pp. 79–85 (quotations from pp. 79, 82).

6. Jim Langley and Suchada Langley, *State-Level Wheat Statistics, 1949–88*, USDA, ERS, *Statistical Bulletin* no. 779 (Washington, D.C., 1989), pp. 42 table 38, 56 table 52, 67 table 63; Bailey and Myrick, "Adapting Farming," p. 63-12; M. L. Upchurch, "Problems in Adjusting Land Use and Production of Great Plains Farms," GPC, *Proceedings, 1958*, pp. 60 table 1, 62; *Williston Daily Herald*, September 6, 1958; *Dickinson Press*, August 22, December 18, 1958.

7. Computations from data in Gary Lucier, Agnes Chesley, Mary Ahearn, *Farm Income Data: A Historical Perspective*, USDA, ERS, *Statistical Bulletin* no. 740 (Washington, D.C., 1986), pp. 180 table 47, 220 table 55, 255 table 62. Cf. L. D. Loftsgard and N. A. Dorow, "The Economics and Complexities of Land Use Adjustments," GPC, *Proceedings, 1963*, pp. 75–82.

8. Taylor and Anderson, *Wheat Statistics*, sec. C, p. 2.

9. John A. Schnittker, "A Government Policy for Agriculture, Especially the Great Plains," GPC, *Proceedings, 1960*, pp. 172–73; Taylor and Anderson, *Wheat Statistics*, figs. C-1, C-2; computations from GPC, *Proceedings, 1957*, p. 6 table 1, *1959*, p. 4 table 1, *1964*, appendix table 1.

10. Douglas E. Bowers, Wayne D. Rasmussen, Gladys L. Baker, *History of Agricultural Price-Support and Adjustment Programs, 1933–84*, USDA, ERS, *Agricultural Information Bulletin* no. 484 (Washington, D.C., 1984), pp. 21–24; Edward L. Schapsmeier and Frederick H. Schapsmeier, *Ezra Taft Benson and the Politics of Agriculture: The Eisenhower Years, 1953–1961* (Danville, Ill.: The Interstate Printers and Publishers, 1975), pp. 161–65; *Williston Daily Herald*, June 11, 1955, February 10, March 9, 10, May 19, 29, 1956, May 2, 1958, and May 12, 1960; *Dickinson Press*, August 18, 1954, and August 9, 1958; GPC, *Proceedings, 1963*, p. 24.

11. L. Orlo Sorenson and Donald Anderson, *The Grain Marketing Operations of the Commodity Credit Corporation through 1962*, NDAES, *Bulletin* no. 458; *North Central Regional Research Bulletin* no. 167 (Fargo, N.D., 1965), p. 7 table 1, Bowers, Rasmussen, and Baker, *History of Agricultural Price-Support*, pp. 24–25.

12. *Dickinson Press*, April 2, 1959; Schapsmeier and Schapsmeier, *Ezra Taft Benson*, pp. 51–52; Wayne D. Rasmussen, "History of Soil Conservation, Institutions, and Incentives," Harold G. Halcrow, Earl O. Heady, and Melvin L. Cotner, eds., *Soil Conservation Policies, Institutions, and Incentives* (Ankeny, Iowa: Published for North Central Research Committee 111: Natural Resource Use and Environmental Policy by the Soil Conservation Society of America, 1982), p. 10; M. P. Leaming, "Report of the Agricultural Conservation Program," GPC, *Proceedings, 1960*, statistical p. 1. The cited expenditures included the cost-share payments under the Soil Bank program discussed below on page pp. 230–32.

Tensions in administration of the overlapping programs have been frequently noted, but at the local level in the northern Plains there was considerable identity of membership between the ACP/PMA committees and the SCD boards of supervisors. In 1961 the linkage of soil conservation with price supports was signalized in renaming the administering agency the Agricultural Stabilization and Conservation Service.

13. 70 *U.S. Stat.* 191–95 (May 28, 1956); GPC, *Proceedings, 1966*, slide 11; *Dickinson Press*, June 6, 1956, May 21, December 3, 4, 6, 1957, and May 8, 1959; *Williston Daily Herald*, April 10, 23, 24, 1958, and May 17, 1960.

14. Leaming, "Report of the Agricultural Conservation Program," p. 248; Thomas E. Hamilton, "Report on the Conservation Reserve," GPC, *Proceedings, 1960*, pp. 276–78, 286–88; Raymond P. Christensen and Ronald O. Aines, *Economic Effects of Acreage Control Programs in the 1950's*, USDA, ERS, *Agricultural Economic Report*

no. 18 (Washington, D.C., 1962), p. 2; U.S. House of Representatives, 86th Cong., 2d sess., *Hearings before the Subcommittee on Appropriations*, pt. 1, p. 19 (February 15, 1960), and pt. 3, pp. 388, 390. The inclusion of noncropland in Bottineau, Divide, and Williams counties of North Dakota was cited specifically as requiring corrective action.

The western Dakotas as here cited encompassed counties east to the James River Valley, in accordance with the definition under the Great Plains Program, discussed below on pp. 232–35.

Nationally the "conservation reserve" program cost a total of $2.48 billion for rental payments and $162 million for cost sharing—a total cost during the life of the program averaging $86.43 an acre placed in the reserve (W. A. Laycock, "History of Grassland Plowing and Grass Planting on the Great Plains," John E. Mitchell, ed., *Impacts of the Conservation Reserve Program in the Great Plains, Symposium Proceedings, September 16–18, 1987*, USDA, Forest Service, *General Technical Report* RM-158 [Fort Collins, Col.: Rocky Mountain Forest and Range Experiment Station, 1987], p. 6). Combined with supplemental price-support payments, the annual government expenditure for conserving practices on agricultural lands increased from $1.7 billion in 1960 to $3.7 billion in 1970 (Earle J. Bedenbaugh, "History of Cropland Set Aside Programs in the Great Plains," ibid., p. 16).

15. Christensen and Aines, *Economic Effects of Acreage Control Programs*, pp. 2, 5, 12–15; National Planning Association Agriculture Committee, *A Statement, Special Report* no. 59 (Washington, D.C.: National Planning Association, 1961), p. 17; *Dickinson Press*, June 2, 1959; *Williston Daily Herald*, April 14, 1960.

16. Loftsgard and Dorow, "Economics and Complexities of Land Use Adjustments," pp. 75–82; Neil Sampson, *Farmland or Wasteland, A Time to Choose: Overcoming the Threat to America's Farm and Food Future* (Emmaus, Pa.: Rodale Press, 1981), p. 60; Evans and Lemon, "Conserving Soil Moisture," pp. 349–52. Developing research was indicating that fallow provided inefficient moisture conservation after the first few months, as compacted topsoil retarded moisture penetration and increased the amount of hygroscopic evaporation. See below, p. 272.

17. Hamilton, "Report on the Conservation Reserve," p. 284; Forest W. Beall, "Retirement of Low Grade Cropland to Grass under Conservation Reserve and Agricultural Conservation Programs," GPC *Proceedings, 1958*, pp. 29–30; Leroy C. Rude, *Land Use Alternatives for Dryland Grain-Livestock Operators in South-Central Montana*, MAES, *Bulletin* no. 570 (Bozeman, Mont., 1962), pp. 6–10; Leroy C. Rude, *Land Use Alternatives for Dryland Grain-Livestock Operators, North-Central Montana*, MAES, *Bulletin* no. 571 (Bozeman, Mont., 1962), pp. 7–10; Leroy C. Rude, *Land Use Alternatives for Dryland Grain-Livestock Operators, Northeastern Montana*, MAES, *Bulletin* no. 572 (Bozeman, Mont., 1962), pp. 7–10; *Dickinson Press*, September 16, 1958, and June 25, 1959; *Williston Daily Herald*, June 13, 1956. See also Bailey and Myrick, "Alternative Farming Programs," pp. 63–13, 63–14; Ronald D. Krenz, LeRoy W. Schaffner, and Enrique Valdivia, *Seeding Cropland to Grass in Southwestern North Dakota*, NDAES, *Bulletin* no. 470 (Fargo, N.D., 1967), pp. 14, 19, 21–22.

18. M. L. Upchurch, "New Programs to Meet Farm Problems of the Plains," GPC, *Proceedings, 1963*, p. 26; *Dickinson Press*, August 19, 1959; *Williston Daily Herald*, June 13, 1956, October 23, 24 (quoted), November 8, 1958, February 27, November 4, 1959, and February 19, September 22, 1960; U.S. House of Representatives, 86th Cong., 2d sess., *Hearings before the Subcommittee on Appropriations*, pt. 3, pp. 359–61 (March 7, 1960); 74 *U.S. Stat.* 239 (June 29, 1960); USDA, *Agricultural Statistics, 1972*, p. 637 table 755. The Soil Bank Act was formally repealed in 1965 but was replaced on a much more limited scale by a similar program of "cropland adjustment."

That the program increased depopulation in the dry-land districts is questionable. This was a period when sharply declining farm parity ratios were depressing agriculture generally. In the northern Plains the proportional rate of increased decline in the number of farms was greater for the states as a whole than for the dryland districts during the fifties. See Table 6.2.

19. 70 *U.S. Stat.* 1115–17 (August 7, 1956); *Dickinson Press*, March 6, 1957.

20. Detail in this and the following paragraph is taken from Cyril Luker, "Progress under Public Law 1021," GPC, *Proceedings, 1957*, pp. 36–41; Paul Griffith, "Report of the Great Plains Committee on State and Local Action," ibid., pp. 46–47 (phrases quoted).

21. E. J. Haslerud, GPC, *Proceedings, 1956*, pp. 39–40; Luker, "Progress under Public Law 1021," pp. 37–38.

22. *Dickinson Press*, September 7, 1957; D. A. Williams, "Progress of the SCS in the Great Plains," GPC, *Proceedings, 1956*, p. 28; R. Burnell Held and Marion Clawson, *Soil Conservation in Perspective* (Baltimore, Md.: Published for Resources for the Future, Inc., by The Johns Hopkins Press, 1965), p. 81. Mont H. Saunderson, a range specialist, had warned in 1952 that 10 million acres were being cropped in the northern Plains, alone, which were marginally productive and should be permanently committed to grazing utilization ("Range Problems of Marginal Farm Lands," *Journal of Range Management* 5 [January 1952]: 13–15). Held and Clawson linked Saunderson's recommendation as an addition to the maximum of 14 million acres in the SCS estimate, thus raising the projected area for retirement to 24 million acres.

23. Data in this and the following paragraph drawn from Held and Clawson, *Soil Conservation*, p. 84 table 3; GPC, "Report of the Executive Committee," *Proceedings, 1966*, p. 121; U.S. House of Representatives, 84th Cong., 2d sess., *Great Plains Conservation Program: Hearings before the Committee on Agriculture . . . on H.R. 11831 and H.R. 11833, June 28, 1956* (Washington, D.C.: U.S. Government Printing Office, 1956), pp. 12–13; Luker, "Progress under Public Law 1021," p. 41; K. William Easter and Melvin L. Cotner, "Evaluation of Current Soil Conservation," Halcrow, Heady, and Cotner, eds., *Soil Conservation Policies*, pp. 293–94; Douglas Helms, "Conserving the Plains: The Soil Conservation Service in the Great Plains," *Agricultural History* 64 (Spring 1990): 69. See also Sandra S. Batie, "Policies, Institutions, and Incentives for Soil Conservation," Halcrow, Heady, and Cotner, eds., *Soil Conservation Policies*, pp. 27–28.

24. GPC, *Proceedings, 1955*, p. 48; Williams, "Progress of the SCS," pp. 27–28; Philip F. Aylesworth, "Progress in the Program for the Plains," GPC, *Proceedings, 1961*, p. 91. Cf. True D. Morse, "Progress with the Great Plains Program," GPC, *Proceedings, 1957*, p. 26. Much larger areas of survey are reported for the states as a whole at the end of 1955 in USDA, *Agricultural Statistics, 1956*, p. 537 table 726: for Montana, 12.5 million acres; for North Dakota, 20.4 million; and for South Dakota, almost 22.2 million; but cf. below, pp. 262, 267–68.

25. Theodore C. Green, "The National Inventory of Soil and Water Conservation Needs," GPC, *Proceedings, 1959*, pp. 108–12.

26. Held and Clawson, *Soil Conservation*, pp. 132 (quoted)–34; Roy D. Hockensmith, "Opportunities for Adjustment in the Great Plains: Land Use Adjustments for Conservation—The Situation," GPC, *Proceedings, 1963*, p. 114. Hockensmith conceded that new agricultural technologies in the future might modify the evaluation.

27. Hockensmith, "Opportunities for Adjustment," pp. 112 table 2, 113 table 3, 132 fig. 8; 133 fig. 9. Data for the northern Plains were calculated from dot maps. The boundaries of the western Dakotas as presented here included the area east to the James River Valley.

28. See pp. 220–23; GPC, *Proceedings, 1956*, p. 79; Roy Freeland, "Wheat—the Troubled Giant," GPC, *Proceedings, 1959*, p. 18. For continuing expressions of confidence in the availability of measures to combat a recurrent "dust bowl," see comments by the Stark County extension agent and the superintendent of the Dickinson, North Dakota, experimental substation, *Dickinson Press*, August 7, 1959; John A. Schnittker and M. L. Upchurch, "Alternative Farm Programs and the Wheat Situation," GPC, *Proceedings, 1961*, p. 63-3; Upchurch, even in arguing for land retirement, in GPC, *Proceedings, 1963*, pp. 24–25, and E. D. Hunter, "Report on Survey of Priority Problems in the Great Plains," GPC, *Proceedings, 1968*, p. 145.

29. 49 *U.S. Stat.* 274–75 (August 24, 1935); Sorenson and Anderson, *Grain Marketing Operations*, p. 46; Nick Kotz, *Let Them Eat Promises: The Politics of Hunger in America* (Englewood Cliffs, N.J.: Prentice-Hall, Inc., 1969), p. 50; Norwood Allen Kerr, "Drafted into the War on Poverty: USDA Food Stamp and Nutrition Programs, 1961–1969," *Agricultural History* 64 (Spring 1990): 160.

30. Except as noted, this and the following four paragraphs are based on Kerr, "Drafted into the War on Poverty," pp. 157–66 passim.

31. Taylor and Anderson, *Wheat Statistics*, secs. G2, G5.

32. Barbara A. Claffey and Thomas A. Stucker, "The Food Stamp Program," Don F. Hadwiger and Ross B. Talbot, eds., *Food Policy and Farm Programs, Proceedings of the Academy of Political Science* 34 (no. 3, 1982): 44–45.

33. U.S. Senate, 86th Cong., 1st sess., *Hearings before a Subcommittee of the Committee on Agriculture and Forestry . . . on . . . Bills to Facilitate the Distribution of Surplus Food Products to Needy Families in the United States, to Safeguard the Health, Efficiency, and Morale of the American People, to Promote the Full Use of Agricultural Resources, and for Other Purposes, June 4, 5, and 8, 1959* (Washington, D.C.: U.S. Government Printing Office, 1959), pp. 74–75 (quoted); U.S. Senate, 88th Cong., 2d sess., *Hearings before the Committee on Agriculture and Forestry, on H.R. 10222, An Act to Strengthen the Agricultural Economy . . . June 18 and 19, 1964* (Washington, D.C.: U.S. Government Printing Office, 1964), pp. 44, 46. See also Kotz, *Let Them Eat Promises*, pp. 51, 56; Claffey and Stucker, "Food Stamp Program," p. 45; Don F. Hadwiger, "The Freeman Administration and the Poor," *Agricultural History* 45 (January 1971): 21–32.

34. USDA, *Agricultural Statistics, 1948*, pp. 500–501 table 603, *1950*, pp. 530–31 table 614, and *1952*, pp. 585–86 table 631.

35. Allen J. Matusow, *Farm Policies and Politics in the Truman Years*, Harvard Historical Studies, vol. 80 (Cambridge, Mass.: Harvard University Press, 1967), pp. 82–109; W. E. Hamilton and W. M. Drummond, *Wheat Surpluses and Their Impact on Canada–United States Relations* (Washington, D.C.: Canadian-American Committee, sponsored by National Planning Association, U.S.A., and Private Planning Association of Canada, 1959), pp. 18–19; Helen C. Farnsworth, "International Wheat Agreements and Problems, 1949–56," *Quarterly Journal of Economics* 70 (May 1956): 217–48.

36. Hamilton and Drummond, *Wheat Surpluses*, pp. 17–18, 24, 45 appendix table 12; but cf. USDA, *Agricultural Statistics, 1952*, p. 12 table 9, which places the amount somewhat lower; Sherman E. Johnson, "Great Plains Agriculture in Relation to World Markets," GPC, *Proceedings, 1962*, p. 10.

37. Computed from Table 1.1; USDA, *Agricultural Statistics, 1952*, p. 12 table 9, and *1962*, p. 12 table 9. See also Farnsworth, "International Wheat Agreements," pp. 217, 226–41, 248; *Dickinson Press*, April 24, 1953; Ovid A. Martin (AP), *Lexington Herald-Leader* (Ky.), February 4, 1968.

38. Computations based on Table 8.1. *See also* USDA, Technical Committee on Grain Exports, *Analysis of Grain Export Programs, A Report*, USDA, *Miscellaneous Publication* no. 905 (Washington, D.C., 1962), p. 7.

39. 68 *U.S. Stat.* 454-59 (July 10, 1954), and 833-64 (August 26, 1954); USDA, *Agricultural Statistics, 1973,* p. 8 table 9; Sorenson and Anderson, *Grain Marketing Operations,* pp. 43-45.

40. Canadian-American Committee, *Wheat Surpluses and the U.S. Barter Program, A Statement* (Washington, D.C.: Sponsored by National Planning Association, U.S.A., and Private Planning Association of Canada, 1960), pp. 1, 3-13; Hamilton and Drummond, *Wheat Surpluses,* pp. 26, 32-33; 72 *U.S. Stat.* 1791 (quoted; September 6, 1958); USDA, *Agricultural Statistics, 1973,* p. 8 table 9, *1986,* p. 11 table 13. Statutory authorization for barter exchanges under the stipulated restraints continued, however. Cf. 104 *U.S. Stat.* 3676-77 (November 28, 1990).

Complaints that U.S. trade under Public Law 480 in effect represented "dumping" were lodged with the State Department by Argentina, New Zealand, Denmark, Mexico, Uruguay, Australia, Burma, Italy, and Peru, as well as Canada, during the 1950s (Dan Morgan, *Merchants of Grain* (New York, N.Y.: The Viking Press, 1979), p. 104).

41. USDA, *Agricultural Statistics, 1973,* p. 8 table 9; and the tables of U.S. wheat and flour exports by country of destination, reported annually in this statistical series; *Dickinson Press,* August 30, 1956, under New Delhi dateline; *Grain Market News,* February 18, June 3, 1966; Leo V. Mayer, "Farm Exports and Soil Conservation," Hadwiger and Talbot, eds., *Food Policy and Farm Programs,* p. 100; Horace J. Davis, "Institutions and Conditions of World Trade," GPC, *Proceedings, 1966,* pp. 50-51. On the development of the Japanese market, cf. Morgan, *Merchants of Grain,* pp. 104-5.

42. *Dickinson Press,* March 12, 1959; *Williston Daily Herald,* June 30, 1959, and June 30, 1960.

43. K. A. Gilles, "Wheat as Human Food," Carle C. Zimmerman and Seth Russell, eds., *Symposium on the Great Plains of North America* (Fargo, N.D.: The North Dakota Institute for Regional Studies, 1967), p. 85; Ivon Ulrey, "The Development of Transportation Rates and Rate Cost Relationships with Particular Reference to the Plains," GPC, *Proceedings, 1964,* pp. 89-97 passim; Leonard W. Schruben, "Grain Market Structure for Foreign and Domestic Trade," ibid., pp. 128-29, 136, 138; Clive R. Harston, *Marketing High Protein Wheat in the Northern Great Plains,* MAES, *Bulletin* no. 527 (Bozeman, Mont., 1957), pp. 33-35; Ted Worrall interview, "Montanans at Work," OH 246, Reel 2, MHS; Everett Snortland, ASCS, Montana State Office, interview with author, November 9, 1984.

44. Table 8.1; USDA, *Agricultural Statistics, 1963,* p. 12 table 12, *1975,* p. 4 table 4; Bernard Brenner (UPI), *Lexington Herald-Leader,* July 3, 1966; Sterling F. Green (AP), *Louisville Courier-Journal* (Ky.), January 16, 1966; Jean M. White (*Los Angeles Times–Washington Post* Service), ibid., October 20, 1966; Ovid A. Martin (AP), ibid., July 17, 1967; 79 *U.S. Stat.* 1199-1206 (November 3, 1965).

45. Carle C. Zimmerman, "Socio-Cultural Changes in the Plains," Zimmerman and Russell, eds., *Symposium,* pp. 207-208; *New York Times* News Service, in *Louisville Courier-Journal,* December 19, 1968; M. L. Upchurch, "Changes in the Economic Climate Affecting Great Plains Agriculture," GPC, *Proceedings, 1967,* pp. 29-34a; L. O. Fine, "Impacts of Change on Land Use and Conservation," ibid., pp. 35-36, reporting Paarlberg speech.

46. 80 *U.S. Stat.* 1526-38 (November 11, 1966; quotations from pp. 1526, 1534-35, 1536).

47. USDA, *Foreign Agriculture* 7 (no. 16; April 21, 1969): 16; Edwin L. Dale, Jr., *New York Times,* January 6, 1967; USDA, *Agricultural Statistics, 1974,* p. 11 table 12.

48. Morgan, *Merchants of Grain,* p. 129.

49. Bruce L. Gardner, *The Governing of Agriculture* (Lawrence, Kans.: Published for the International Center for Economic Policy Studies and the Institute for the

Study of Market Agriculture by the Regents Press of Kansas, 1981), p. 37; *New York Times*, September 11, 1977, July 8, 1985, and November 27, 1990; *Lexington Herald-Leader* (AP), December 8, 1990.

50. Morgan, *Merchants of Grain*, pp. 108–20.

51. Langley and Langley, *State-Level Wheat Statistics*, p. 5 table 1; Lucier, Chesley, and Ahearn, *Farm Income Data*, p. 29 table 8; Norbert A. Dorow, "Possible Implications to the Plains States of Changes in Farm Programs," GPC, *Proceedings, 1973*, p. 15 table 4; Bowers, Rasmussen, and Baker, *History of Agricultural Price-Support*, pp. 27–28; *Lexington Leader* (AP), July 18, 1972.

52. Table 8.1; A. H. Boerma, "The World Could Be Fed," *Journal of Soil and Water Conservation* 30 (January-February, 1975): 4–11 passim; Clayton K. Yeutter, "Food for an Anxious World," ibid., 12–14; Phil Newsom (UPI), *Lexington Leader*, July 5, 1973; Gladwin Hill, "Population Boom and Food Shortage: World Losing Fight for Vital Balance," *New York Times*, August 14, 1974; William Robbins, "Rise in Farm Output Said to Falter as Need Grows," *New York Times*, January 13, 1975; USDA, *Agricultural Statistics, 1976*, p. 5 table 5, *1986*, pp. 4 table 4, and 11 table 13. The next to the last reference reports somewhat larger stocks, but a similar trend for the period.

53. Gerald W. Thomas, "World Hunger: The New Global Challenge," *Journal of Soil and Water Conservation* 30 (January-February, 1975): 8-10; Yeutter, "Food for an Anxious World," pp. 13–14.

54. Quoted remarks from Thomas, "World Hunger," pp. 9, 11.

55. Thomas, "World Hunger," p. 11; Tom Wicker (*New York Times* News Service), *Louisville Courier-Journal*, August 31, 1976.

56. Bowers, Rasmussen, and Baker, *History of Agricultural Price-Support*, p. 33; Gardner, *Governing of Agriculture*, pp. 35–40.

57. Bowers, Rasmussen, and Baker, *History of Agricultural Price-Support*, pp. 29–31; Langley and Langley, *State-Level Wheat Statistics*, pp. 5, 42–45, 56, 58–59, 67–68, 70.

58. Lucier, Chesley, and Ahearn, *Farm Income Data*, pp. 29 table 8, 180 table 47, 220 table 55, 255 table 62; Langley and Langley, *State-Level Wheat Statistics*, pp. 42 table 38, 56 table 52, 67 table 63.

59. Keith F. Myers, "The Plains Today," GPC, *Proceedings, 1974*, pp. 5, 12; regional calculations based on Table 1.1 of this volume.

60. *Farm Real Estate: Historical Series Data, 1950–85*, USDA, ERS, *Statistical Bulletin* no. 738 (Washington, D.C., 1985), pp. 23 table 21, 24 table 22, 41 table 39; Luther Tweeten, "Impacts of Monetary, Fiscal, and International Trade Policies on Great Plains Agriculture," GPC, *Proceedings, 1985*, pp. 24 (quoted), 30–31; John Oster, "Inflation and Its Effect on Agriculture and Cash Flow Problems," GPC, *Proceedings, 1981*, pp. 48–49; *Agricultural Income and Finance, Situation and Outlook Report*, USDA, ERS, AFO-36 (February 1990): 34 table B-1, 35 table B-2.

61. Langley and Langley, *State-Level Wheat Statistics*, p. 5 table 1; USDA, *Agricultural Statistics, 1988*, p. 4 table 5; Lucier, Chesley, and Ahearn, *Farm Income Data*, p. 29 table 8; Bowers, Rasmussen, and Baker, *History of Agricultural Price-Support*, pp. 31–32, 35, 36; *Williston Daily Herald*, January 10, 12, 13, 18, 20, 1978.

62. On the political use of P.L. 480, see Morgan, *Merchants of Grain*, pp. 258–61; on the effect of the 1980–1981 embargo, see Sampson, *Farmland or Wasteland*, pp. 41–43. Computations based on USDA, *Agricultural Statistics, 1975*, p. 5 table 5, *1978*, p. 4 table 6, *1982*, p. 4 table 6, and *1986*, p. 4 table 6.

63. Harston, *Marketing High Protein Wheat*, pp. 14–15; Langley and Langley, *State-Level Wheat Statistics*, pp. 42–44, 56–58, 67–69, tables 38–40, 52–54, 63–65; computations based on data from USDA, *Agricultural Statistics*, tables on "Wheat: Supply and Disappearance, by Classes," for the extended period.

64. Harston, *Marketing High Protein Wheat*, pp. 7–8, 14–15, 63, 80, 87–89, 93, 97–

98; Schruben, "Grain Market Structure," GPC, *Proceedings, 1964*, pp. 137–38; Floyd W. Smith, GPC, *Proceedings, 1976*, pp. 60–70, 133–34; Hamilton and Drummond, *Wheat Surpluses*, p. 29; oral interviews of author with James B. Johnson, Farm Management Specialist, MExtS, and Everett Snortland, November 8, 9, 1984. There was a demand for high protein wheat in South America, Cuba, South Africa, and much of Europe. China, however, preferred soft wheats (Harston, *Marketing High Protein Wheat*, pp. 93, 98).

65. Bowers, Rasmussen, and Baker, *History of Agricultural Price-Support*, p. 40.

66. Keith Schneider, *New York Times*, September 9, 1986; Reed Karaim (Knight-Ridder News Service), *Lexington Herald-Leader*, July 26, 1987; Tweeten, *Impacts of Monetary, Fiscal, and International Trade Policies*, p. 37; computation from USDA, *Agricultural Statistics, 1989*, p. 4 table 6. Twenty-five countries, including former net importers such as Finland, Saudi Arabia, and China, had farm surpluses in 1985 (USDA, *The Second RCA Appraisal: Soil, Water, and Related Resources on Nonfederal Land in the United States, Analysis of Conditions and Trends* [Washington, D.C.: U.S. Government Printing Office, 1989], pp. 145–46).

Scandals, well documented in the mid-seventies, showing short weights and contaminated grain in American shipments contributed to the trade decline. Continued foreign complaints of low quality finally led to passage of the Grain Quality Improvement Act in 1986 (Lowell D. Hill, *Grain Grades and Standards: Historical Issues Shaping the Future* [Urbana, Ill.: University of Illinois Press, 1990], pp. 141–42, 145, 166–69).

67. Langley and Langley, *State-Level Wheat Statistics*, pp. 5, 42, 44, 56, 58, 67, tables 1, 38, 40, 52, 54, 63; Lucier, Chesley, and Ahearn, *Farm Income Data*, p. 29 table 8.

68. Glenn D. Pederson, "Economic Strategies for Coping with Financial Problems Faced by Farmers and Ranchers in the Great Plains," GPC, *Proceedings, 1985*, pp. 45–46, 48–49, 60 n. 3; Tweeten, *Impacts of Monetary, Fiscal, and International Trade Policies*, p. 26 table 2; Glenn Hertzler, Jr., "Farmers Home Administration Policies Which Deal with Control of Inflation in Great Plains Agriculture," GPC, *Proceedings, 1981*, pp. 56–57; [USDA, ERS] *Agricultural Income and Finance*, AFO-36 (February 1990): 30. On the role of the Farm Credit System, see above, p. 207.

69. Langley and Langley, *State-Level Wheat Statistics*, pp. 42 table 38, 56 table 52, 67 table 63; Teigen and Singer, *Weather in United States Agriculture*, pp. 145, 177, 205; *Lexington Herald-Leader*, June 21, 1989; USDA, ERS, *Agricultural Income and Finance*, AFO-30 (September, 1988): 13 table 5, AFO-33 (May 1989): 8 table 1, 9; AFO-38 (August 1990): 15; Lucier, Chesley, and Ahearn, *Farm Income Data*, pp. 180 table 47, 220 table 55, 255 table 62.

Drought was identified as damaging to American agriculture nationally in 1955, 1981, 1984, and 1988 (William Robbins, *New York Times*, national edn., August 7, 1988). Neither of the first two cited years was marked by severe drought in the northern Plains. Montana experienced greatly reduced wheat production in 1985, perhaps as a consequence of short rainfall in 1983 and 1984, but neither of the Dakotas recorded low yields through the period. Only the drought in 1988 seriously affected wheat production throughout the region during the eighties.

70. Bowers, Rasmussen, and Baker, *History of Agricultural Price-Support*, pp. 40, 41, 43.

71. Louis Uchitelle, "As Grain Reserves Drop, Prices May Rise," *New York Times*, May 31, 1990; USDA, *Agricultural Statistics . . . 1989*, pp. 12–14 tables 14–16; and Table 8.1 of this volume. See also Lester R. Brown, "Securing Food Supplies," Lester R. Brown and others, *State of the World 1984: A Worldwatch Institute Report on Progress*

toward a Sustainable Society (New York, N.Y.: W. W. Norton & Co., 1984), pp. 188–92. The rivalry for the Algerian trade was so intense that the United States reportedly subsidized a recent sale to that traditional market of the European Economic Community by nearly 50 percent of the purchase price (*London Economist*, August 24, 1991).

72. Mike Robinson (AP), *Lexington Herald-Leader*, June 5, 1988; also ibid., December 20, 1988, January 14, 1989, and March 27, 1990; *New York Times*, May 31, 1990; 104 *U.S. Stat.* 3633 (November 28, 1990).

73. 104 *U.S. Stat.* 3382–85, 3390, 3633–45, 3670.

74. Earl O. Heady and John F. Timmons, "United States Land Needs for Meeting Food and Fiber Demands," *Journal of Soil and Water Conservation* 30 (January-February, 1975): 16 (quoted)–17.

75. See Morgan, *Merchants of Grain*, pp. 227–30, 345–55 passim.

CHAPTER NINE: POLICY DILEMMAS

1. Don F. Hadwiger and Ross B. Talbot, eds., *Food Policy and Farm Programs, Proceedings of the Academy of Political Science* 34 (no. 3; New York, N.Y.: 1982): 10 (quoted).

2. On the American Agricultural Movement, see Gilbert C. Fite, *American Farmers: The New Minority* (Bloomington, Ind.: Indiana University Press, 1981), pp. 209–21 (quotations on p. 216).

3. Warren R. Bailey and D. C. Myrick, "Alternative Farming Programs—Effects on Individual Farms," Great Plains Agricultural Council (hereafter cited as GPC), *Proceedings, 1973*, pp. 63/12–63/14. In the central Plains Kansas and Colorado would have shifted production from wheat to sorghum if wheat prices had declined to a feed equivalent.

4. Gary Lucier, Agnes Chesley, and Mary Ahearn, *Farm Income Data: A Historical Perspective*, USDA, ERS, *Statistical Bulletin* no. 740 (Washington, D.C., 1986), pp. 180 table 47, 220 table 55, 255 table 62; Norbert A. Dorow, "Possible Implications to the Plains States of Changes in Farm Programs" GPC, *Proceedings, 1973*, pp. 9 table 1, 10 tables 2 and 3.

5. Lester R. Brown and others, *State of the World 1984: A Worldwatch Institute Report on Progress toward a Sustainable Society* (New York, N.Y.: W. W. Norton and Co., 1984), pp. 62, 185 (quoted); Earl O. Heady, "Trade-Offs among Soil Conservation, Energy Use, Exports, and Environmental Quality," Harold G. Halcrow, Earl O. Heady, and Melvin L. Cotner, eds., *Soil Conservation, Policies, Institutions, and Incentives* (Ankeny, Iowa: Published for North Central Research Committee III: Natural Resources Use and Environmental Policy, by Soil Conservation Society of America, 1982), p. 268; Ann Crittenden, "Soil Erosion Threatens U.S. Farm Output," *New York Times*, October 26, 1980.

6. R. Burnell Held and Marion Clawson, *Soil Conservation in Perspective* (Baltimore, Md.: Published for Resources for the Future, Inc., by Johns Hopkins University Press, 1965), pp. 58 (quoted), 70 (quoted)–74 (quoted), 75 (quoted) et passim.

7. 86 *U.S. Stat.* 657–77, 816–903 (August 30, October 18, 1972); Harold M. Price, "New Directions in Land-Use Planning," reprinted from *Montana Business Quarterly*, Summer, 1973, GPC, *Proceedings, 1973*, pp. 3, 6–7; Maurice Baker, "New Developments in Land-Use Legislation," GPC, *Proceedings, 1974*, pp. 57–61.

8. George R. Dawson, comments, GPC, *Proceedings, 1973*, pp. 90–91, 93 (all quoted).

9. U.S. House of Representatives, 94th Cong., 1st sess., *Hearings before the Subcommittee on Energy and the Environment of the Committee on Interior and Insular Affairs . . . on H.R. 3510 and Related Bills to Encourage Conservation of Natural Resources . . . March 17, 18, 24, 25, and April 14, 1975,* Serial no. 94-7 (Washington, D.C., 1975), pp. 172, 266–68, 351–55, 370–74, 574–76.

10. Leo V. Mayer, "Farm Exports and Soil Conservation," Hadwiger and Talbot, eds., *Food Policy and Farm Programs,* p. 102; R. Neil Sampson, *Farmland or Wasteland: A Time to Choose, Overcoming the Threat to America's Farm and Food Future* (Emmaus, Pa.: Rodale Press, 1981), p. 78; Raymond D. Vlasin, "Food Production and Its Implications for Resource Conservation," *Journal of Soil and Water Conservation* 30 (January-February, 1975): 2–4 (quoted); Heady, "Trade-Offs," p. 266.

The premise that land formerly retired was at this time being brought back into production in the northern Plains is questionable. There the land in farms had increased 79 percent between 1929 and 1969 but had actually decreased by 4 percent between 1969 and 1982. The acreage in cropland had decreased in both periods, 7 percent during the first forty years and 1 percent during the last thirteen. The statement might more accurately have noted that the acreage in *harvested* cropland had increased 12 percent during the earlier period but 15 percent during the seventies. Why that acceleration had occurred very largely resulted from falling livestock prices at a time when grain prices were rising, but it was also a consequence of the introduction of dry-land methodology that lessened the practice of alternating summer-fallow. From an environmental standpoint, the development of "no-till" farming reduced the potential for wind erosion. (Computations based on Table 1.1.) On the problems of the livestock industry, see *Williston Daily Herald,* February 14, 22, 1978; *Lexington Herald-Leader* (Ky.), June 10, 1988, June 1, 1989. On the introduction of "no-till" methodology, see below, p. 272.

11. 88 *U.S. Stat.* 476–80 (August 17, 1974): Lawrence W. Libby, "Interaction of RCA with State and Local Conservation Programs," Halcrow, Heady, and Cotner, eds., *Soil Conservation Policies,* pp. 112–13; Sandra S. Batie, "Policies, Institutions, and Incentives for Soil Conservation," ibid., pp. 27–28, 31; Christopher Leman, "Political Dilemmas in Evaluating and Budgeting Soil Conservation Programs: The RCA Process," ibid., pp. 51–52; K. William Easter and Melvin L. Cotner, "Evaluation of Current Soil Conservation Strategies," ibid., pp. 290–94.

12. W. Wendell Fletcher and Emma Blacken, "Land Use Policy," U.S. Senate, 95th Cong. 1st sess., Senate Committee on Interior and Insular Affairs, *Congress and the Nation's Environment: Energy and Natural Resources Actions of the 94th Congress,* Prepared by the Environment and Natural Resources Policy Division, Congressional Research Service, Library of Congress (Committee Print, *Publication* no. 95-5, 1977), pp. 642, 685–89; Leman, "Political Dilemmas," pp. 58–60; Mayer, "Farm Exports," p. 102; *National Agricultural Lands Study: Final Report, 1981* (Washington, D.C., U.S. Government Printing Office, 1981), pp. 13 (quoted), 15, 17, 25–26, 38 (quoted), 52–62. Participating agencies in the study were the Council on Environmental Quality; the federal departments of Agriculture, Commerce, Defense, Energy, Housing, and Urban Development, Interior, State, Transportation, and Treasury; the Environmental Protection Agency; and the Water Resources Council.

Assistant Secretary of Agriculture M. Rupert Cutler, departing from the official version, published a signed article in the *New York Times,* July 1, 1980, entitled "The Peril of Vanishing Farmlands," declaring: "America is on the brink of a crisis in the loss of agricultural land that may soon undermine our ability to produce sufficient food for ourselves and other nations of our hungry world."

13. 95 *U.S. Stat.* 1407–11 (November 18, 1977). Detail of this and the following paragraph drawn from Leman, "Political Dilemmas," pp. 61–68, 72, 82.

14. Leman, "Political Dilemmas," p. 55; Easter and Cotner, "Evaluations of Current Soil Conservation Strategies," pp. 290, 292; David J. Allee, "Implementation of RCA: A Problem Accommodating Economics in Soil and Water Conservation," Halcrow, Heady, and Cotner, eds., *Soil Conservation Policies*, p. 97.

15. USDA, *A National Program for Soil and Water Conservation: 1982 Final Program Report and Environmental Impact Statement* (Washington, D.C.: U.S. Government Printing Office, 1982), p. 43; comments reviewed and the USDA responses, pp. 52–61. For criticism of the role of public participation and specifically of the report's failure to analyze programs, see also Leman, "Political Dilemmas," pp. 63–83 passim; Herbert H. Stoevener, critique in Libby, "Interaction of RCA with State and Local Conservation Programs," p. 129.

16. USDA, *The Second RCA Appraisal: Soil, Water, and Related Resources on Nonfederal Land in the United States, Analysis of Condition and Trends* (Washington, D.C.: U.S. Government Printing Office, 1989), pp. vi, 141–54, esp. p. 151 fig. 82. Note the regional definitions, which in the RCA reports differ from those used generally in this study.

17. Kenneth A. Cook, "Problems and Prospects for the Agricultural Conservation Program," *Journal of Soil and Water Conservation* 36 (January-February, 1981): 27; Sample, *Farmland or Wasteland*, Appendix B, pp. 345–49; National Cattlemen's Association comments in USDA, *National Program for Soil and Water Conservation: 1982*, Appendix A, p. 87; National Association of Conservation Districts, *Soil Degradation: Effects on Agricultural Productivity*, National Agricultural Lands Study, *Interim Report* no. 4 (Washington, D.C.: 1980), p. 9 (quoted); Heady, "Trade-Offs," p. 263, and attached critique by Lloyd K. Fischer, p. 274; William L. Miller, "The Farm Business Perspective and Soil Conservation," Halcrow, Heady, and Cotner, eds., *Soil Conservation Policies*, pp. 153, 158–59; Marion Clawson, "Conserving the Soil," Hadwiger and Talbot, eds., *Food Policy and Farm Programs*, pp. 97–98.

18. USDA, *Second RCA Appraisal*, pp. 27, 32 figs. 23 and 24, 33, 218–219 appendix table 9, 224–25 appendix table 12; USDA, *A National Program for Soil and Water Conservation: The 1988–97 Update* (Washington, D.C., 1989), p. 10. Cf. the wind erosion data reported during the recent drought, *New York Times*, national edn., August 7, 1988.

19. Computations based on data in Table 9.1. The distinction between a high rate of erodibility and a low rate of erosion was specifically noted relative to the northern Plains and explained as "primarily from differences in the crops and types of farming operation typical of the regions" (USDA, *Second RCA Appraisal*, pp. 28–30 [quoted]).

20. USDA, *National Program, 1982*, pp. 25–27, 45.

21. USDA, *National Program, 1982*, Appendix A, pp. 66, 84–87 (quoted), 89–90 (quoted), 135–36, 143, 148.

22. 95 *U.S. Stat.* 1328–45 (December 22, 1981). The Special Areas Program was not funded and not implemented (Emma R. Corcoran, SCS, to author, January 28, 1991).

23. Anthony Grano and others, *Analysis of Policies to Conserve Soil and Reduce Surplus Crop Production*, USDA, ERS, *Agricultural Economic Report* no. 534 (Washington, D.C., 1985), p. 8 table 3. The options were defined on p. vi. "Highly erodible" soils had an "Erodibility Index" of 15 or greater; EI rates from 5 to 8 were "Moderate"; and those from 8 to 15 were "High" USDA, *Second RCA Appraisal*, p. 33.

24. Grano and others, *Analysis of Policies to Conserve Soil*, p. 11 table 7, p. 16 table 12. The difference centers in the lesser cost for long-term rental, with presumably a one-time shared-cost payment for establishing grass cover, compared to annual pay-

ments for crop diversion or "set-aside," entailing also repeated payments for "conservation practices" on the idled land.

25. 99:2 *U.S. Stat.* 1378–95, 1454 (December 23, 1985) (quotations on pp. 1392, 1393, 1454).

26. 99:2 *U.S. Stat.* 1504–07, 1509–16. On the considerations involved in the secretary's formulation of the choices on definition of high erodibility, acreage allotment accorded individual regions, states, or districts, and bid selection criteria, see Michael B. Dicks and Katherine Reichelderfer, *Issues in Agricultural Policy: Choices for Implementing the Conservation Reserve*, USDA, ERS, *Agricultural Information Bulletin* no. 507 (Washington, D.C., 1987).

27. 99:2 *U.S. Stat.* 1506–07, 1509 (both quoted). Under the legislation cropland classified by the SCS as classes IV, VI, VII, or VIII was to be considered "highly erodible." The Great Plains Agricultural Council had announced completion in 1972 of base mapping of the Plains according to SCS land capability classes (GPC, *Proceedings, 1972*, p. 107, but see pp. 267–68 in this volume).

28. Leman, "Political Dilemmas," p. 83; E. Wayne Chapman, "Rationale and Legislation for the Creation of the Conservation Reserve Program," John E. Mitchell, ed., *Impacts of the Conservation Reserve Program in the Great Plains, Symposium Proceedings, September 16–18, 1987, Denver, Colorado*, USDA, Forest Service, *General Technical Report* RM-158 (Fort Collins, Col.: Rocky Mt. Forest and Range Experiment Station, 1987), p. 13 (quoted). Secretary Bergland was reported as being particularly concerned about the effect of American "set-asides" on world needs (Raymond F. Hopkins, "Food Policymaking," Hadwiger and Talbot, eds., *Food Policy and Farm Programs*, p. 17).

On the needed research basis for the program, see Pierre R. Crosson, "Environmental Considerations in Expanding Agricultural Production," *Journal of Soil and Water Conservation* 30 (January-February 1975): 27–28; Pierre R. Crosson, ed., *The Cropland Crisis: Myth or Reality?* (Baltimore, Md.: Johns Hopkins University Press for Resources for the Future, 1982), pp. 7–16, 18; Robert N. Shulstad and Ralph D. May, "Conversion of Noncropland to Cropland: The Prospects, Alternatives, and Implications," *American Journal of Agricultural Economics* 62 (December 1980): 1082; Allee, "Implementation of RCA," pp. 107–8 and comment by Roger W. Strohbehn, p. 110; Easter and Cotner, "Evaluation of Current Soil Conservation Strategies," pp. 297–300; Sterling Brubaker and Emery N. Castle, "Alternative Policies and Strategies to Achieve Soil Conservation," Halcrow, Heady, and Cotner, eds., *Soil Conservation Policies*, pp. 311–12; John E. Mitchell and Gary R. Evans, "A Prospectus for Research Needs Created by Passage of the Conservation Reserve Program," Mitchell, ed., *Impacts of the Conservation Reserve Program . . . Symposium*, pp. 128–34.

29. See Neil W. Johnson, *Farm Adjustments in Montana: Graphic Supplement* (Washington, D.C.: USDA in cooperation with MAES and with the assistance of WPA, 1940), passim; John R. Lacey, "Status of Plowout in Montana from an Extension Viewpoint," GPC, *Proceedings, 1983*, pp. 65–66. Lacey did not identify the base of his percentages, but his source citation was the RCA, a cropland survey. To have rested the calculations on land area or land in farms would have produced far greater distortions.

30. See p. 201; USDA, *Second RCA Appraisal*, pp. 158, 160–61; USDA, *National Program for Soil and Water Conservation: 1988–97 Update*, p. 11.

31. On the estimated cost of the program, see W. A. Laycock, "History of Grassland Plowing and Grass Planting on the Great Plains," Mitchell, ed., *Impacts of the Conservation Reserve Program*, p. 7; and cf. USDA, ERS, *Farm Real Estate: Historical Series Data, 1950–85, Statistical Bulletin* no. 738 (Washington, D.C., 1985), p. 41. Proce-

dural detail related in this and the following two paragraphs is based on Michael R. Dicks, Felix Llacuna, and Michael Linsenbigler, *The Conservation Reserve Program: Implementation and Accomplishments, 1986–87*, USDA, ERS, *Statistical Bulletin* no. 763 (Washington, D.C., 1988), pp. 7–8.

32. USDA, *Second RCA Appraisal*, pp. 41 table 8, 224–25 appendix table 12; James B. Newman, "Overview of the Present Land-Use Situation and the Anticipated Ecological Impacts of Program Implementation," Mitchell, ed., *Impacts of the Conservation Reserve Program*, p. 57 table 2; Linda A. Joyce and Melvin D. Skold, "Implications of Changes in the Regional Ecology of the Great Plains," Mitchell, ed., *Impacts of the Conservation Reserve Program*, p. 118 table 1.

33. C. Tim Osborn, Felix Llacuna, Michael Linsenbigler, *The Conservation Reserve Program: Enrollment Statistics for 1987–88 and Signup Periods 1–7*, USDA, ERS, *Statistical Bulletin* no. 785 (Washington, D.C., 1989), pp. 42, 98 table 8B, and 116–18 table 1. Repeated signups, including three the first year, indicated the official disappointment at the response (Reed Karaim, "Response Light to Federal Plan to Remove Land from Production," Knight-Ridder News Service, *Lexington Herald-Leader*, March 3, 1986; Don Kendall, "U.S. Seeks to Idle More Land in Farm Conservation Plan," Associated Press, *Lexington Herald-Leader*, May 4, 1986).

34. E. T. Bartlett, "Social and Economic Impacts of the Conservation Reserve Program," Mitchell, ed., *Impacts of the Conservation Reserve Program*, pp. 52–53; 104 *U.S. Stat.* 3394–96 (November 28, 1990).

During the crop year 1987 1,294 contracts idled 244,209 base acres of wheat in Montana, 1,849 contracts idled 157,011 acres in North Dakota, and 978 contracts idled 78,612 acres in South Dakota. The number of contracts during the drought year 1988 rose to 1,929, idling 369,569 wheat acres, in Montana; to 3,652, covering 352,665 such acres, in North Dakota; and to 1,322, for 136,395 acres, in South Dakota. Yields on the idled acreage had been low, averaging twenty-one to twenty-four bushels an acre for the wheat land retired in 1987 and twenty-two to twenty-six bushels an acre in 1988.

35. Harry A. Steele, Erling D. Solberg, and Howard L. Hill, in GPC, *Proceedings, 1958*, pp. 40–58, esp. p. 54 table 2; Dov M. Grunschlag, "Administering Federal Programs of Production Adjustment," *Agricultural History* 49 (January 1975): 139–44; Batie, "Policies, Institutions, and Incentives for Soil Conservation," pp. 33–35; Libby, "Interaction of RCA with State and Local Conservation Programs," pp. 116–24; Philip M. Glick, "The Coming Transformation of the Soil Conservation District," *Journal of Soil and Water Conservation* 22 (March-April, 1967): 47; William D. Anderson, Gregory C. Gustafson, and Robert F. Boxley, "Perspectives on Agricultural Land Policy," *Journal of Soil and Water Conservation* 30 (January-February, 1975): 41; John Opie, *The Law of the Land: Two Hundred Years of American Farmland Policy* (Lincoln, Nebr.: University of Nebraska Press, 1987), pp. 179–80; Clawson, "Conserving the Soil," pp. 90–91; Sampson, *Farmland or Wasteland*, pp. 290–91.

36. Libby, "Interaction of RCA with State and Conservation Programs," p. 125; Leman, "Political Dilemmas," pp. 62–66, 68–69, 81.

37. USDA, *Second RCA Appraisal*, p. 36 table 5; Miller, "Farm Business Perspective," pp. 155–57; Opie, *Law of the Land*, pp. 193–99; Sampson, *Farmland or Wasteland*, pp. 14, 48–61, 81–94, 292; Crosson, "Environmental Considerations in Expanding Agricultural Production," pp. 23–28; Mayer, "Farm Exports and Soil Conservation," pp. 99–111 passim; "Plan for Preserve Aired in Montana," *New York Times*, February 22, 1987; Deborah Epstein Popper and Frank J. Popper, "The Great Plains from Dust to Dust," *Planning*, December 1987, pp. 11–18 (quotation, p. 17); Anne Matthews, "The Poppers and the Plains," *New York Times Magazine*, June 24, 1990, pp. 24–26, 41,

48–49, 53; "Hugh Sidey's America: Where the Buffalo Roamed," *Time,* September 24, 1990, pp. 53–54, 56; Tad Szulc, "One Person Too Many?" *Parade Magazine,* April 29, 1984, pp. 16–17, 19; Brown, "Securing Food Supplies," pp. 183 table 10-4, 184; Lester R. Brown and others, *State of the World 1985: A Worldwatch Institute Report on Progress toward a Sustainable Society* (New York, N.Y.: W. W. Norton and Co., 1985), p. 24; John Tierney, "Betting the Planet," *New York Times Magazine,* December 2, 1990, pp. 52–53, 74, 76, 78, 80–81; Don Hinrichsen and Alex Marshall, "Population and the Food Crisis," *Populi: The Journal of the United Nations Population Fund* 18 (no. 2; June 1991): 26.

38. USDA, *Agricultural Statistics, 1989,* p. 5 table 7; *1987 Census of Agriculture,* vol. 1, pt. 26, pp. 204–11, pt. 34, pp. 204–11, pt. 41, pp. 212–20—table 7.

39. USDA, ERS, *Economic Indicators of the Farm Sector: Production and Efficiency Statistics, 1987,* ECIFS 7-5 (Washington, D.C., 1989), pp. 30 table 28, 31 table 29; USDA, ERS, *Economic Indicators of the Farm Sector: Costs of Production, Major Field Crops, 1988,* ECIFS 8-4 (Washington, D.C., 1990), pp. 69–77 tables 51–59; Jim Langley and Suchada Langley, *State-Level Wheat Statistics, 1949–88,* USDA, ERS, *Statistical Bulletin* no. 779 (Washington, D.C., 1989), pp. 42 table 38, 56 table 52, 67 table 63.

40. USDA, ERS, *Economic Indicators . . . Production and Efficiency Statistics, 1987,* pp. 22 table 17 and 25 table 22; Norman A. Berg, "Potential for All-Out Production and the Resulting Critical Problems," GPC, *Proceedings, 1975,* p. 24; B. W. Greb, D. E. Smika, and J. R. Welsh, "Technology and Wheat Yields in the Central Great Plains," *Journal of Soil and Water Conservation* 34 (November-December 1979): 264–68; Sampson, *Farmland or Wasteland,* pp. 232–36; D. G. Hanway, "Enroute to Conservation Production Farming Systems," GPC, *Proceedings, 1982,* pp. 49–52; Gail A. Wicks, "Update on Ecofallow in the Winter Wheat-Sorghum-Fallow Rotation," ibid., pp. 53–54; C. R. Fenster, "Potential and Problems of Ecofarming in Drier Environments," ibid., pp. 55–58; Roger Schroeder, "Value of Residue," ibid., pp. 60–61; Norman L. Klocke, "Planting Winter Wheat into Wheat Residue—Dilemma or Challenge for Great Plains Wheat Production," ibid., pp. 74–75, 77; Norman J. Rosenberg, "Adaptations to Adversity: Agriculture, Climate, and the Great Plains of North America," *Great Plains Quarterly* 6 (Summer 1986): 206; Crosson, ed., *Cropland Crisis,* p. 15.

41. A. D. Halvorson and A. L. Black, "Saline-Seep Development in Dryland Soils of Northeastern Montana," *Journal of Soil and Water Conservation* 29 (no. 2; 1974): 77–81; A. L. Black, T. L. Brown, A. D. Halvorson, and F. H. Siddaway, "Dryland Cropping Strategies for Efficient Water Use to Control Saline Seeps in the Northern Great Plains, U.S.A.," *Agricultural Water Management* 4 (1981): 295–311; P. L. Brown, A. D. Halvorson, F. H. Siddaway, H. F. Mayland, and M. R. Miller, *Saline-Seep Diagnosis, Control and Reclamation,* USDA, ARS, *Conservation Research Report* no. 30 (Washington, D.C., 1983), pp. 6 fig. 4, 21; Hayden Ferguson and Tom Bateridge, "Salt Status of Glacial Till Soils of North Central Montana as Affected by the Crop-Fallow System of Dryland Farming," *Soil Science Society of America Journal* 46 (July-August, 1982): 807–10; "Farm Policy Is Destroying Cropland Study Warns," *New York Times,* February 11, 1979; K. J. Dalsted, B. K. Worster, and L. J. Brun, "Interpretation of Remote Sensing Products to Detect Saline Seeps," GPC, *Proceedings, 1979,* p. 89; J. L. Heilman and D. G. Moore, "Thermography for Estimating Soil Water Content," ibid., p. 91; Paul M. Seevers, "Farmer Use of Remote Sensing," ibid., p. 95. On recent research trends see John F. Stone and Wayne O. Willis, eds., *Plant Production and Management under Drought Conditions, Papers Presented at the Symposium, 4–6 October, 1982, Held at Tulsa . . . , Agricultural Water Management* 7 (1983): 89, 97–99, 143–55, 265–80.

42. USDA, *Second RCA Appraisal,* pp. 57, 248–49 appendix table 30.

43. See Table 1.1; Sampson, *Farmland or Wasteland,* pp. 177, 180–81. USDA restrictions on cost-sharing for production-oriented measures may retard further de-

velopment of the long-awaited breakthroughs in hybridization of wheat, as well as other technological advances such as have heretofore provided the basis for American leadership in agricultural productivity. Cf. Earl O. Heady, "The Adequacy of Agricultural Land: A Demand-Supply Perspective," in Crosson, ed., *Cropland Crisis*, pp. 40–42; Sterling Brubaker, "Agricultural Land: Policy Issues and Alternatives," in ibid., p. 219; Sylvan Wittwer, "New Technology, Agricultural Productivity, and Conservation," Halcrow, Heady, and Cotner, eds., *Soil Conservation Policies*, pp. 205, 208; USDA, *Second RCA Appraisal*, p. 146; Wayne D. Rasmussen, "History of Soil Conservation, Institutions," Halcrow, Heady, and Cotner, eds., *Soil Conservation Policies*, p. 10; Sampson, *Farmland or Wasteland*, p. 292.

44. See Table 6.2.

45. See Table 9.2. Production percentages were computed from state reports for 1987, *USDA Agricultural Statistics, 1988*, p. 5, table 7, and Langley and Langley, *State-Level Wheat Statistics*, pp. 43, 45, 57, 59, 68, 70, tables 39, 41, 53, 55, 64, 66.

46. Leman, "Political Dilemmas," pp. 72 (quoting), 85; USDA, *National Program for Soil and Water Conservation*, pp. 156 (quoted)–57, Appendix C; USDA, *Second RCA Appraisal*, p. 143. See also Allee, "Implementation of RCA," p. 108.

47. USDA, *Second RCA Appraisal*, p. 33 table 4; p. 234 in this volume; Sampson, *Farmland or Wasteland*, p. 292 (quoted).

48. Takashi Sato, "Finding the Balance," *Populi: The Journal of the United Nations Population Fund* 18 (no. 2; June, 1991): 48; Don Hinrichsen and Alex Marshall, "Population and the Food Crisis," ibid., p. 26; Nafis Sadik, "Choice or Chance?" ibid., pp. 5–6; USDA, *Second RCA Appraisal*, pp. 12–13 (quoted), 155; Martin E. Abel, "Growth in Demand for U.S. Crops and Animal Production by 2005," and the critique by Leroy Quance in Crosson, ed., *Cropland Crisis*, pp. 65–69, 88, 91.

49. USDA, *Second RCA Appraisal*, pp. 142 table 49, 213 appendix table 6; Laycock, "History of Grassland Plowing," p. 7 (quoted). The *Second RCA Appraisal* listed for Montana and the Dakotas about 3,182,000 acres with "high" and 12,079,000 acres with "medium" potential for conversion from nonfederal pasture, range, or forest to cropland.

50. USDA, *Second RCA Appraisal*, p. 143 (quoted).

51. USDA, *Second RCA Appraisal*, p. 201 appendix table 3a; computations from *1982 Census of Agriculture*, vol. 1, pt. 26, pp. 120–27, county table 1.

BIBLIOGRAPHICAL NOTE

This bibliography indicates the kinds of materials utilized and their value to the analysis; it encompasses only a fraction of the titles consulted. The reader is urged also to check the Notes for other authors cited.

SOURCE WORKS

Among the most helpful materials available are the long-term statistical compilations by the Economic Research Service of the United States Department of Agriculture, data that until very recently were available chiefly as highly generalized summations or fragmentary references in newspapers. For the early years of the study, runs of price and production data were pieced together from the USDA *Yearbook* publications and *Agricultural Statistics* with some supplementation from census reports and the Bureau of the Census, *Historical Statistics of the United States, 1789–1945: A Supplement to the Statistical Abstract of the United States* (Washington, D.C.: U.S. Government Printing Office, 1949). An experiment station bulletin, *Annual and Seasonal Precipitation at Six Representative Locations in Montana*, MAES, *Bulletin* no. 447 (Bozeman, Mont., 1947), provided some local differentiation. Publications now include Lloyd D. Teigen and Florence Singer, *Weather in United States Agriculture: Monthly Temperature and Precipitation by State and Farm Production Region, 1950–86*, USDA, ERS, *Statistical Bulletin* no. 765 (Washington, D.C., 1988); Jim Langley and Suchada Langley, *State-Level Wheat Statistics, 1949–88*, USDA, ERS, *Statistical Bulletin* no. 779 (Washington, D.C., 1989); Gary Lucier, Agnes Chesley, and Mary Ahearn, *Farm Income Data: A Historical Perspective*, USDA, ERS, *Statistical Bulletin* no. 740 (Washington, D.C., 1986); *Farm Real Estate: Historical Series Data, 1950–85*, USDA, ERS, *Statistical Bulletin* no. 738 (Washington, D.C., 1985); and Robert G. McElroy and Cole Gustafson, *Costs of Producing Major Crops, 1975–81*, USDA, ERS, *ERS Staff Report* no. AGES 850329 (Washington, D.C., 1985), supplemented since the early eighties by the serial USDA, ERA, *Economic Indicators of the Farm Sector: Costs of Production.*

The vast number of experiment station bulletins, too numerous to list here separately, were invaluable, especially for reporting varietal development, testing of tillage methods, and research on cost of production. Particularly helpful were for Montana the county soil surveys by L. F. Gieseker from the twenties into the early fifties and the reports by Roland R. Renne on farm and county finance through the thirties; for North Dakota, the early type-of-farming studies by Rex E. Willard, the work by E. A. Willson as an early rural sociologist, and studies in agricultural economics by Rainer Schickele and the statistical handbooks by Baldur H. Kristjanson and C. J. Heltemes in the fifties; for South Dakota, the farm credit studies by Gabriel Lundy and Henry A. Steele and the focus on town-country community adjustments in the research of Douglas Chittick, John P. Johansen, W. F. Kumlien, and Paul H. Landis. The numerous surveys during the thirties by the Federal Emergency Relief Administration, the Resettlement Administration, and the Division of Social Research of the WPA reinforced these sociological and economic studies with particular relevance to the needs for relief. Among the most useful were P. G. Peck and M. C. Forster, *Six Rural Problem Areas: Relief—Resources—Rehabilitation*, U.S. FERA, Division of Research Statistics, *Research Monograph* no. 1 (Washington, D.C., 1935); Francis D. Cronin and Howard W. Beers, *Areas of Intense Drought Distress, 1930–1936*, U.S. WPA, Division of Social Research, *Research Bulletin*, series 5, no. 1 (Washington, D.C., 1937); Berta Asch and A. R. Mangus, *Farmers on Relief and Rehabilitation*, U.S. WPA, Division of Social Research, *Research Monograph* no. 8 (Washington, D.C.: U.S. Government Printing Office, 1937); R. S. Kifer and H. L. Stewart, *Farming Hazards in the Drought Area*, U.S. WPA, Division of Social Research, *Research Monograph* no. 16 (Washington, D.C.: U.S. Government Printing Office, 1938; H. L. Stewart, *Natural and Economic Factors Affecting Rural Rehabilitation Problems in Southwestern North Dakota (as Typified by Hettinger County)*, Resettlement Administration, *Research Bulletin* K-4 (Washington, D.C., mimeographed, 1936); and H. L. Stewart, *Natural and Economic Factors Affecting Rural Rehabilitation Problems in Northwestern North Dakota and Northeastern Montana (as Typified by Divide County, North Dakota)*, Resettlement Administration, *Research Bulletin* no. K-7 (Washington, D.C., mimeographed, 1936).

Federal and state statutes, congressional hearings, and agency reports, both annual and special committee reports—some as chronological runs for long periods, others as topical searches—were basic research. The following documents, with related supplements, are central to the policy discussions of this study:

U.S. National Resources Board. *A Report on National Planning and Public Works in Relation to Natural Resources and Including Land Use and Water Resources with Findings and Recommendations, December 1, 1934*. Washington, D.C.: U.S. Government Printing Office, 1934.

U.S. National Resources Board, Land Planning Committee. *Maladjustments in Land Use in the United States, Supplementary Report*, pt. 6. Washington, D.C.: U.S. Government Printing Office, 1935.

U.S. Great Plains Drought Area Committee. *Report.* Washington, D.C., 1936.

U.S. Great Plains Committee. *The Future of the Great Plains, Report.* Washington, D.C.: U.S. Government Printing Office, 1936.

Missouri Basin Interagency Committee. *The Missouri: A Great River Basin of the United States, Its Resources and How We Are Using Them.* Public Health Service Publication no. 604. Washington, D.C.: U.S. Government Printing Office, 1958.

USDA. *A National Program for Soil and Water Conservation: 1982 Final Program Report and Environmental Impact Statement.* Washington, D.C.: U.S. Government Printing Office, 1982.

———. *The Second RCA Appraisal: Soil, Water, and Related Resources on Nonfederal Land in the United States, Analysis of Condition and Trends.* Washington, D.C., U.S. Government Printing Office, 1989.

The Great Plains Agricultural Advisory Council *Proceedings*, were surveyed as a file running annually from 1947 through 1985, affording the most complete review available on the regional planning process throughout the postwar period. Organization of the programs reveals the shifting focus of problems year by year, and routine reports extending through series of years summarize crop and range conditions, plow-up of "restoration" lands, growing use of fertilizers, and other similar developments. Brief histories of the Great Plains Council through the early years are to be found in A. M. Eberle, "Achievements and Activities of the Great Plains Council, 1935–1953," GPC, *Proceedings . . . August 2–5, 1953*, pp. 36–37; and Henry J. Tomasek, "The Great Plains Agricultural Council" (Ph.D. diss., University of Chicago, 1959).

Newspapers also afforded sweeping review, more varied although not always as relevant to the purposes of the inquiry. They included publication of county agent reports, information on localized adjustment activities and program response, as well as surveys of events related to community development. Files were searched from settlement through 1925 for the *Fergus County Argus* (Lewistown), the *Hardin Tribune,* the *Havre Plaindealer* or *Havre Promoter,* the *Scobey Sentinel* or the *Plentywood Herald,* and the *Wibaux Pioneer*—all in Montana; the *Dickinson Recorder-Post* and the *Williston Herald* in North Dakota; and the *Kadoka Press* in South Dakota. Microfilm runs continued the coverage for central Montana through the *Lewistown Democrat-News,* with occasional issues of the *Fergus County Argus,* through 1950; for the Havre area through the *Hill County Democrat,* succeeded by the *Hill County Journal,* through 1937; for northwestern North Dakota through the *Williston Herald* through 1960; and for southwestern North Dakota through the *Dickinson Press* through 1959 and the *Adams County Record* (Hettinger) through 1956. Difficulties earlier experienced in gaining library access to South Dakota journals were unfortunately repeated in obtaining microfilm through interlibrary loan, although master copies are listed as available. The coverage given to neighboring Perkins and Corson counties through the *Adams County Record* in North Dakota was consequently of particular value.

Manuscript research was centered on the USDA files covering "State and Local Planning," RG 83, at the National Archives; the regional files relating to reclamation program development at the Federal Records Center in Denver, Colorado; miscellaneous files at the Montana Historical Society, most notably the oral history series taped by Laurie K. Mercier on "Montanans at Work"; and scattered materials in the library of Montana State University. Manuscript searches as published by William D. Rowley in *M. L. Wilson and the Campaign for the Domestic Allotment* (Lincoln, Nebr.: University of Nebraska Press, 1970) and conversations reported by Russell Lord in *The Wallaces of Iowa* (Boston, Mass.: Houghton Mifflin Co., 1947) provided insight on the views of M. L. Wilson. Elmer A. Starch, in a letter dated November 21, 1977, very generously supplied a nine-page manuscript statement on the locally initiated land adjustment effort in Montana during the thirties. The recently announced acquisition of the papers of Thomas D. Campbell and the Campbell Farming Corporation by the Montana Historical Society offers exciting prospects for much useful material not yet consulted.

Numerous published papers by O. E. Baker, L. C. Gray, Elmer Starch, Rexford Tugwell, and M. L. Wilson have been cited and discussed at some length in the volume and notes, as indexed. Sherman E. Johnson's *From the St. Croix to the Potomac—Reflections of a Bureaucrat* (Bozeman, Mont.: Big Sky Books, 1974) is particularly useful for its reporting of USDA actions and concerns in program development at the operational level during the thirties and at a level which he describes as "second and third tier research" (p. 197) on planning through the forties and fifties. Two documentary compilations also include much relevant material: Richard Lowitt and Maurine Beasley, eds., *One Third of a Nation: Lorena Hickok Reports on the Great Depression* (Urbana, Ill.: University of Illinois Press, 1981); and Edgar B. Nixon, comp. and ed., *Franklin D. Roosevelt and Conservation, 1911–1945*, 2 vols. (New York, N.Y.: Arno Press, 1972).

SECONDARY WORKS

A few works cover relatively broad periods of this study. Excellent state histories provide general background: Michael P. Malone and Richard B. Roeder, *Montana: A History of Two Centuries* (Seattle, Wash.: University of Washington Press, 1976); Elwyn B. Robinson, *History of North Dakota* (Lincoln, Nebr.: University of Nebraska Press, 1966); and Herbert S. Schell, *History of South Dakota* (Lincoln, Nebr.: University of Nebraska Press, 1961).

Douglas E. Bowers, Wayne D. Rasmussen, and Gladys L. Baker, *History of Agricultural Price Support and Adjustment Programs, 1933–84*, USDA, ERS, *Agricultural Information Bulletin* no. 485 (Washington, D.C., 1984) summarizes the major provisions of this legislation, noting particularly the major changes through the years.

Charles M. Studness, "Development of the Great Plains Farming Economy: Frontier to 1953" (Ph.D. diss., Columbia University, 1963), focused on the economic rationale for expansion through that period, concluding that "a long-run rising trend of the revenue-to-cost ratio" dominated the development (p. 2). He found that because of the high social and economic cost of returning cropland to grass, there was little tendency to retire acreage even during periods of depression. Such periods were marked, instead, by stagnation, as expansion evolved in a ratchet pattern with the acreage remaining in production. The "price-support-acreage-restriction mechanism" had made the western extension of wheat acreage permanent as farmers sought to increase their efficiency through mechanization and expanded operations.

Secondary works applicable to this study for the most part, however, relate to topical segments. For the early years (Part One) those most relevant, apart from the Rowley and Lord volumes heretofore mentioned, were Hiram M. Drache, *Beyond the Furrow, Some Keys to Successful Farming in the Twentieth Century* (Danville, Ill.: The Interstate Printers and Publishers, 1976) on Thomas D. Campbell's operations; Thomas D. Isern, *Custom Combining on the Great Plains: A History* (Norman, Okla.: University of Oklahoma Press, 1981); James H. Shidler, *Farm Crisis: 1919–1923* (Berkeley, Calif.: University of California Press, 1957); and an article by Albert Z. Guttenberg, "The Land Utilization Movement of the 1920s," *Agricultural History* 50 (July 1976): 477–90.

On the Depression and related adjustment of the thirties (Part Two) Richard S. Kirkendall, *Social Scientists and Farm Politics in the Age of Roosevelt* (Columbia, Mo.: University of Missouri Press, 1966), offers particularly useful insights. Richard Lowitt, *The New Deal and the West* (Bloomington, Ind.: Indiana University Press, 1984) is strong in noting the role of interdepartmental rivalry as a factor in program development. In emphasizing the geographic division effected as a settlement of this controversy, however, the bulk of the study relates to the reclamation work of the Department of the Interior rather than the activities of independent agencies such as the Resettlement and Farm Security Administrations and the work of the Department of Agriculture through the AAA or the SCS in the Plains.

Several monographs on the specific programs of the AAA provide useful background: Murray R. Benedict, *Farm Policies of the United States: A Study of Their Origins and Development* (New York, N.Y.: The Twentieth Century Fund, 1953); Edwin G. Nourse, Joseph S. Davis and John D. Black, *Three Years of the Agricultural Adjustment Administration* (Washington, D.C.: The Brookings Institution, 1937); Van L. Perkins, *Crisis in Agriculture: The Agricultural Adjust-*

ment and the New Deal, 1933, University of California Publications in History, vol. 81 (Berkeley, Calif.: University of California Press, 1969); and an article by C. Roger Lambert, "Want and Plenty: The Federal Surplus Relief Corporation and the AAA," *Agricultural History* 46 (July 1972): 390–400. Leonard J. Arrington has compiled the most complete accounting of the New Deal expenditures for relief on a state basis in "The New Deal in the West: A Preliminary Statistical Inquiry," *Pacific Historical Review* 38 (August 1969): 311–16, and "Western Agriculture and the New Deal," *Agricultural History* 44 (October 1970): 337–53.

On the agricultural discontent that precipitated relief efforts, see Theodore Saloutos and John D. Hicks, *Agricultural Discontent in the Middle West, 1900–1939* (Madison, Wisc.: University of Wisconsin Press, 1951); Lowell K. Dyson, *Red Harvest: The Communist Party and American Farmers* (Lincoln, Nebr.: University of Nebraska Press, 1982); Larry Remele, "The North Dakota Farm Strike of 1932," *North Dakota History* 41 (Fall 1974): 5–18; John E. Miller, "Restrained, Respectable Radicals: The South Dakota Farm Holiday," *Agricultural History* 59 (July 1985); 429–47; Charles Vindex, "The Agonizing Years: Radical Rule in Montana," *Montana: The Magazine of Western History* 18 (January 1968): 3–18; William C. Pratt, "Rethinking the Farm Revolt of the 1930s," *Great Plains Quarterly* 8 (Summer 1988): 131–44; Allan J. Mathews, "Agrarian Radicals: The United Farmers League of South Dakota," *South Dakota History* 3 (Fall 1973): 408–21; John L. Shover, *Cornbelt Rebellion: The Farmers' Holiday Association* (Urbana, Ill.: The University of Illinois Press, 1965); Arthur M. Schlesinger, Jr., *The Age of Roosevelt: The Coming of the New Deal* (Boston, Mass.: Houghton Mifflin Co., 1959).

On policy regulating settlement of public lands, the fullest account of federal legislation is Paul W. Gates, *History of Public Land Law Development* (Washington, D.C.: U.S. Government Printing Office, 1968). Discussion of early efforts at state regulatory action is provided by William A. Hartman, "State Policies in Regulating Land Settlement Activities," *Journal of Farm Economics* 13 (April 1931): 259–69; W. A. Hartman, *State Land-Settlement Problems and Policies in the United States*, USDA, *Technical Bulletin* no. 357 (Washington, D.C., 1933). The New Deal land purchase program is treated in L. C. Gray, "Federal Purchase and Administration of Submarginal Land in the Great Plains," *Journal of Farm Economics* 21 (February 1939): 123–31; H. H. Wooten, *The Land Utilization Program, 1934 to 1964: Origin, Development, and Present Status*, USDA, ERS, *Agricultural Economic Report* no. 85 (Washington, D.C.: U.S. Government Printing Office, 1965). Local projects were discussed in James H. Marshall and Stanley Voelker, *Land-Use Adjustment in the Buffalo Creek Grazing District, Yellowstone County, Montana*, USDA, BAE, *Land Economics Report* no. 6 (Washington, D.C.: mimeographed, 1940); M. B. Johnson, "Land Use Adjustment in North Dakota," *Soil Conservation* 5 (March 1940): 221–22, 226; Loyd [sic] Glover, *The Future of Federal Land Use Purchase Projects in South Dakota,* SDAES, *Bulletin* no. 464 (Brookings, S.D., 1957); Roy E. Huff-man and James L. Paschal, "Integrating the Use of Irrigated and Grazing Land in the Northern Great Plains," *Journal of Land and Public Utility Economics* 18 (February 1942): 17–27; and H. L. Lantz, "Readjustment of Population to Land Resources in Northern Montana," *Soil Conservation* 5 (February 1940); 210–13.

Irrigation projects developed as resettlement areas under the land-retire-ment program are described in Gladys R. Costello, "Irrigation History and Resettlement on Milk River Project, Montana," *Reclamation Era* 30 (1940): 136–38, 142, 170–71, 231–33; Ernest Eugene Melvin, "Problems Related to the De-velopment of the Greenfields Irrigation District" (Ph.D. diss., Northwestern University, 1952); Clyde E. Stewart and D. C. Myrick, *Control and Use of Re-sources in the Development of Irrigated Farms: Buffalo Rapids and Kinsey, Montana,* MAES, *Bulletin* no. 476 (Bozeman, Mont., 1951); Marvin P. Riley, W. F. Kum-lien, and Duane Tucker, *Fifty Years Experience on the Belle Fourche Irrigation Project,* SDAES, *Bulletin* no. 450 (Brookings, S.D., 1955); and Roy E. Huffman, *Irrigation Development and Public Water Policy* (New York, N.Y.: Ronald Press Co., 1953). Several MAES *Bulletins,* written by P. L. Slagsvold in the mid-thirties reviewed the status of agriculture on the irrigation projects in Mon-tana unfavorably. The best account found on the resettlement experience is the contemporary, in-house report to the USDA, published many years later, by Marion Clawson, "Resettlement Experience on Nine Selected Resettle-ment Projects," *Agricultural History* 52 (January 1978): 1–92.

Useful critiques of the FSA "rehabilitation-in-place" program are provid-ed in Olaf F. Larson, "Lessons from Rural Rehabilitation Experience," *Land Policy Review* 9 (Fall 1946): 13–18, and Olaf F. Larson, ed., *"Ten Years of Rural Rehabilitation in the United States," Summary of a Report* (Bombay, India: Indian Society of Agricultural Economics, 1950).

Several articles by George S. Wehrwein published in the *Journal of Farm Economics* during the latter half of the thirties developed proposals for rural zoning legislation, but the concept was most fully explored in relation to the northern Plains by R. J. Penn, W. F. Musbach, and W. C. Clark, *Possibilities of Rural Zoning in South Dakota: A Study in Corson County,* SDAES, *Bulletin* no. 345 (Brookings, S.D., 1940). Charles M. Hardin, *The Politics of Agriculture: Soil Conservation and the Struggle for Power in Rural America* (Glencoe, Ill.: The Free Press, 1952) provides a survey of the early efforts to utilize such authority in Colorado. The legal issues involved in land-use regulation were discussed by Edwin E. Ferguson, in "Nation-Wide Erosion Control: Soil Conservation Dis-tricts and the Power of Land-Use Regulation," *Iowa Law Review* 34 (January 1949): 166–87.

The county plannig program was explained by the director of the pro-gram, Bushrod W. Allin, with accompanying commentary, in "County Plan-ning Project—A Cooperative Approach to Agricultural Planning," *Journal of*

Farm Economics 22 (February 1940); 292–316; and by Ellery A. Foster and Harold A. Vogel, "Cooperative Land Use Planning—A New Development in Democracy," USDA, *Farmers in a Changing World, Yearbook of Agriculture: 1940*, pp. 1138–56; and John D. Lewis, "Democratic Planning in Agriculture," *American Political Science Review* 35 (April and June 1941): 244–49, 454–63. Its demise was discussed by Neil C. Gross, "A Post Mortem on County Planning," *Journal of Farm Economics* 25 (August 1943): 652–55, 659; Christiana McFadyen Campbell, *The Farm Bureau and the New Deal: A Study of the Making of National Farm Policy* (Urbana, Ill.: University of Illinois Press, 1962); Grant McConnell, *The Decline of Agrarian Democracy* (Berkeley, Calif.: University of California Press, 1953); and Kirkendall, *Social Scientists and Farm Politics*, ch. 11.

For the post-depression period (Part III) the study written as a project for the Great Plains Agricultural Advisory Council by Howard W. Ottoson, Eleanor M. Birch, Philip A. Henderson, and A. H. Anderson, *Land and People in the Northern Plains Transition Area* (Lincoln, Nebr.: University of Nebraska Press, 1966) affords a comprehensive survey of the eastern borderland section of the region in the late fifties. The authors summarize the agricultural history, adaptive responses to the dry years as evidenced through localized analysis of central Nebraska, and contemporary views on desirable adjustments and the outlook for the future. Socio-economic data on prevalent conditions in the focus area are particularly noteworthy.

Policy in relation to agricultural programming during the war years was analyzed by Bela Gold, *Wartime Economic Planning in Agriculture: A Study in the Allocation of Resources* (New York, N.Y.: Columbia University Press, 1949), with strong criticism on the basis of economic considerations. Walter W. Wilcox, *The Farmer in the Second World War* (Ames, Iowa: The Iowa State College Press, 1947), viewed the effort more favorably, with particular praise for the development of agricultural innovations during the period. While these achievements pointed the way to major agricultural adjustments in the postwar years, they were not generally apparent in the northern Plains until the mid-fifties.

Background surveys on the politics of administrative policy may be found in Allen J. Matusow, *Farm Policies and Politics in the Truman Years*, Harvard Historical Studies, vol. 80 (Cambridge, Mass.: Harvard University Press, 1967); and Edward L. Schapsmeier and Frederick H. Schapsmeier, *Ezra Taft Benson and the Politics of Agriculture: The Eisenhower Years, 1953–1961* (Danville, Ill.: The Interstate Printers and Publishers, 1975).

The above-noted Bowers, Rasmussen, and Baker, *History of Agricultural Price Support and Adjustment Programs* and the GPC, *Proceedings*, remain basic for overview of the policy trends. Bruce L. Gardner, *The Governing of Agriculture* (Lawrence, Kans.: Published for the International Center for Economic Policy Studies and the Institute for the Study of Market Agriculture by The Regents Press of Kansas, 1981), provides an excellent critique of the effects of those policies in application; "It seems likely that the considerable entrepreneurial skills that farmers possess have undergone a substantial shift from outsmarting nature and the markets to outsmarting the government" (p. 31). See also Raymond P. Christensen and Ronald O. Aines, *Economic Effects of Acreage Control Programs in the 1950's*, USDA, ERS, *Agricultural Economic Report* no. 18 (Washington, D.C.: 1962); and Earle J. Bedenbaugh, "History of Cropland Set Aside Programs in the Great Plains," and W. A. Laycock, "History of Grassland Plowing and Grass Planting on the Great Plains," papers in John E. Mitchell, ed., *Impacts of the Conservation Reserve Program in the Great Plains, Symposium Proceedings, September 16–18, 1987, Denver, Colorado*, USDA, Forest Service, *General Technical Report* RM-158 (Fort Collins, Col.: Forest Service Rocky Mountain Forest and Range Experiment Station, 1987, pp. 3–8, 14–17. The GPC, *Proceedings*, are also particularly helpful for such analysis.

On changes developing in livestock production, useful discussion is found in Mont H. Saunderson, *Western Stock Ranching* (Minneapolis, Minn.: University of Minnesota Press, 1950); John T. Schlebecker, *Cattle Raising on the Plains, 1900–1961* (Lincoln, Nebr.: University of Nebraska Press, 1963); and Charles L. Wood, *The Kansas Beef Industry* (Lawrence, Kans.: The Regents Press of Kansas, 1980), as well as in experiment station bulletins.

On crop varietal introductions, the station bulletins are especially important. Cf., however, Jack Ralph Kloppenburg, Jr., *First the Seed: The Political Economy of Plant Biotechnology, 1492–2000* (Cambridge, Eng.: Cambridge University Press, 1988), which presents comment on the transfer of such research to commercial seed houses as United States agricultural policy shifts away from production technology.

Postwar planning for irrigation development in the region is discussed in Huffman, *Irrigation Development*; U.S. Congress, House Committee on Interior and Insular Affairs, *Garrison Diversion Unit Reformulation Act of 1986*, 99th Cong., 2d. sess., 1986, *House Report* 99-525, pt. 1, pp. 8–15; Harry A. Steele, "The Missouri River Development Program," *Journal of Farm Economics* 31 (February 1949): 1010–16; Harry A. Steele and John Muehlbeier, "Land and Water Development Programs in the Northern Great Plains," *Journal of Farm Economics* 32 (August 1950): 431–44; Marc Reisner, *Cadillac Desert: The American West and Its Disappearing Water* (New York, N.Y.: Viking Penguin Inc., 1986); Robert G. Athearn, *High Country Empire: The High Plains and Rockies* (Lincoln, Nebr.: University of Nebraska Press, 1960); and Schell, *History of South Dakota*.

On domestic food distribution programs see Nick Kotz, *Let Them Eat Promises: The Politics of Hunger in America* (Englewood Cliffs, N.J.: Prentice-Hall, 1969); Barbara A. Claffey and Thomas A. Stucker, "The Food Stamp Program," in Don F. Hadwiger and Ross B. Talbot, eds., *Food Policy and Farm Programs: Proceedings of the Academy of Political Science* 34 (no. 3; 1982): 40–53; Norwood Allen Kerr, "Drafted into the War on Poverty: USDA Food Stamp and Nutrition Programs, 1961–1969," *Agricultural History* 64 (Spring 1990): 154–66; Don F. Hadwiger, "The Freeman Administration and the Poor," *Agricultural History* 45 (January 1971): 21–32.

On foreign trade in wheat, more studies are greatly needed. Dan Morgan, *Merchants of Grain* (New York, N.Y.: The Viking Press, 1979) is excellent on the developing trade of the seventies. It should be used in conjunction with Lowell D. Hill, *Grain Grades and Standards: Historical Issues Shaping the Future* (Urbana, Ill.: University of Illinois Press, 1990). Two pamphlets sponsored by the National Planning Association are helpful on the developing programs under Public Law 480: W. E. Hamilton and W. M. Drummond, *Wheat Surpluses and Their Impact on Canada-United States Relations* (Washington, D.C.: Canadian-American Committee, 1959); Canadian-American Committee, *Wheat Surpluses and the U.S. Barter Program, A Statement* (Washington, D.C.: Sponsored by National Planning Association, U.S.A., and Private Planning Association of Canada, 1960). R. Neil Sampson, *Farmland or Wasteland, A Time to Choose: Overcoming the Threat to America's Farm and Food Future* (Emmaus, Pa.: Rodale Press, 1981), pp. 41–42, discusses the effect of the embargo policy in 1980–1981. Leo V. Mayer, "Farm Exports and Soil Conservation," in Hadwiger and Talbot, eds., *Food Policy and Farm Programs*, pp. 99–111, provides a brief for maintenance of the farm export program; Robert L. Paarlberg, "Food as an Instrument of Foreign Policy," ibid., pp. 25–39, warns of the dangers involved. Helen C. Farnsworth, "International Wheat Agreements and Problems, 1949–56," *Quarterly Journal of Economics* 70 (May 1956): 217–48, analyzes one phase of such arrangements, but a comprehensive discussion of these programs has not been found.

Scholarly publication is also needed on the problems of world hunger. The pioneering population studies of Warren S. Thompson and P. K. Whelpton, *Population Trends in the United States* (New York, N.Y.: McGraw-Hill Book Co., 1933) were challenged by Joseph S. Davis, *The Population Upsurge in the United States, War-Peace Pamphlets* no. 12 (Stanford, Calif.: Food Research Institute, Stanford University, 1949). The *Journal of Soil and Water Conservation* 30 (January-February, 1975), presented a collection of papers on the food crisis of the seventies. The United Nations journal *Populi* is devoted to consideration of population growth trends, and U.N. Food and Agriculture Organization reports on food needs are extensive. The Resources for the Future volume, Pierre R. Crosson, ed., *The Cropland Crisis: Myth or Reality?* (Baltimore, Md.: Johns Hopkins University Press, 1982), also brought together a series of papers on the topic, notably Martin E. Abel, "Growth in Demand for U.S. Crop and Animal Production by 2005," with questioning comment by Leroy Quance, pp. 63–91. Lester R. Brown and others for the Worldwatch Institute have reported on world needs in annual publications since 1984. The *Second RCA Appraisal* offers some comparison of USDA data with alternative projections.

The reassertion of the conservation theme which precipitated much of the current legislation on agricultural planning was expressed by R. Burnell Held and Marion Clawson in the volume *Soil Conservation in Perspective* (Baltimore, Md.: Published for Resources for the Future by The Johns Hopkins University Press, 1965). Sampson's *Farmland or Wasteland* (1981) provides a generally well-balanced exposition of the environmental argument, given the assumptions which the computerized modeling of the RCI has established.

Two excellent collections of papers explore the problems of the policy development represented in the RCA: Hadwiger and Talbot, eds., *Food Policy and Farm Programs*, cited above; and Harold G. Halcrow, Earl O. Heady, and Melvin L. Cotner, eds., *Soil Conservation Policies, Institutions, and Incentives* (Ankeny, Iowa: Published for North Central Research Committee 3: Natural Resource Use and Environmental Policy, by the Soil Conservation Society of America, 1982), the proceedings of a symposium at Illinois State Park, near Zion, Illinois, May 19–21, 1981. Both raise considerable question concerning the RCA. Two research papers developed in USDA give insight on the considerations underlying the program: Anthony Grano and others, *Analysis of Policies to Conserve Soil and Reduce Surplus Crop Production*, USDA, ERS, *Agricultural Economic Report* no. 534 (Washington, D.C., 1985); and Michael R. Dicks and Katherine Reichelderfer, *Issues in Agricultural Policy: Choices for Implementing the Conservation Reserve*, USDA, ERS, *Agricultural Information Bulletin* no. 507 (Washington, D.C., 1987). Michael R. Dicks, Felix Llacuna, and Michael Linsenbigler, *The Conservation Reserve Program: Implementation and Accomplishments, 1986–87*, USDA, ERS, *Statistical Bulletin* no. 763 (Washington, D.C., 1988), provides data on the progress under this program. James B. Newman, "Overview of the Present Land-Use Situation and the Anticipated Ecological Impacts of Program Implementation," and Linda A. Joyce and Melvin D. Skold, "Implications of Changes in the Regional Ecology of the Great Plains," both in Mitchell, ed., *Impacts of the Conservation Reserve Program*, pp. 55–59 and 115–27, respectively, provide previously unpublished data on the extent of acreage considered eligible for land retirement under the CRP by states in the Great Plains.

Norman J. Rosenberg, "Adaptations to Adversity: Agriculture, Climate, and the Great Plains of North America," *Great Plains Quarterly* 6 (Summer 1986): 202–17, presents a summary of current methods of managing semiarid lands and relates such agriculture to recent concerns on climatological change. He concludes that the region is unlikely to develop into either a desert or a paradise (p. 215).

INDEX

AAA (PMA, ACP, ASCS) program, 72, 111, 256; acreage allotments under, 92–95, 142–43, 156, 160, 168, 174–76, 183, 224, 227–32, 234, 242–43, 246, 248–49, 251–54, 256, 264, 265; conservation provisions in, 95, 99, 139, 142–44, 196–99, 201, 211–12, 229–32, 234, 248, 257 (*see also* Conservation reserve program; Great Plains Conservation Program; Soil Bank program); cost of, 198, 229–30, 257, 261, 278, 347n14; criticism of, 93, 160, 257, 261; crop insurance under, 93, 95, 131, 208–10; local planning in, 94, 144–45, 257; marketing quotas under, 95, 166, 175, 183, 234, 265; popularity of, 94, 209; price supports under, 82, 92–93, 174, 175, 182, 183, 228–30, 246, 248–49, 251–55, 257, 265 (*for dated measures see* Price supports: legislation for)

Adams County, N.D., 20, 35, 47, 81, 94, 137, 148, 193, 208, 337n18

Adams County (N.D.) Soil Conservation District, 197, 201

Africa, 250, 253. *See also* Algeria; Libya; Morocco

Agricultural Credit Corporation, 88

Agricultural discontent, 39, 79–85, 89–90, 101, 122–23, 125, 128, 154–55, 161, 249, 251–52, 255, 308n71

Agricultural Marketing Act (1929), 88

"Agricultural Outlook" reports, 69, 103

Agricultural Stabilization and Conservation Service. *See* AAA program

Agricultural Trade Development and Assistance Act. *See* Public Law 480

Agriculture and Consumer Protection Act. *See* Crop insurance

Agriculture and Food Security Act. *See* Conservation reserve program

Alfalfa: in conservation programs, 140, 143, 144; marketing problems of, 39, 221; moisture requirements of, 51, 226; production of, 4, 12–13, 15, 23, 43, 139, 160, 168, 180, 186, 273–74; profitability of, 189; in salinity treatment, 273; variety development of, 191–92

Algeria: and criticism of U.S. agricultural policy, 247; wheat exports to, 253, 352n71

American Agricultural Movement, 249, 251–52, 255

American Farm Bureau Federation, 32, 80; divergent views of, 40, 109, 216, 258; hostility toward Bureau of Agricultural Economics, 156–57; opposition to National Farmers' Union, 155; opposition to resettlement program, 122; opposition to Resources Conservation Appraisal, 264

American Farm Credit Administration, 207, 252. *See also* Credit, availability of

American Red Cross, 73–74, 87–88, 100

Anderson, Clinton P., 181

Angostura: dam, 214; irrigation project, 128, 134, 135, 200

Armstrong County, S.D., 76

Association of Agricultural Colleges and Experiment Stations, 40

Association of Land Grant Colleges and Universities, 149

Atkinson, Alfred, 33

Australia, 10, 251

Automobiles, 56–57, 202

Ayers, Roy E., 119

Bailey, William R., 226–27

Baker, O. E., 29, 106–7

Bankhead-Jones Farm Tenancy Act (1937), 115, 116, 121, 129, 131

Bank of North Dakota, 9, 34, 84, 170, 330n16

Banks, commercial: as credit source, 7, 8, 41, 207; funded by government, 90–92, 207; liquidate land holdings, 170

Barley: crop insurance on, 341n52; moisture requirements of, 51; pelleting development of, 190; production of, 4, 15, 23, 43, 139, 160, 168, 178, 180, 186; regulated under domestic allotment program, 142, 227; varietal development of, 44, 191, 273

Beans, 175, 176

Belgium, 239

Belle Fourche irrigation project, 39, 40, 131, 134

373